Bark Anatomy of Trees and Shrubs in the Temperate Northern Hemisphere

Fritz H. Schweingruber • Peter Steiger
Annett Börner

Bark Anatomy of Trees and Shrubs in the Temperate Northern Hemisphere

Springer

Fritz H. Schweingruber
Institute of Forest, Snow
and Landscape Research - WSL
Birmensdorf, Switzerland

Peter Steiger
Landscape architect
Rodersdorf, Switzerland

Annett Börner
Calyptra Pty Ltd.
Melrose Park, Australia

Fritz Schweingruber	all microscopic content
Peter Steiger	all macroscopic content
Annett Börner	editing and layout

Image source: all microscopic photos by Fritz Schweingruber,
all macroscopic photos by Peter Steiger, except:
p. 298 lower right, p. 386 lower right by Fritz Schweingruber
p. 3 top left & lower center, p. 6 top right,
p. 18/19, p. 322/323 by Annett Börner
p. 3 lower left, p. 286/287 by Danny Wild

ISBN 978-3-030-14055-7 ISBN 978-3-030-14056-4 (eBook)
https://doi.org/10.1007/978-3-030-14056-4

Library of Congress Control Number: 2019934760

Cover illustration
Spine: Xylem and phloem in polarized light
Frontcover, top: Crystal druses in the phloem of *Paulownia tomentosa* (polarized light)
Middle left: Xylem and phloem of *Malva moschata*
Middle right: Phloem of *Salix foetida*

This Springer imprint is published by the registered company Springer Nature Switzerland AG
The registered company address is: Gewerbestrasse 11, 6330 Cham, Switzerland

Content

1 | Introduction. 1
Species selection & biomes featured in this book . 2
Macroscopic bark aspects 4
Microscopic bark aspects. 6
Principal bark elements and their function 7
Illustrated overview of microscopic bark features . 8
Microscopic preparation techniques 16

2 | The boreal taiga. 18
Juniperus communis 20
Larix sibirica 22
Picea abies 24
Picea glauca 26
Picea mariana 28
Pinus pumila 30
Pinus sylvestris 32

Arctostaphylos uva-ursi 34
Betula nana 36
Betula papyrifera 38
Hippophae rhamnoides 40
Myrica gale 42
Populus tremula 44
Rhododendron dauricum 46
Rhododendron tomentosum 48
Vaccinium uliginosum 50

3 | Subalpine and subarctic dwarf shrubs 52
Arctostaphylos alpina 54
Empetrum nigrum. 56
Erica carnea 58
Loiseleuria procumbens 60
Phyllodoce caerulea. 62
Rhamnus pumila 64
Rhododendron ferrugineum. 66
Rhodothamnus chamaecistus. 68
Salix foetida 70
Salix reticulata 72
Salix retusa 74
Vaccinium myrtillus. 76
Vaccinium vitis-idaea 78

4 | Mountain ranges in Eurasia 80
Abies alba. 82
Cedrus atlantica 84
Juniperus sabina. 86
Larix decidua. 88
Pinus cembra. 90
Pinus mugo subsp. *uncinata*. 92
Pinus nigra subsp. *nigra* 94
Alnus viridis 96
Amelanchier ovalis 98
Myricaria germanica 100
Rhamnus alpina. 102
Ribes alpinum 104
Ribes petraeum 106
Salix appendiculata. 108
Sorbus aria 110
Sorbus chamaemespilus 112

5 | Deciduous forest of temperate Europe 114
Taxus baccata 116

Acer campestre 118
Acer platanoides. 120
Acer pseudoplatanus 122
Alnus glutinosa. 124
Alnus incana 126
Berberis vulgaris 128
Betula pendula. 130
Carpinus betulus. 132
Cornus sanguinea 134
Corylus avellana. 136
Cotoneaster integerrimus 138
Crataegus monogyna. 140
Euonymus europaeus. 142
Euonymus verrucosus. 144
Fagus sylvatica 146
Fraxinus excelsior 148
Hedera helix 150
Hippocrepis emerus. 152
Ilex aquifolium. 154
Ligustrum vulgare 156
Lonicera xylosteum 158
Populus nigra. 160
Prunus avium 162
Prunus padus. 164
Prunus spinosa. 166
Pyrus pyraster 168
Quercus petraea 170
Quercus robur 172
Rhamnus cathartica 174
Rosa arvensis. 176
Rosa elliptica 178
Salix alba 180
Salix purpurea 182
Sambucus nigra 184
Sambucus racemosa 186
Sorbus torminalis 188
Tilia cordata 190

Ulmus glabra . . . 192
Viburnum lantana . . . 194
Viburnum opulus . . . 196
Viscum album . . . 198

6 | Deciduous forest of submediterranean Europe . 200
Acer monspessulanum . . . 202
Acer opalus . . . 204
Aesculus hippocastanum . . . 206
Buxus sempervirens . . . 208
Castanea sativa . . . 210
Celtis australis . . . 212
Cotinus coggygria . . . 214
Cydonia oblonga . . . 216
Daphne laureola . . . 218
Fraxinus ornus . . . 220
Juglans regia . . . 222
Laburnum anagyroides . . . 224
Lonicera caprifolium . . . 226
Malus x domestica . . . 228
Ostrya carpinifolia . . . 230
Prunus laurocerasus . . . 232
Prunus mahaleb . . . 234
Quercus cerris . . . 236
Quercus pubescens . . . 238
Sorbus domestica . . . 240
Staphylea pinnata . . . 242
Syringa vulgaris . . . 244
Vitis vinifera . . . 246

7 | Deciduous forest of Eastern Asia . . . 248
Cryptomeria japonica . . . 250
Ginkgo biloba . . . 252
Larix kaempferi . . . 254
Metasequoia glyptostroboides . . . 256
Sciadopitys verticillata . . . 258
Thujopsis dolabrata . . . 260

Ailanthus altissima . . . 262
Cercidiphyllum japonicum . . . 264
Decaisnea insignis . . . 266
Diospyros lotus . . . 268
Edgeworthia chrysantha . . . 270
Hamamelis mollis . . . 272
Hydrangea aspera . . . 274
Koelreuteria paniculata . . . 276
Morus alba . . . 278
Paulownia tomentosa . . . 280
Stachyurus praecox . . . 282
Trochodendron aralioides . . . 284

8 | Deciduous forest of Eastern North America . 286
Pinus strobus . . . 288
Tsuga canadensis . . . 290

Acer rubrum . . . 292
Asimina triloba . . . 294
Calycanthus floridus . . . 296
Carya myristiciformis . . . 298
Gleditsia triacanthos . . . 300
Halesia carolina . . . 302
Hamamelis virginiana . . . 304
Liquidambar styraciflua . . . 306
Liriodendron tulipifera . . . 308
Magnolia grandiflora . . . 310
Nyssa sylvatica . . . 312
Quercus rubra . . . 314
Rhus typhina . . . 316
Robinia pseudoacacia . . . 318
Sassafras albidum . . . 320

9 | Coniferous forest of Pacific North America . . 322
Calocedrus decurrens . . . 324
Pinus ponderosa . . . 326
Pseudotsuga menziesii . . . 328
Sequoia sempervirens . . . 330
Sequoiadendron giganteum . . . 332
Thuja plicata . . . 334
Torreya californica . . . 336
Xanthocyparis nootkatensis . . . 338

Gaultheria shallon . . . 340
Mahonia aquifolium . . . 342
Umbellularia californica . . . 344

10 | Mediterranean hard-leaved forest . . . 346
Cupressus sempervirens . . . 348
Juniperus oxycedrus . . . 350
Pinus halepensis . . . 352

Arbutus unedo . . . 354
Ceratonia siliqua . . . 356
Cercis siliquastrum . . . 358
Cistus laurifolius . . . 360
Erica arborea . . . 362
Ficus carica . . . 364
Laurus nobilis . . . 366
Myrtus communis . . . 368
Nerium oleander . . . 370
Olea europaea . . . 372
Phillyrea latifolia . . . 374
Pistacia lentiscus . . . 376
Platanus x hispanica . . . 378
Punica granatum . . . 380
Quercus coccifera . . . 382
Quercus ilex . . . 384
Quercus suber . . . 386
Spartium junceum . . . 388
Tamarix gallica . . . 390

References and recommended reading . . . 393

1 | Introduction

Bark is an essential part of all plants with secondary growth. Young bark, especially the cortex, plays an active role in plant metabolism. Parenchyma cells store photosynthetic products and young bark is photosynthetically active. Barks protect sensitive meristems mechanically and chemically against fire and frost, and defend essential physiological tissues such as the xylem, phloem and cortex against pathogens. Barks are also a habitat for fungi, algae, mosses, lichens, vascular plants and invertebrates. Since prehistoric times, bark has been used by humans for fibers, resins, rubber, medicinal products, dyes, tannins and cork.

Why a new book about the anatomy of bark? There is an obvious lack of books about taxonomic bark anatomy. The reference list of studies on general bark anatomy is long (see Angyalossy *et al.* 2016), and many books present the macroscopic aspect of barks (e.g. Vaucher 1990), but most of those publications focus on general phloem anatomy or on the characterization of a few genera or species (e.g. Ravaomanalina *et al.* 2017). However, Holdheide 1951 and Crivellaro & Schweingruber 2013 are the only publications with regional anatomical inventories. The former concentrates on Central European species, the latter on Eastern Mediterranean species.

In contrast, there is a wealth of publications and atlases describing the anatomical structures of the xylem, featuring many species in large and diverse regions (e.g. Grosser 1977, Schweingruber 1978, 1990 and 2005). Recently, two studies described a comprehensive number of bark features occurring in the phloem and periderm. Crivellaro & Schweingruber 2015 chose a pragmatic way, describing all bark features observed in double-stained slides (Astrablue/Safranin) of trees, shrubs and herbs from all climatic zones of the Northern Hemisphere. Angyalossy *et al.* 2016 take a more academic approach, describing anatomical bark features observed in differently stained slides of trees from a worldwide collection. Unfortunately, the two approaches are at least partially difficult to compare. The masterpiece of bark characterization remains the book by Esau (1969).

In this new book, we try to give an overview over the variability of the macroscopic and microscopic aspect of bark, especially in phloem structures, from the temperate zone of the Northern Hemisphere.

F. H. Schweingruber et al., *Bark Anatomy of Trees and Shrubs in the Temperate Northern Hemisphere*, https://doi.org/10.1007/978-3-030-14056-4_1

Species selection and the biomes featured in this book

The original concept of this book only included a wide variety of Central European and Mediterranean trees and shrubs. Because many readers may be of East Asian or North American origin, the selection of species was extended to these regions, also giving the book a broader view of the whole Northern Hemisphere. Because of the inclusion of other regions beyond Europe, the species were grouped by their geographical origins, such as Central Europe, the Mediterranean, Eastern Asia and Northern America. In each biome chapter, the species are separated into conifers and angiosperms. Within these two groups, the species follow in alphabetic order of their Latin names.

The number of species in this book is not reflecting the much greater biodiversity of Eastern Asia and Northern America compared to Europe, simply because the majority of the sampled bark material and photos compiled over the years by the two Swiss authors is of European origin.

Two of the chosen biomes are forming circumpolar belts, stretching from Europe continuously to Northeastern Asia and Northern America. There is the forest biome with the far biggest surface on earth—the **boreal taiga**. Consequently, the species selected for this chapter in the book are of Eurasian or North American origin. The boreal taiga is characterized by vast areas of species-poor coniferous forests, mostly of the genera *Picea*, *Pinus* or *Larix*, with a few deciduous pioneer trees such as *Populus tremula* or *Betula papyrifera*, and some widespread shrubs such as *Pinus pumila* or *Arctostaphylos uva-ursi*.

The **subarctic biome** is a circumpolar, treeless belt north of the taiga. This book includes only dwarf shrub communities, with most of the species having a circumpolar or at least Eurasian range. Ecologically similar conditions can be found in **subalpine dwarf shrub** communities of high mountain ranges above the tree line, also south of the subarctic, such as in the Alps, the Caucasus, the Altai Mountains, Siberian and Alaskan mountain ranges, the Northern Rockies or mountain tops in Maine.

The following biomes are more or less limited to one continent, with the exception of the Mediterranean Basin, where Northern Africa comes into close contact with Europe across the Strait of Gibraltar, in the same way as the Caucasus itself is sitting on the border of Europe and Asia, and the Greek and Turkish mountain ranges are close to one another. The **mountain systems of Eurasia** are represented in this book only by European species, with the exception of *Cedrus atlantica* from the Mediterranean mountains. The dominant conifers of the pine family in this biome are accompanied by various deciduous shrubs. Central Asian and Himalayan mountains are not included in this book, simply because of a lack of available bark material.

The main focus of this book, due to the huge variety of sampled bark material, is the **deciduous forest of temperate Europe**, represented by a variety of deciduous trees and shrubs. In many areas of Central Europe, *Fagus sylvatica* is able to form vast, pure stands, whereas in Atlantic or more continental areas, mixed oak forests dominate. Geographically and ecologically very close and difficult to separate is the biome of the **deciduous forest of submediterranean Europe**, centered in the higher, frost-influenced areas of the Mediterranean, the Balkans and the Southern Alps, with a few thermically favored islands in Central Europe, such as the Pannonian Basin. These forests are usually more species-rich then those in Central Europe, and are often dominated by deciduous oaks, *Castanea sativa*, *Ostrya carpinifolia* and *Fraxinus ornus*.

The **deciduous forest of Eastern Asia** with its enormous biodiversity and subtropical influences reaching far north, is represented in this book only by a small selection of typical species, with the purpose to include some of the plant families centered in Eastern Asia, and also some species with a worldwide distribution and familiarity as planted trees and shrubs.

A similar, limited representation is given to the **deciduous forest of Eastern America**, although its level of biodiversity and species richness lies somewhere in between Eastern Asia and Europe. Due to a marked continental or Mediterranean summer drought, the **coniferous forests of Pacific North America** are represented by a unique variety of conifers, including the tallest known trees, marked not only by many members of the pine family but also by outstanding members of the cypress family, such as the giant sequoia and coast redwood.

Finally, the **Mediterranean hard-leaved forest** is well represented by hard-leaved trees and shrubs, naturally often dominated by *Quercus ilex*, growing in a sometimes only narrow zone below the frost-influenced areas around the Mediterranean Basin, together with a few conifers and deciduous species such as *Platanus* x *hispanica* or *Punica granatum*. With the exception of some circum-Mediterranean species, species are often centered either in the western or eastern half of the basin.

Boreal taiga

Subalpine and subarctic dwarf shrubs

Mountain systems in Eurasia

Deciduous forest of temperate Europe

Deciduous forest of submediterranean Europe

Deciduous forest of Eastern Asia

Deciduous forest of Eastern America

Coniferous forest of Pacific North America

Mediterranean hard-leaved forest

Macroscopic bark aspects

The contemplation of tree and shrub bark is as fascinating as the study of individual human skin, both reflecting life stories that are getting richer with age. A species' typical bark textures and colors tell—besides an often impressive process of aging—an individual story for each tree and shrub. Its bark is reflecting an often century-long process influenced by the specific locality and soil, light and shadow, competition with neighboring trees, average climate, weather extremes, catastrophic events such as fire, flood, avalanches or attacks by animals and fungi. As a result of all these influences, a bark may be instantly recognizable as a certain species but may also show an impressive variety of individual colors and structures, often additionally characterized by lichen, mosses or higher plant epiphytes.

Some barks, such as this bark of *Pinus cembra*, are very distinct with age.

This *Juniperus thurifera* is at least several hundred years old.

Some barks are unmistakable, such as the orange red, papery scaling bark of the genus *Arctostapylos*.

In species-rich genera, such as the Australian *Eucalyptus*, the bark can be very helpful to distinguish species or groups of closely related species.

Fruticose lichen on a branch of a subalpine *Picea abies*. Bark characteristics are no longer visible because the surface is covered with various lichen species.

Crustaceous lichens on an old *Acer* stem, a food resource for snails. Bark characteristics can be recognized but the original color is falsified by the color of the lichens.

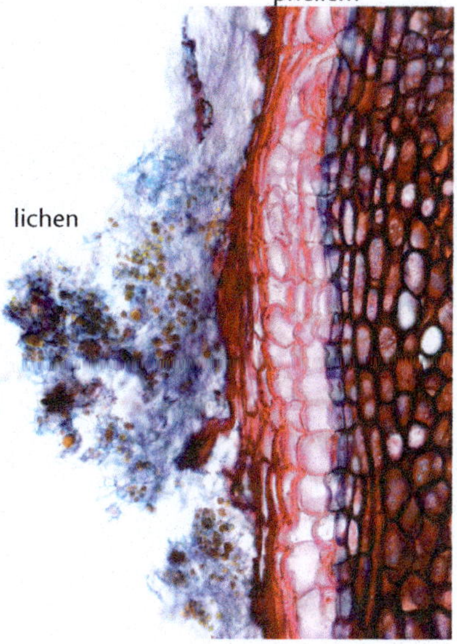

Microscopic aspect of the crustaceous lichen *Phlyctis argena* on a young *Fraxinus* twig. The lichen consists of hyphae (blue cells) with chloroplasts. It sticks to the cork layer (red phellem cells) without penetrating it.

Microscopic bark aspects

Bark consists of the following principal parts: The outermost juvenile part of the bark are the **epidermis** and the **cortex**. The innermost part outside of the cambium is the **living phloem**. Between the cortex and the phloem is the **periderm**. The adult phloem consists of a living and a dead part. The **phellogen** is a meristem separating the living from the dead phloem, and, in earlier stages, the phloem from the cortex. The phellogen is a tertiary cambium that produces the periderm; the **phelloderm** towards the inside and the **phellem** towards the outside.

Bark is a general term and includes all tissues outside of the cambium. The **inner bark** includes the living part of the phloem, and the **outer bark** the dead part of the phloem. The outer bark is also called **rhytidome**. **Bast** is the colloquial term for the living part of the phloem.

Wood and bark terminology

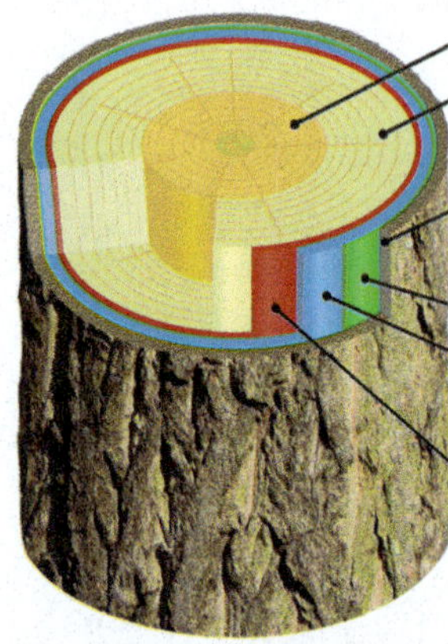

Young bark

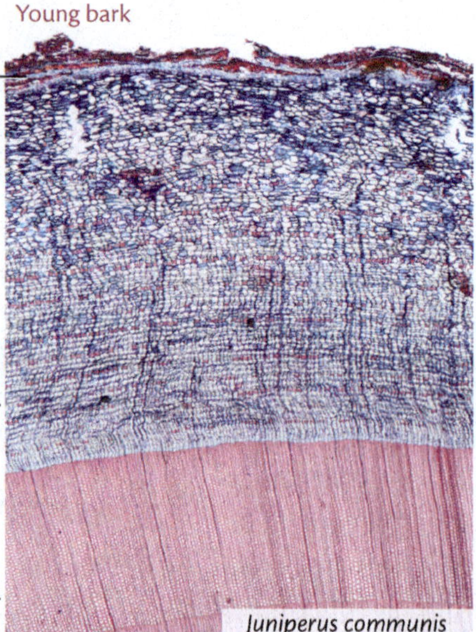

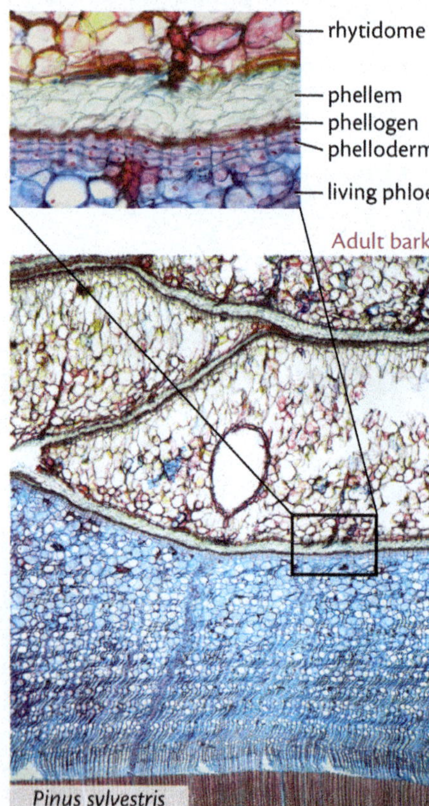

Principal bark elements and their function

Xylem and phloem are principally composed of similar tissues. The only qualitative anatomical difference in the phloem is the presence of sieve elements instead of vessels and tracheids. Several cell types characterize the xylem and the bark. All of them have been previously described, e.g. by Evert 2006, Crivellaro & Schweingruber 2015 and Angyalossy *et al.* 2016. The following descriptions are based on these sources.

Parenchyma cells are long-lived and perform their function while living. Parenchyma cells store, transport and remobilize assimilates, and produce various excretes such as resin and gum. Parenchyma cells are easily recognizable because they contain nuclei in living tissues. Cell-wall thickness, cell shape and size vary. Axial parenchyma cells occur in the bark (cortex, phloem) and in the xylem. Radially oriented parenchyma cells form the rays which occur in the phloem as well as in the xylem. Parenchyma cells are stem cells and are able to differentiate into meristems, parenchyma cells, sclereids and excretion cells (Eschrich 1995, Evert 2006, Schweingruber *et al.* 2013).

Sieve elements (sieve cells or sieve tubes) are short-lived and transport assimilates from all plant parts with chlorophyll to storage elements (parenchyma cells) elsewhere in the plants. Sieve elements are only active in the innermost, conductive part of the phloem near the cambium. They are large cells, but rapidly change their form due to cell collapse while they lose their functionality.

Sieve cells are axially elongated cells in the phloem of conifers. Their sieve plates occur on radial walls. Sieve cells in the phloem are the equivalent of axial tracheids in the xylem. Small, living **Strasburger cells** occur adjacent to sieve cells.

Sieve tubes are axially elongated cells in the phloem of angiosperms. Their sieve plates occur primarily on transverse walls at different inclinations. Sieve tubes in the phloem are the equivalent of vessels in the xylem. Small, living **companion cells** occur adjacent to sieve tubes.

Sclerenchyma cells (fibers and sclereids) are short-lived stabilizing cells and protect the tissue against physical and biological damage.

Fibers are elongated cells with pointed ends, produced in the cambial differentiation belt. Fibers occur in the phloem and in the xylem. **Gelatinous fibers** are a special type of fiber, characterized by an unlignified secondary layer. Gelatinous fibers occur in phloem, cortex and xylem of specific taxa.

Sclereids are thick-walled cells of various forms. They derive from living parenchyma cells that thicken their cell walls and lose cell content. Sclereids are restricted to the pith, phloem and cortex. They are mostly absent in the xylem. In the bark they occur predominantly outside of the conductive phloem.

Excretion cells are parenchyma cells specialized in the production of various substances such as resin, slime or gum. They occur surrounding ducts or in the form of long cells (laticifers). They are found in cortex, phloem and xylem.

Crystals of various forms (prismatic, acicular, raphids, sand etc.) mostly have the same chemical composition (calcium oxalate). They occur in cells of the cortex, phloem and xylem of various species. Crystals have two functions—calcium regulates the transport of organic molecules, and calcium oxalate is an end product of metabolic processes.

The **periderm** is formed by the tertiary cambium and consists of an external belt of parenchyma and cork cells. The outermost layer of the periderm, the **phellem**, varies in dimension, but it is mostly the dominating tissue.

The **rhytidome** is the dead part of the bark. It largely determines the macroscopic aspect of the bark. The periderm and the rhytidome protect the living phloem from dehydration, solar radiation and mechanical damages.

Illustrated overview of microscopic bark features

The following features are visible in high-quality, double-stained sections. The features and codes in brackets relate to Crivellaro & Schweingruber 2015.

Homogeneous structure
(Cr,Sg B.3, B.14)

Tissue without tangential layering. Sieve tubes and parenchyma cells irregularly or radially arranged. Sclerenchyma absent.

Sieve tubes and parenchyma radially arranged

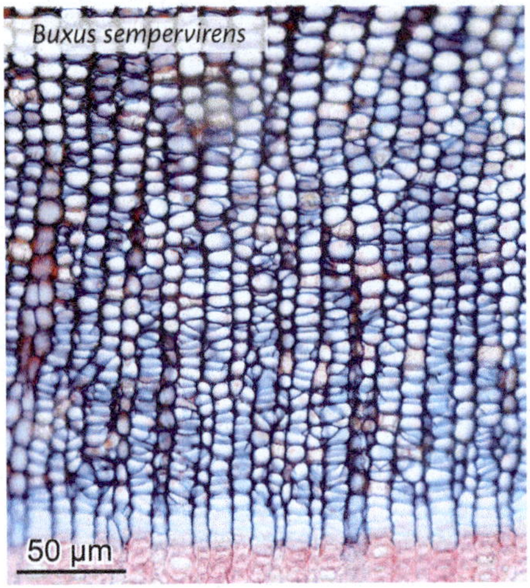

Sieve tubes & parenchyma irregularly alternating

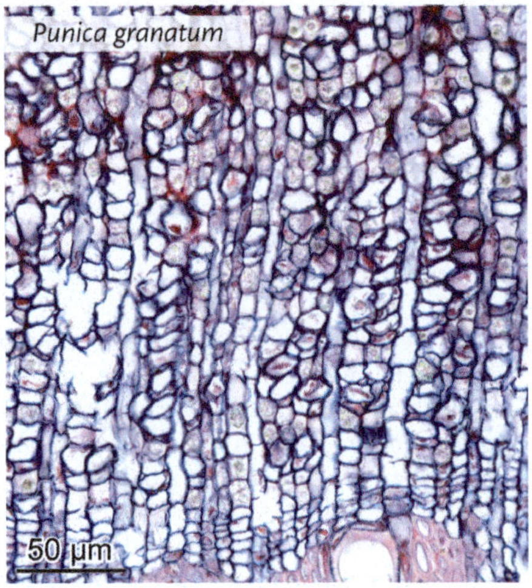

Tangential layering

Multiseriate layers without sclerification
(Cr,Sg B.9.1)

Several rows of parenchyma cells alternating with one or two rows of collapsed sieve tubes.

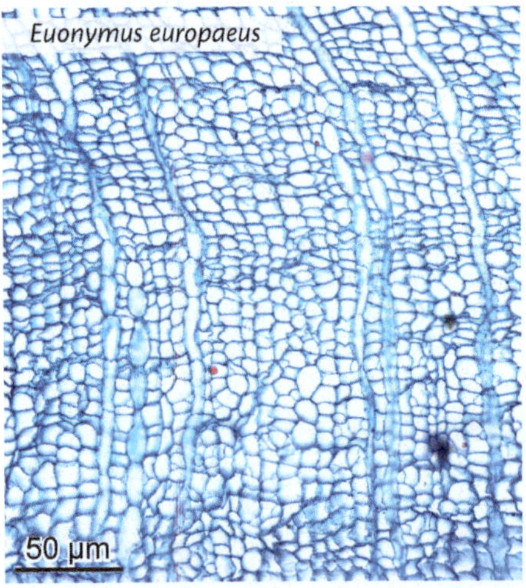

Uniseriate layers without sclerification

Sieve tubes and parenchyma cells alternating.

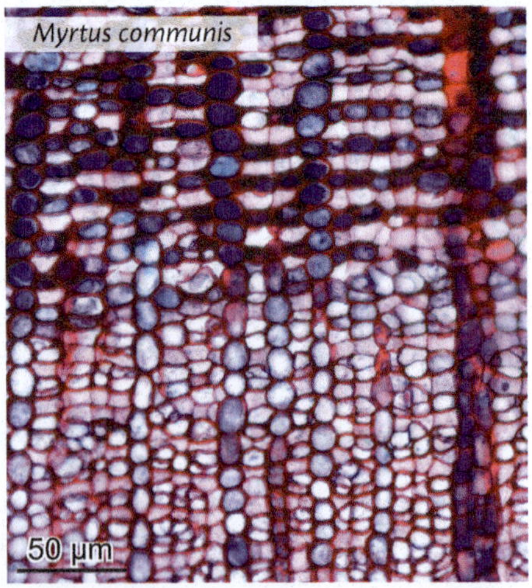

Tangential layering

Fibers uniseriate

Uniseriate layers of fibers alternating with layers of sieve cells and parenchyma cells.

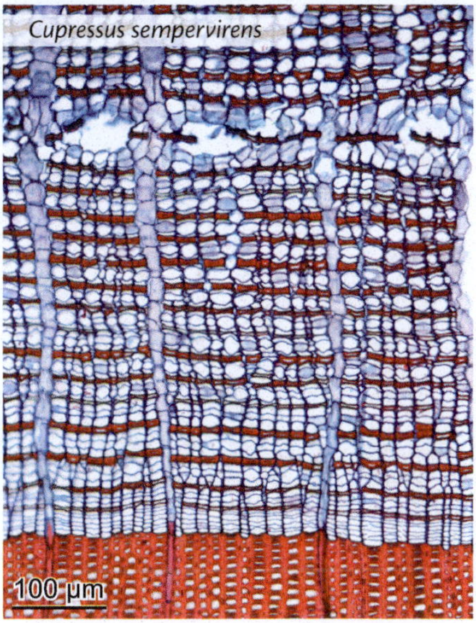

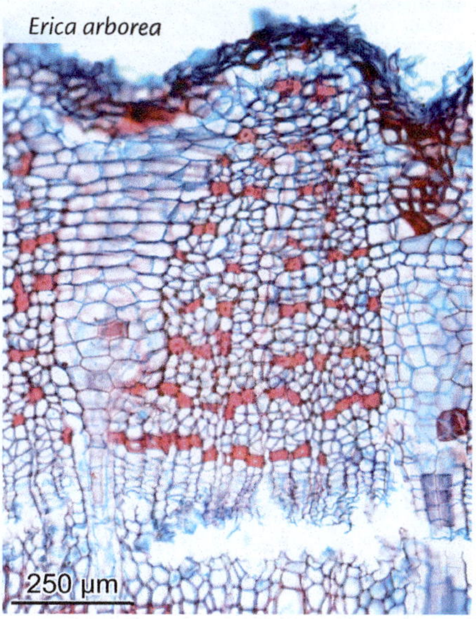

Small layers of fibers with less than 3 cells

(Cr,Sg B.18)

1–3 layers of fibers in tangential bands alternating with layers of sieve tubes and parenchyma cells.

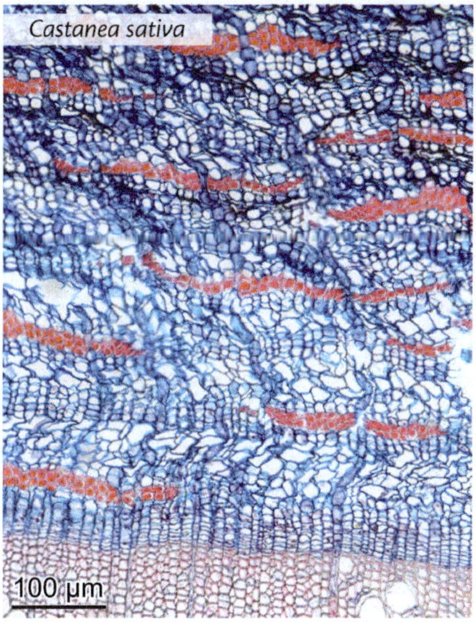

Large layer of fibers with >3 cells

(Cr,Sg B.17.4)

Layers of different widths alternating with tangential bands of sieve tubes and parenchyma cells.

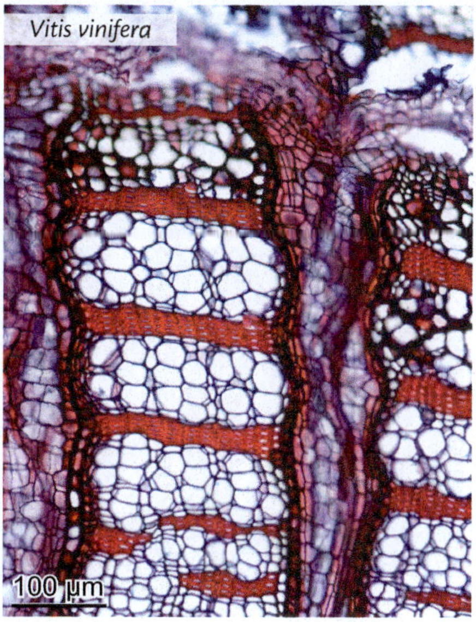

Tangential layering

Small layers of sclereids with less than 3 cells

Small layers of sclereids alternating with a few layers of sieve tubes and parenchyma cells.

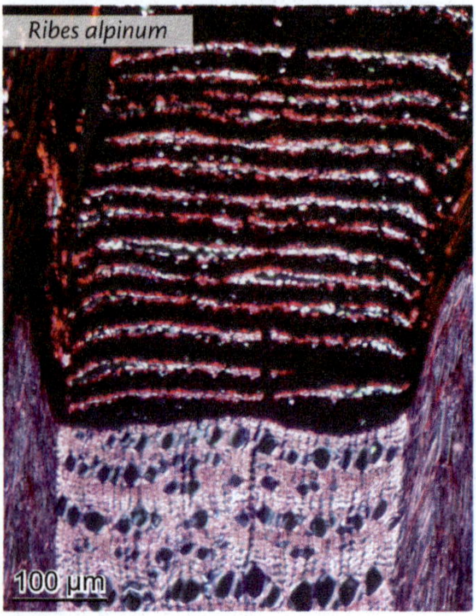

Large layers of sclereids with >3 cells

Small layers of sclereids alternating with a few or many rows of sieve tubes and parenchyma cells.

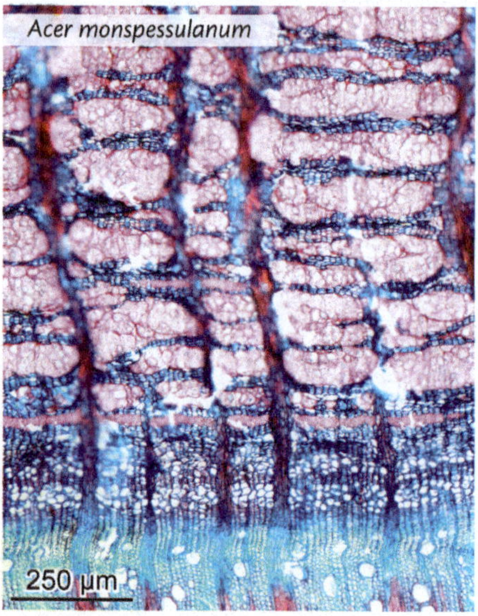

Radial orientation of sclereids

(Cr,Sg B.29.2)

Sclereids arranged in radial stripes or blocks.

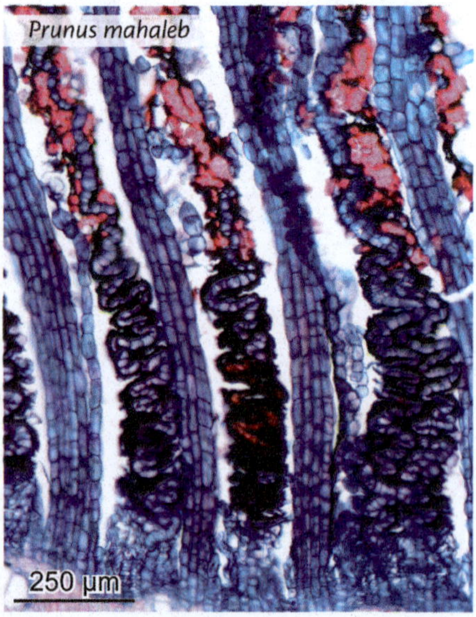

Grouping of sclerenchyma

Fibers solitary

(Cr,Sg B.15)

Individual fibers distributed irregularly within a tissue of sieve tubes and parenchyma cells.

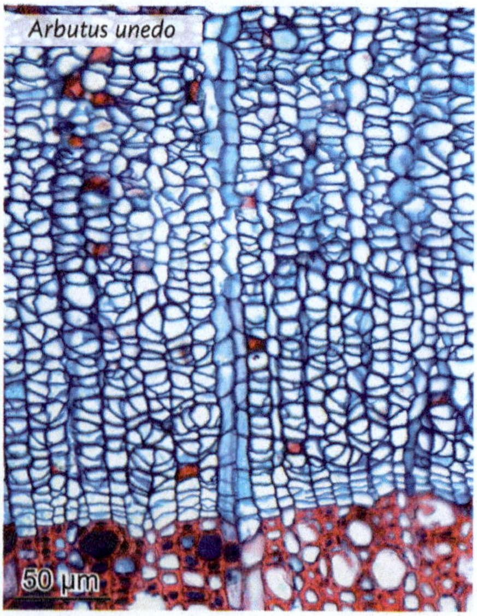

Grouping of sclerenchyma

Fibers in groups

(Cr,Sg B.27)

Fibers in small or large clusters.

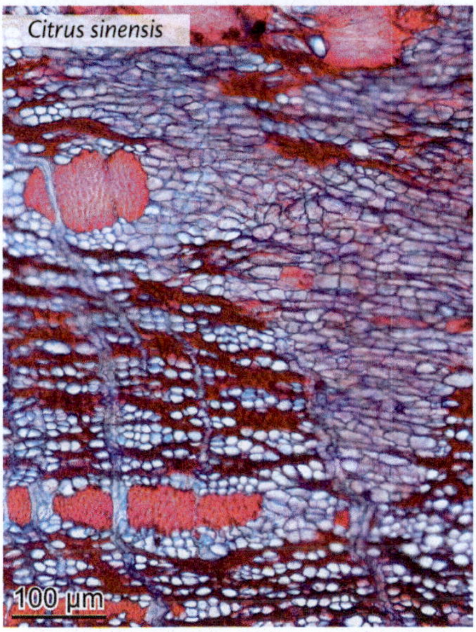

Sclereids solitary

(Cr,Sg B.27)

Sclereids solitary within a tissue of sieve tubes and parenchyma cells.

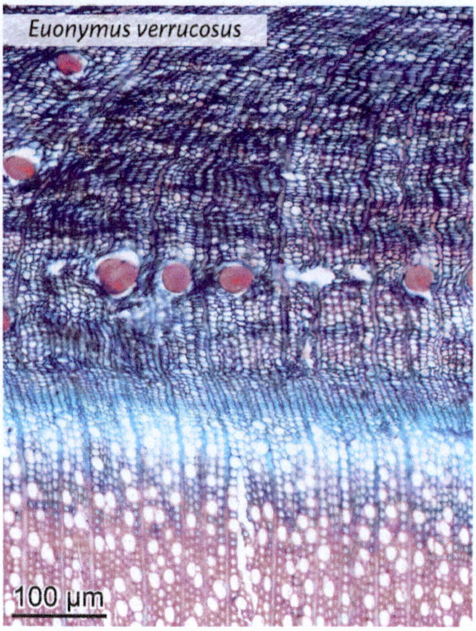

Sclereids in groups

(Cr,Sg B.28,29)

Sclereids in round to oval clusters.

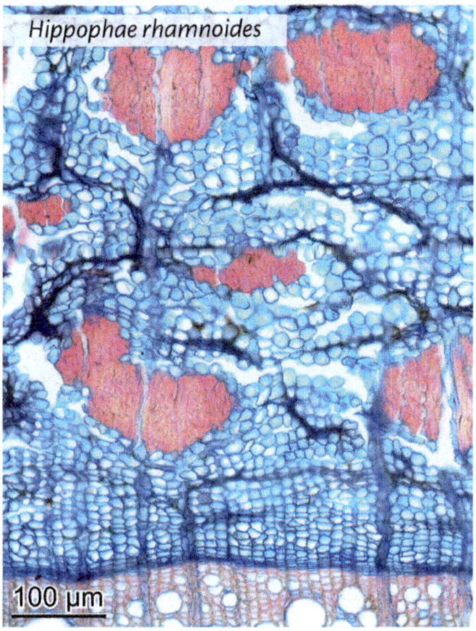

Sclereids in large rays or aggregate rays

(Cr,Sg B.41)

Sclereids of any form restricted to ray zones.

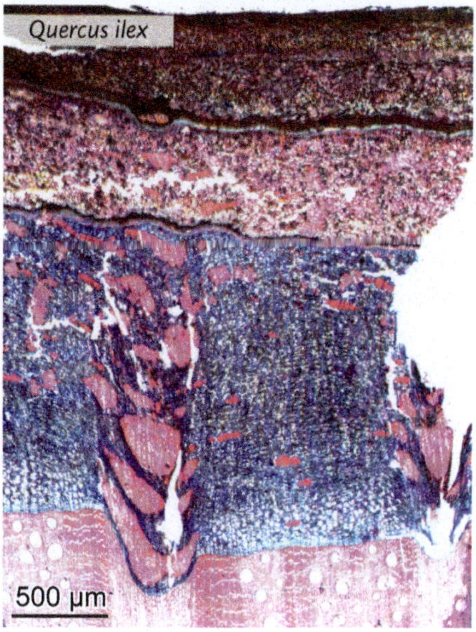

Grouping of sclerenchyma

Sclereids in large, irregularly distributed groups

(Cr,Sg B.29.3)

Groups of sclereids of any form irregularly distributed.

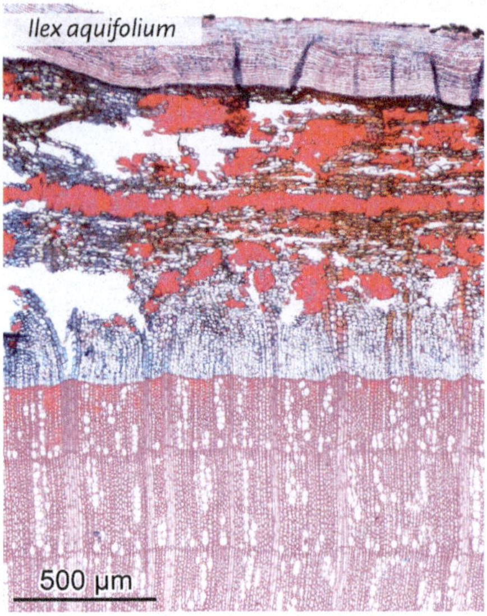

Rays

Rays absent or unrecognizable

(Cr,Sg B.30)

Rays not differentiated from radially arranged tissue of sieve tubes and parenchyma cells.

Rays

Rays straight

(Cr,Sg B.31)

Ray course not influenced by mechanical forces of other elements.

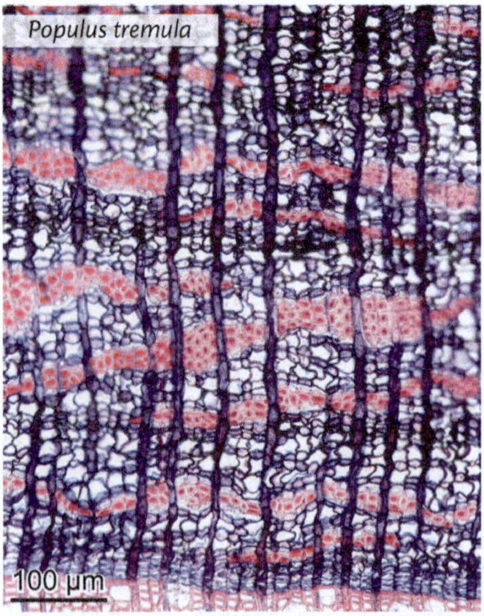

Rays wavy or bent

(Cr,Sg B.32,45)

Rays wavy due to radial and lateral compression.

Dilatation

Ray dilatation

(Cr,Sg B.37–39)

Rays dilating by forming new cells from an intercalary meristem.

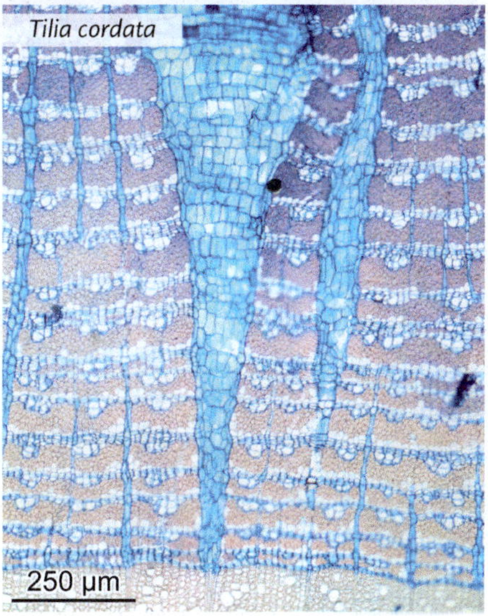

Dilatation origin in axial parenchyma cells

Tissues dilating by forming new cells from axial parenchyma cells.

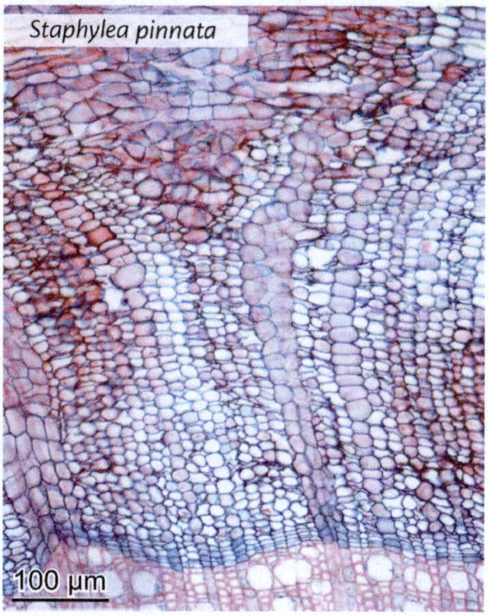

Ducts and laticifers

Ducts

(Cr,Sg B.47)

Ducts are surrounded by small excretion cells.

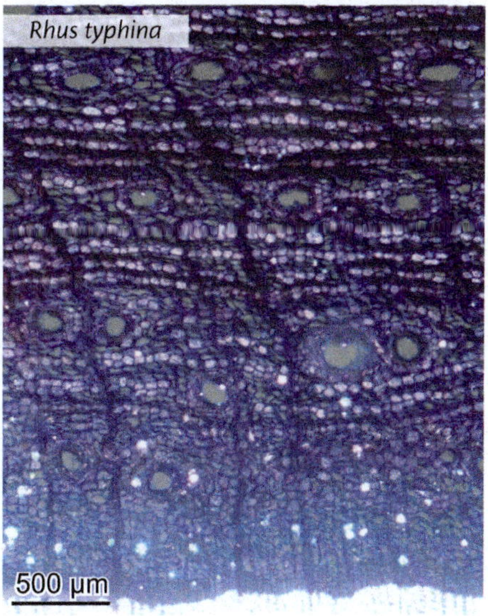

Laticifers

(Cr,Sg B.48)

Laticifers consist of single, axially elongated excretion cells.

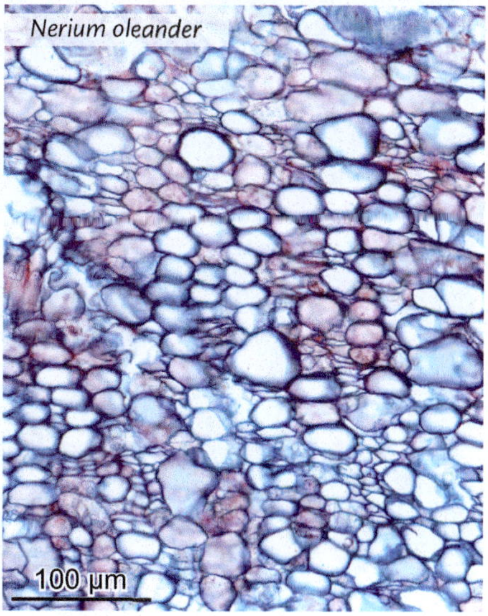

Special cells

Gelatinous fibers

(Cr,Sg B.64)

Unlignified, round fibers resembling tension wood.

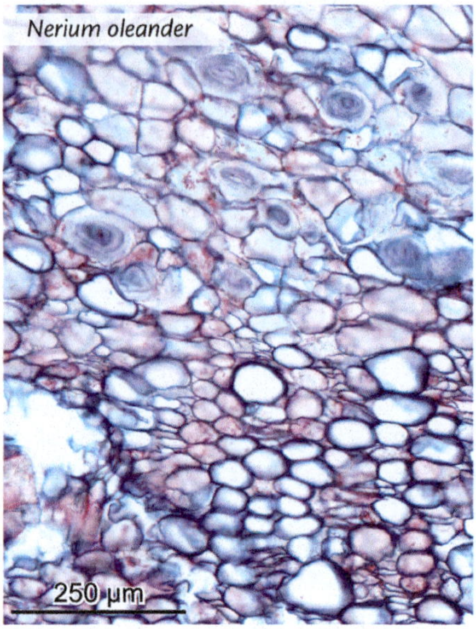

Crystals

Prismatic crystals in sclereids

(Cr,Sg B.53)

Crystals surrounded by dense, sclerotic cells.

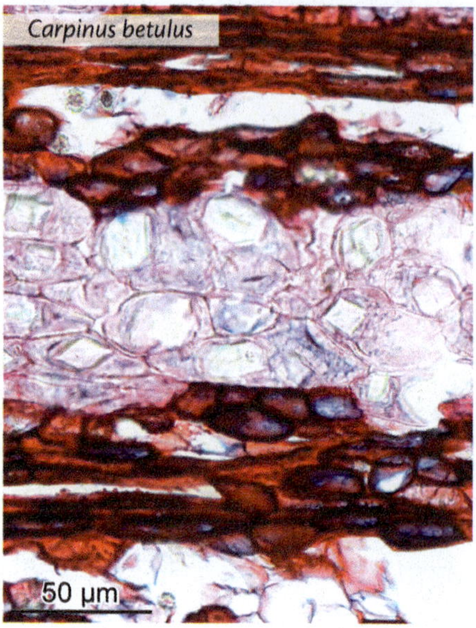

Crystals

Prismatic crystals around sclerenchyma

(Cr,Sg B.55)

Groups of sclereid surrounded by layers of crystals.

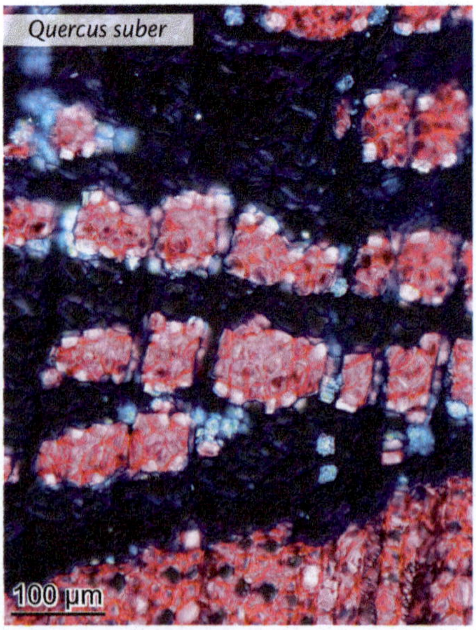

Crystal druses

(Cr,Sg B.63,64)

Druses associated with rays or unlignified axial tissues.

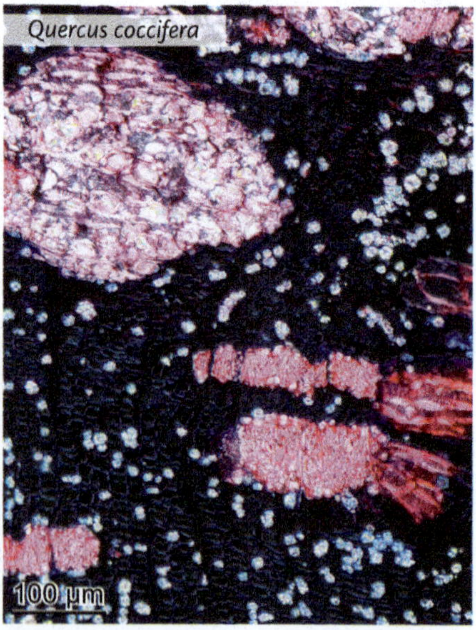

Periderm

Radially arranged rectangular cells

(Cr,Sg B.134,138)

Carpinus betulus

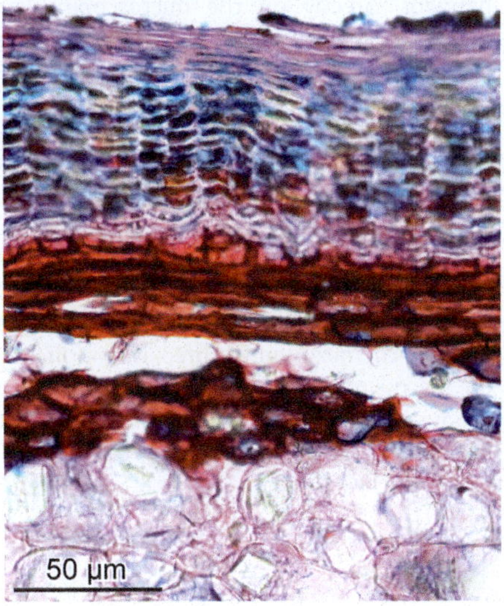

Rhytidome

Regular layers of old phloem compartments

(Cr,Sg B.131,137)

Buxus sempervirens

Rhytidome

Phelloids

(Cr,Sg B.132)

Rows of large square cells alternating with many rows of small cells.

Arbutus unedo

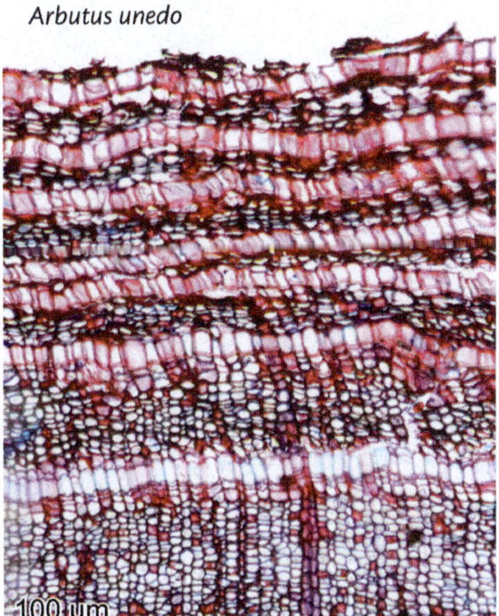

Irregular

(Cr,Sg B.133)

Groups of irregularly shaped cells separating compartments of phloem.

Microscopic preparation techniques

There are many publications on the anatomy of wood, but not many on the anatomy of bark. One likely reason for this discrepancy lies in the technical difficulty of sectioning bark. Bark is a very heterogeneous material—very tough tissues, with a density of around 1 g/cm³, are embedded into soft tissues with a very low density of 0.3 g/cm³. If the microtome knives are just slightly blunt, the denser tissues will tear the more fragile tissues during sectioning, rendering these sections scientifically useless.

All bark sections shown in this atlas were prepared from freshly collected samples. The material was sectioned with paper-knife blades, fixed in a sledge microtome with a special blade holder (Gärtner & Schweingruber 2013). All sections were stained with a one-to-one mixture of Safranin/Astrablue for several minutes. Staining and dehydration of the samples with ethanol 96%, absolute alcohol and xylene happens directly on the microscopic glass slide, reducing the amount of work. All slides were embedded in Canada Balsam for preservation, making them permanent and comparable even after many years. Photographs were taken in transmitting, normal and polarized light with an Olympus BX51 microscope. Sections were documented under 40, 100, 200 and 400 times magnification.

The use of sharp disposable paper-knife blades and the chemical sample preparation directly on the microscopic slide make it possible to produce many high-quality slides in a short amount of time. Detailed preparation techniques are described in Gärtner & Schweingruber 2013.

Microtome designed by Gärtner *et al.* 2015, featuring a blade holder for disposable paper-knife blades.

Sectioning of a sample with a high-quality disposable paper-knife blade.

Staining and dehydration of the samples happens directly on the microscopic glass slide.

Storage of microscopic slides in boxes. All slides are entered into a digital database with their taxonomical, morphological, geographical and anatomical characteristics.

Unstained slide

Stained with Safranin/ Astrablue, normal light

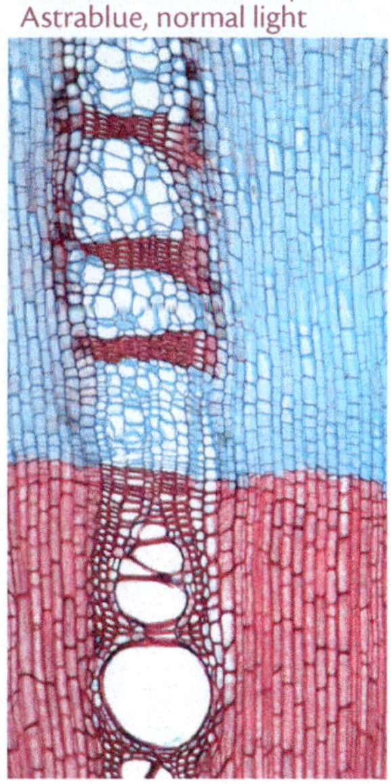

Stained with Safranin/ Astrablue, polarized light

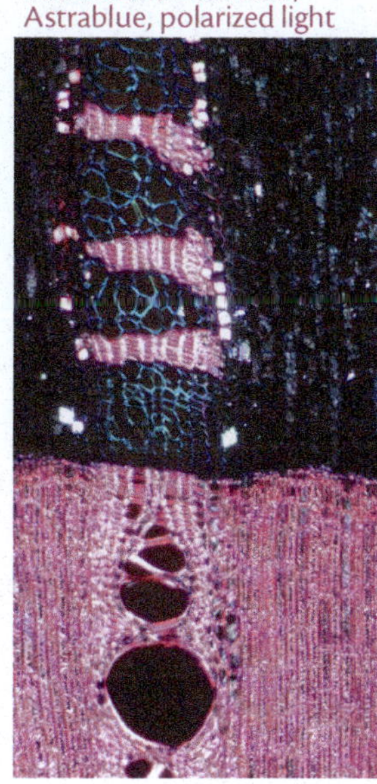

The **boreal taiga** is characterized by vast areas of species-poor coniferous forests, mostly of the genera *Picea*, *Pinus* or *Larix*, with a few deciduous pioneer species such as *Populus tremula* or *Betula papyrifera*, and some wide-spread shrubs, such as *Pinus pumila* or *Arctostaphylos uva-ursi*.

F. H. Schweingruber et al., *Bark Anatomy of Trees and Shrubs in the Temperate Northern Hemisphere*, https://doi.org/10.1007/978-3-030-14056-4_2

2 | The boreal taiga

Juniperus communis L. Cupressaceae

juniper, Gewöhnlicher Wacholder, genévrier commun, ginepro comune, enebro comun

Form: Conifer, usually a very columnar shrub, rarely a small tree, to 12 m tall; the dwarf subsp. *nana* is always a low-spreading dwarf shrub.

Features: Needles evergreen, stiff, short, sharply pointed in subsp. *communis*, to 2 cm long, upper side with a white waxy streak giving the needles a bluish appearance; needles in subsp. *nana* curved, without sharply pointed tip. Flowers dioecious, female and male flowers in tiny yellowish globes, to 5 mm wide. Cones with a berry-like, light bluish structure of coalescing scales.

Habitat: Subspecies *communis* in nemoral and boreal Eurosiberia, from Morocco eastwards to the Ob and Yenisei rivers; subsp. *nana* in subarctic Siberia, Greenland and North America, south only in high mountain ranges.

Subspecies *communis* on dry or wet soils, mostly on rocky, gravelly or sandy soils, also in bogs in full sun, to 1800 m; subsp. *nana* only in alpine or subarctic tundra, to 4200 m.

Bark: Young glossy, silvery or dark red brown, later light gray brown, with fine longitudinal fissures.

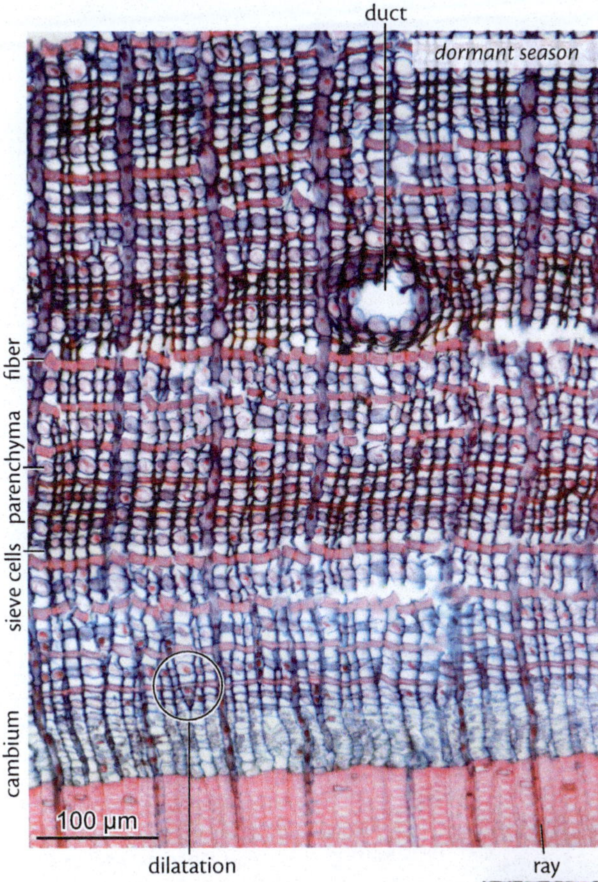

Bark anatomy: Regular tangential layers of sieve cells, fibers, and parenchyma cells are characteristic for Cupressaceae.

Sieve cells thin-walled, rectangular. Parenchyma cells round to oval. Fibers thick-walled, square and rectangular. Sieve cells tending to collapse. Uniseriate rays dilating in later stages. Rays remaining straight. Rarely with large ducts. Crystal sand in radial intracellular layers.

Annual rings mostly indistinct. Ring boundaries of fast-growing trees indicated by thick-walled fibers in the early phloem. Phellem cells thin-walled, irregular shaped.

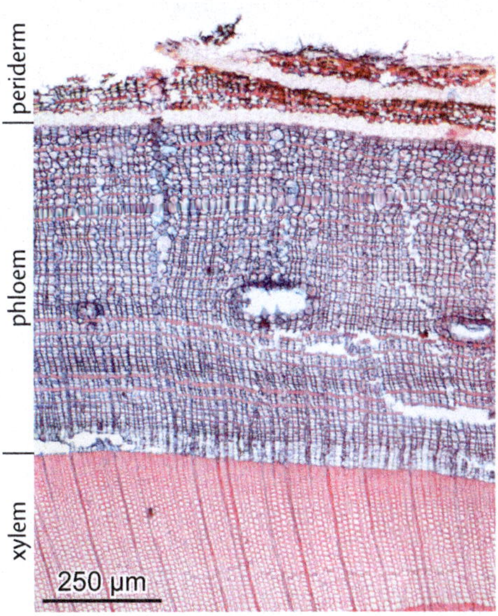

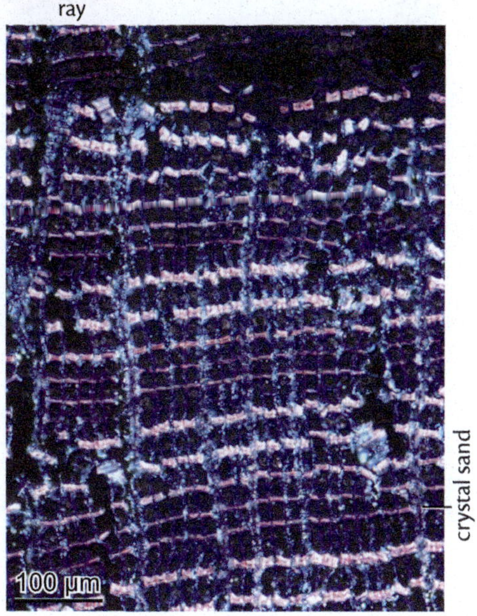

Larix sibirica Ledeb. Pinaceae

Siberian larch, Sibirische Lärche, mélèze de Siberie, larice siberica, alerce de Siberia

Form: Conifer, crown broadly columnar, later rounded, with horizontal branches and hanging twigs, to 45 m tall.

Features: Needles deciduous, soft, to 5 cm long, in star-shaped bundles of 20–40 on short buff shoots. Flowers monoecious, female cone upright, red, 1–2 cm long, male cone hanging, yellow, globular, to 2 cm long. Seed cones soft, globular, red to gray brown, to 5 cm long.

Habitat: Boreal and subarctic Russia from the White Sea eastwards to the Ural Mountains, Yenisei River and Altai Mountains, eastwards gradually being replaced by the east-Siberian *L. dahurica*.

A typical pioneer on all soil types, to 2400 m.

Bark: Thin and flaky, young yellowish or red brown, later gray brown, corky, with deep longitudinal furrows.

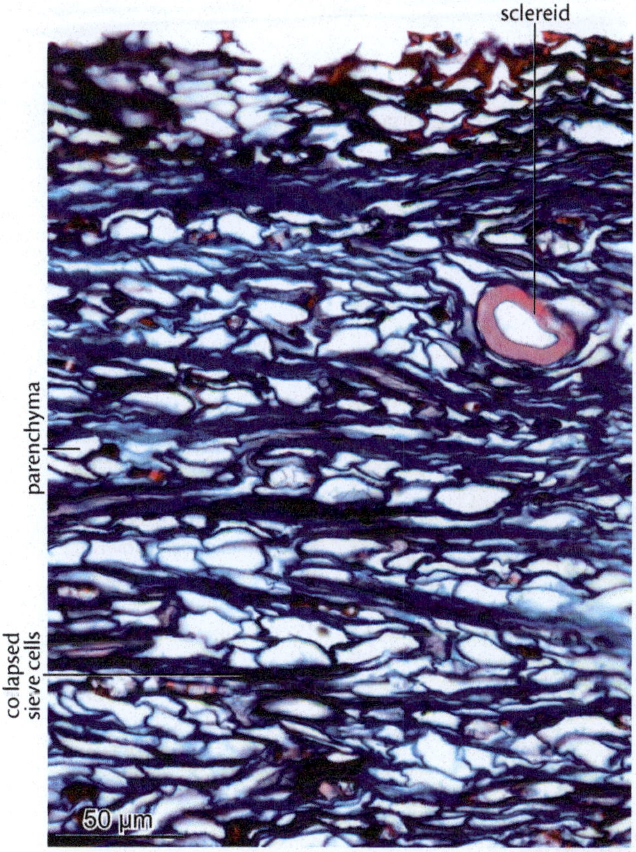

Bark anatomy: Regular tangential layers. The presence of isolated sclereids is characteristic, a rare feature in slow-growing trees.

Sieve cells rectangular, arranged in 3–5-seriate tangential rows, collapsing within a short time after formation. Parenchyma cells small, oval in the juvenile phloem, enlarged in the adult phloem. Fibers absent. Sclereids as isolated, round, very thick-walled, lignified cells. Rays radially wavy. Small prismatic crystals outside of sclereids.

Annual rings indistinct in the present material. Periderm active for several years. Phellem cells rectangular, thin-walled.

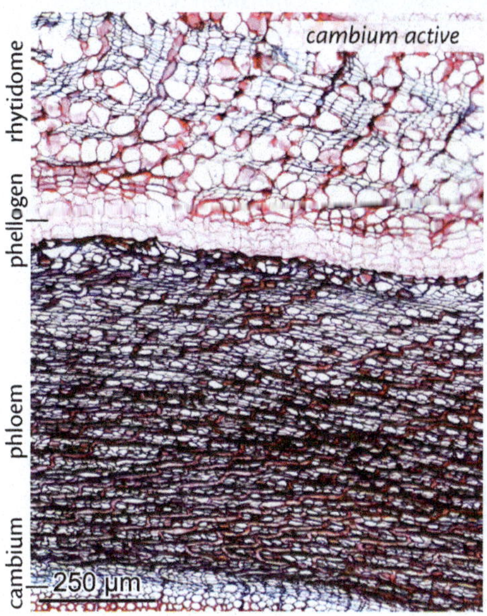

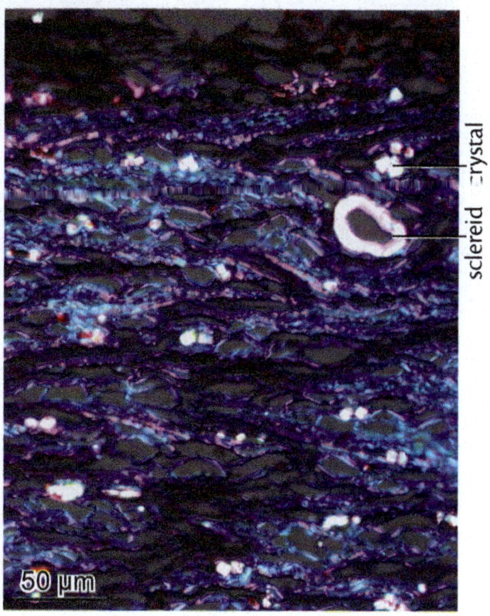

Picea abies (L.) H.Karst

Pinaceae

Norway spruce, Fichte, épicéa, peccio, pícea europaea

Form: Upright conifer, crown narrowly to broadly conical, to 60 m tall.

Features: Needles evergreen, stiff, quadrangular in cross section, to 3 cm long. Flowers monoecious, female cone upright, crimson, to 4 cm long, male cone yellowish red, to 5 cm long. Seed cones brown, hanging, soft, to 15 cm long.

Habitat: Northern Europe and the Alps, from Scandinavia and Northwest Russia south to the Balkans, North Apennines and Massif Central.

On all soil types, often dominant in boreal and subalpine forest, to 2100 m.

Bark: Young red brown, fine scaly, later tessellated-flaky, gray brown.

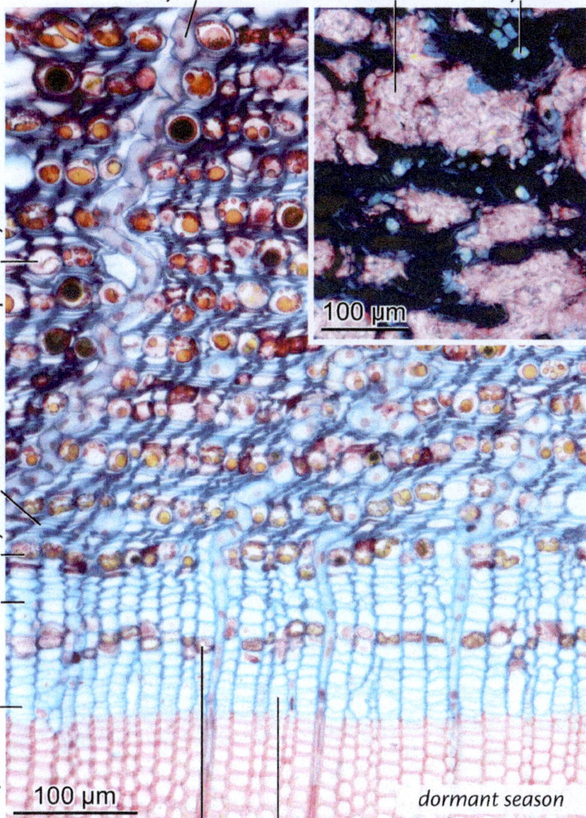

Bark anatomy: Regular tangential layers in the young phloem, adult phloem with many irregular shaped groups of sclereids.

Sieve cells rectangular, collapsing after a few years. Parenchyma cells small and oval in the juvenile phloem, enlarged in the adult phloem. Fibers absent. Sclereids rare in the young phloem, frequent in the old phloem. Rays often dilating. Crystals rare, in form of individual calcium oxalates. With resin ducts.

Annual rings distinct for a few years near the cambium. Rings indistinct in the sclerotic part of the bark.

Periderm distinctly layered, consisting of rectangular, thick-walled cells (sclerotic cork) and thin-walled, irregular shaped cells (*Schwammcork*).

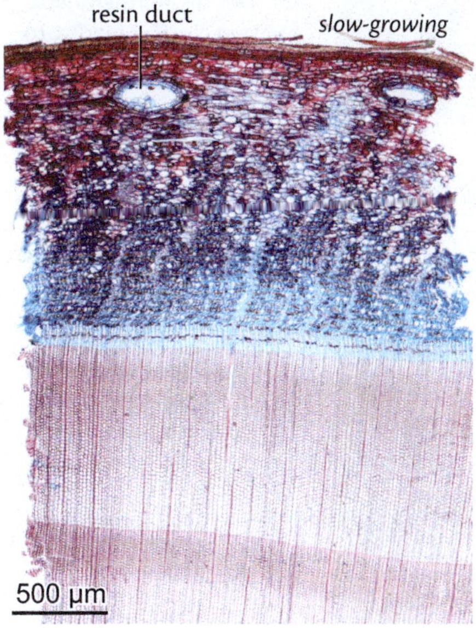

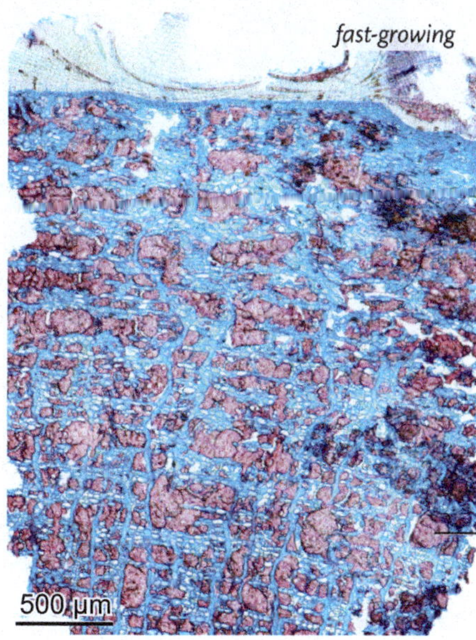

Picea glauca (Moench) Voss — Pinaceae

white spruce, Kanadische Fichte, épinette blanche, peccio bianco, pícea blanca

Form: Upright conifer, crown narrow cylindrical, becoming conical, to 40 m tall.

Features: Needles evergreen, stiff, quadrangular in cross section, glaucous gray green, to 2 cm long. Flowers monoecious, female cone upright, crimson, to 2 cm long, male cone yellowish red, to 3 cm long. Seed cones brown, hanging, soft, to 7 cm long.

Habitat: North America from Central Alaska east to Newfoundland, south to Manitoba, North Dakota, Wisconsin and Maine.

On a variety of soils, preferring river valleys, often dominant in boreal forest, forming the northernmost forest in North America, to 2400 m.

Bark: Young yellowish brown and fine scaly, later circular flaky and ash gray.

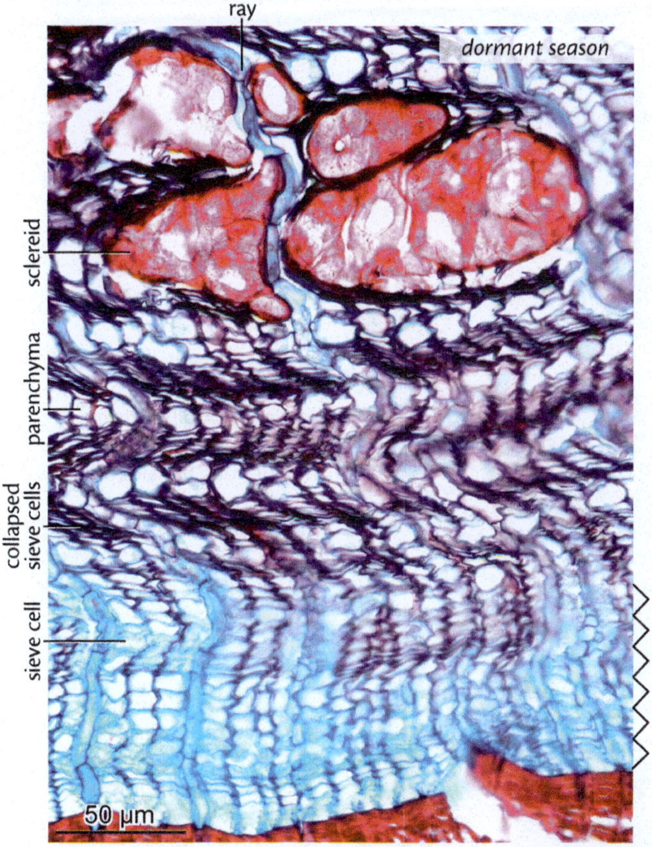

Bark anatomy: Regular tangential layers in the young phloem, adult phloem with many irregular shaped groups of sclereids.

Sieve cells rectangular, collapsing within a few years. Parenchyma cells small, oval. Fibers absent. Sclereids in round groups, rare in the young phloem, frequent in the old phloem. Crystals fairly rare, in form of single calcium oxalates. Resin ducts absent.

Annual rings indistinct. Periderm distinctly layered, consisting of rectangular, thin-walled cells.

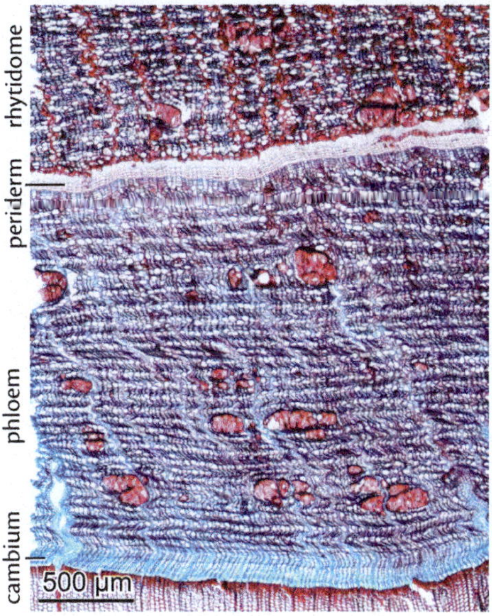

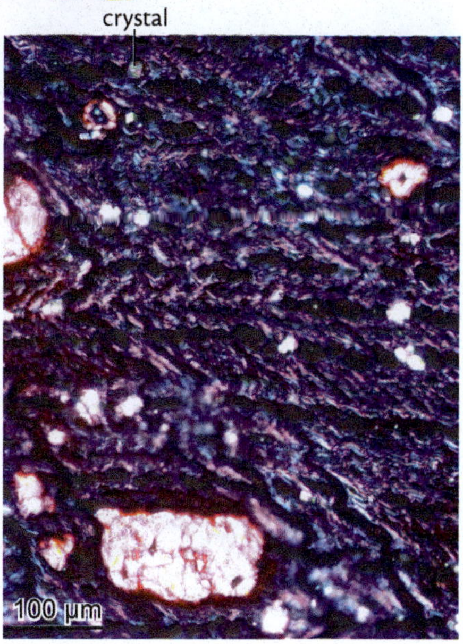

Picea mariana (Mill.) Britton, Sterns & Poggenb. Pinaceae

black spruce, Schwarz-Fichte, épinette noire, peccio nero, pícea negra

Form: Upright conifer, crown narrow and cylindrical, branches compact, drooping, to 20 m tall.

Features: Needles evergreen, stiff, quadrangular in cross section, dark blue green, to 2 cm long. Flowers monoecious, female cone upright, crimson, to 2 cm long, male cone yellowish red, to 3 cm long. Seed cones the shortest of all spruce cones, hanging, deep purple at first, later red brown, to 4 cm long.

Habitat: North America from Central Alaska east to Newfoundland, south to Manitoba, North Dakota, Wisconsin and Maine.

On a variety of soils, preferring bogs, often dominant in boreal forest, often forming even-aged stands after fires, to 2400 m.

Bark: Young red brown, finely scaly, later flaky, gray black.

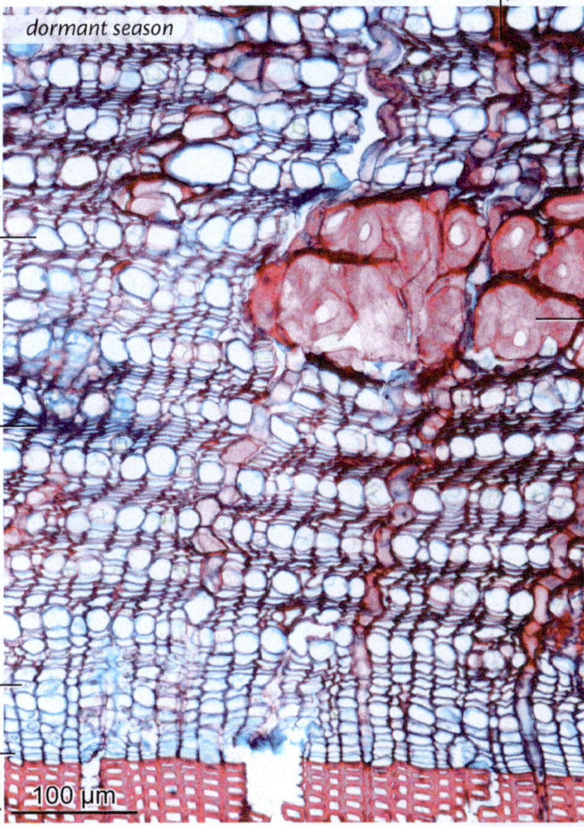

Bark anatomy: Regular tangential layers in the phloem. Adult phloem with some irregular shaped groups of sclereids.

Sieve cells rectangular, collapsing within a few years. Parenchyma cells small, oval in the juvenile phloem, enlarged in the adult phloem. Fibers absent. Sclereids rare. Crystals frequent, in form of single calcium oxalates.

Annual rings distinct. Periderm distinctly layered, consisting of rectangular, thick-walled cells (sclerotic cork) and thin-walled, irregular shaped cells (*Schwammcork*).

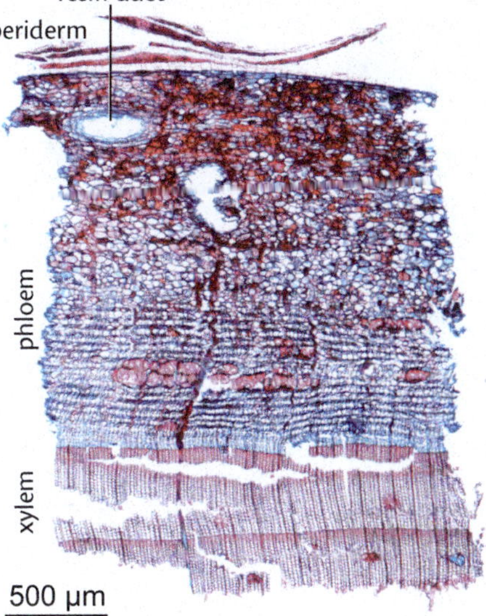

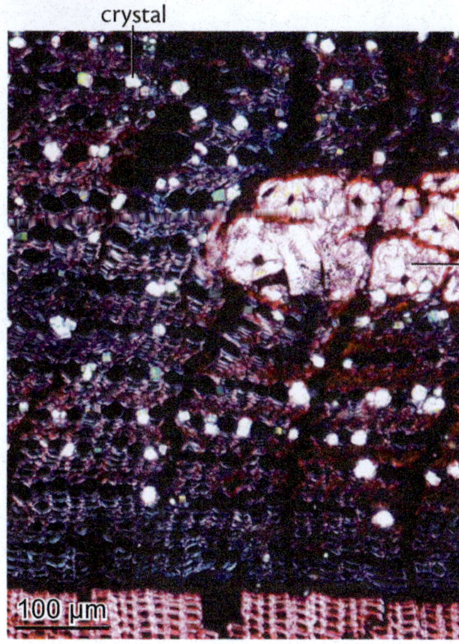

Pinus pumila (Pall.) Regel

Pinaceae

Siberian dwarf pine, Zwerg-Kiefer, pin nain de Sibérie, pino nano siberiano, pino enano siberiano

Form: Multi-stemmed coniferous shrub, shape dense, round, to 3 m tall, often forming pure stands.

Features: Needles evergreen, in bundles of five, soft, green, to 6 cm long. Flowers monoecious, female cone upright, wine red, globular, male cones yellow, crowded together below to top of twigs. Seed cones with large scales, green brown, later yellow brown, to 6 cm long.

Habitat: Northeast Asia, in Siberian Far East often forming gregarious pure stands, south to Mongolia, Korea and northern Japan.

Often on rocky or sandy soils, to 2000 m.

Bark: Young brown, thin and smooth or finely scaly, later gray brown.

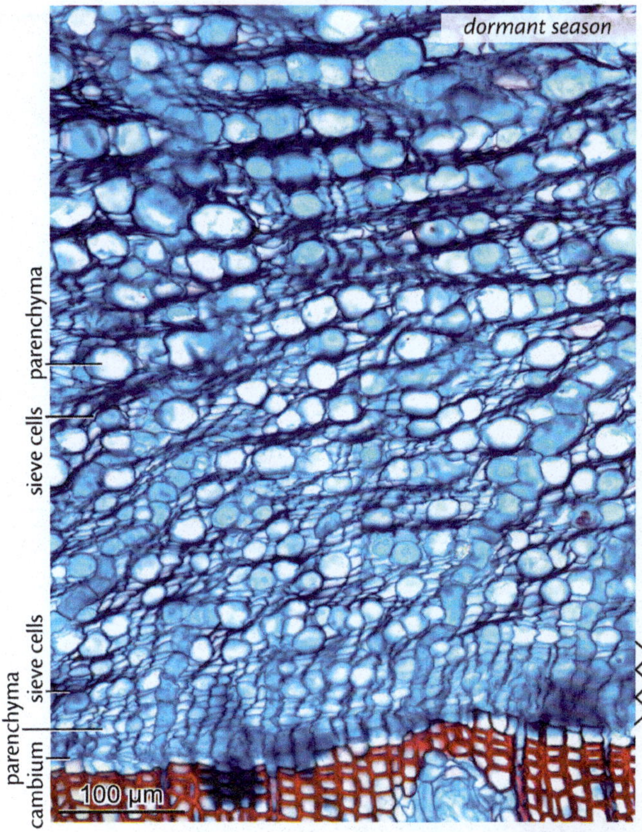

Bark anatomy: Irregular tangential layers in the young phloem. Absence of fibers and isolated sclereids is characteristic.

Sieve cells rectangular, arranged in tangential rows, collapsing within a few years. Parenchyma cells small and oval in the juvenile phloem, enlarged in the adult phloem. Sclereids and fibers absent. Rays radially wavy. Resin ducts rare or absent in the older phloem.

Annual rings indistinct. Phellem irregularly layered, consisting of rectangular thick-walled (sclereids) and thin-walled cells. A few prismatic crystals in the phloem.

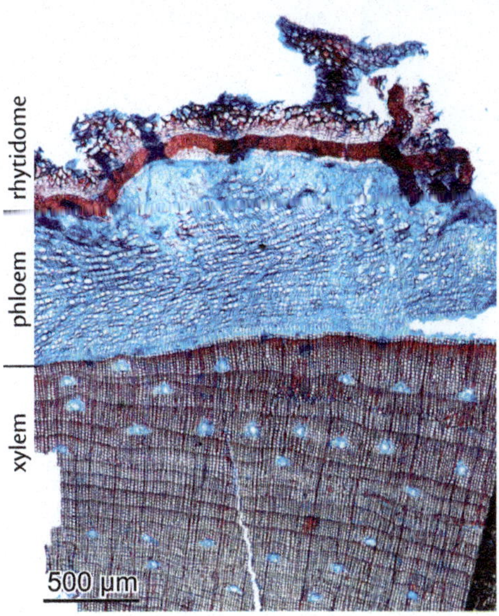

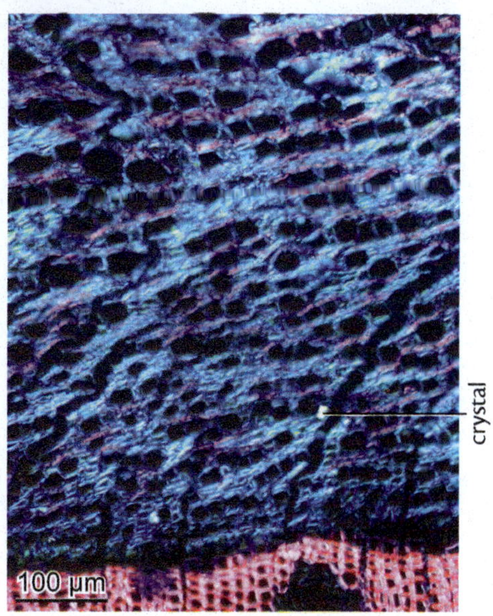

Pinus sylvestris L. Pinaceae

Scots pine, Wald-Kiefer, pin sylvestre, pino silvestre, pino silvestre

Form: Upright conifer, crown conical, becoming umbrella-shaped with age, to 45 m tall.

Features: Needles evergreen, in bundles of two, stiff, dark green, to 7 cm long. Flowers monoecious, female cone upright, carmine, globular, male cones yellow, crowded together below to top of twigs. Seed cones brown, later hanging, to 7 cm long.

Habitat: Eurosiberia, from Spain northeast to Scandinavia and the Siberian Pacific coast, together with aspen the largest distribution area of any tree species.

Usually on rocky, sandy or boggy soils, avoiding competition with other species on deeper soils, to 2600 m.

Bark: Thick, scaly, dark gray brown on the lower trunk, characteristically thin, flaky, orange brown on the upper trunk.

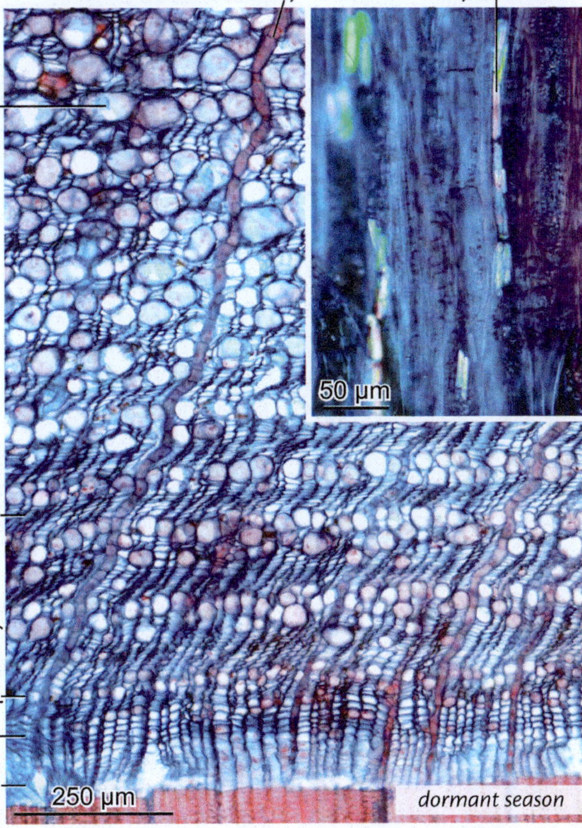

Bark anatomy: Regular tangential layers in the young phloem. Characteristic absence of fibers and isolated sclereids.

Sieve cells rectangular, arranged in 3–5-seriate, tangential rows, collapsing within a few years. Parenchyma cells small and oval in the juvenile phloem, enlarged in the adult phloem. Sclereids and fibers absent. Rays radially wavy. With elongated prismatic crystals outside of rays, mostly in axially elongated cells (crystal tubes); see radial section.

Annual rings distinct for a few years near the cambium. Rings indistinct in the adult phloem due to sieve-cell collapse and dilatation of parenchyma cells.

Phellem irregularly layered, consisting of rectangular thick-walled (sclereids) and thin-walled cells.

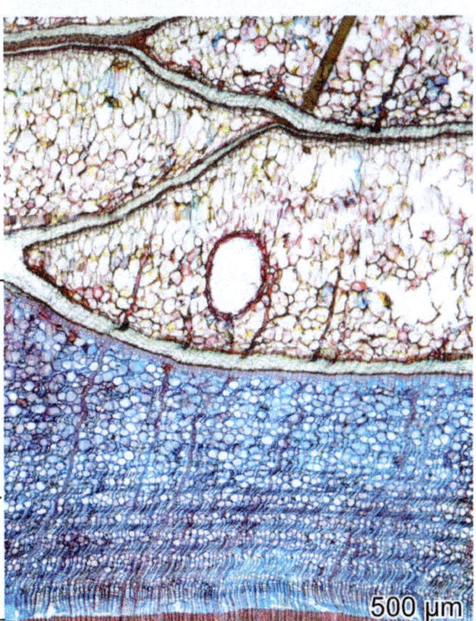

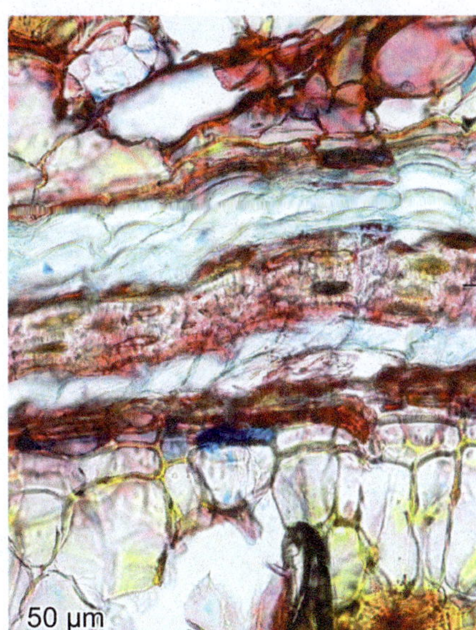

Arctostaphylos uva-ursi (L.) Spreng. Ericaceae

bearberry, Immergrüne Bärentraube, busserole, uva ursina, uva de oso

Form: Low-spreading dwarf shrub with long running twigs, often growing as a hanging carpet over rocks, 10 cm tall.

Features: Leaves evergreen, alternate, flat, ovate and leathery, 1–3 cm long. Flowers hanging white bells with pink petal tips. Fruit a red, berry-like drupe.

Habitat: Subarctic Eurosiberia and North America, south to Spain, Turkey, Mongolia, California and the Great Lakes.

Mostly on dry, sandy or rocky soils in full sun, often gregarious in subarctic tundra or alpine dwarf shrub communities, to 2800 m.

Bark: The papery, shedding bark with its often intense orange brown color is unmistakable.

dormant season

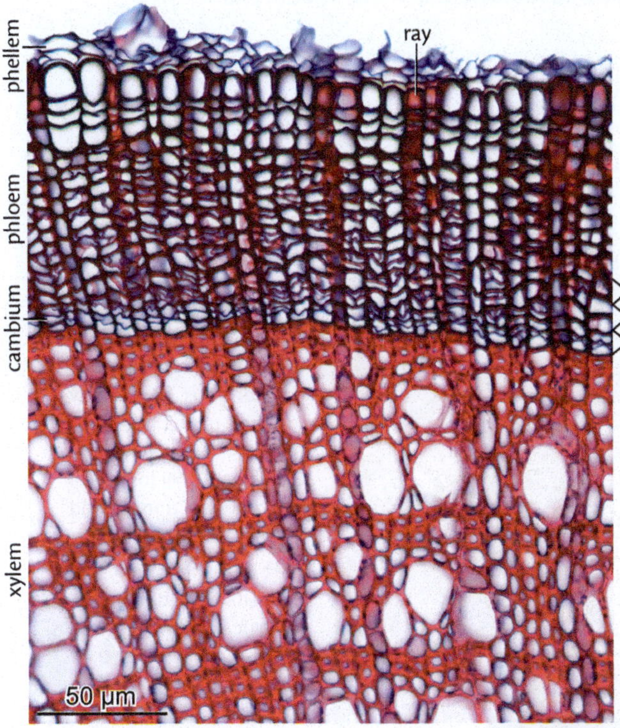

Bark anatomy: Phloem poorly structured. Sieve tubes in small groups. Fibers absent. Crystal druses irregularly distributed.

Cortex consisting of irregular shaped parenchyma. Old phellem cells thick-walled, square (not in the images).

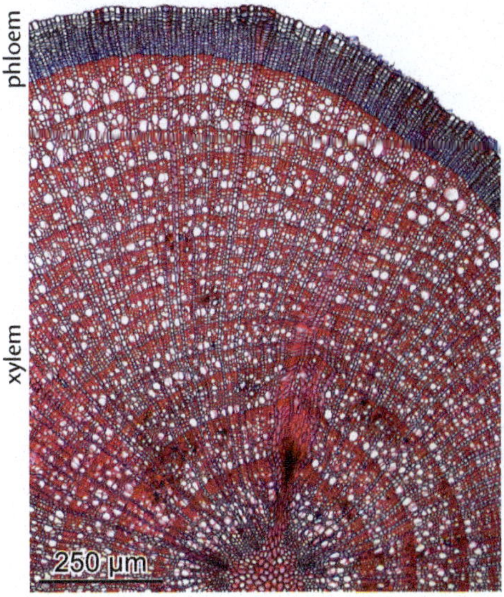

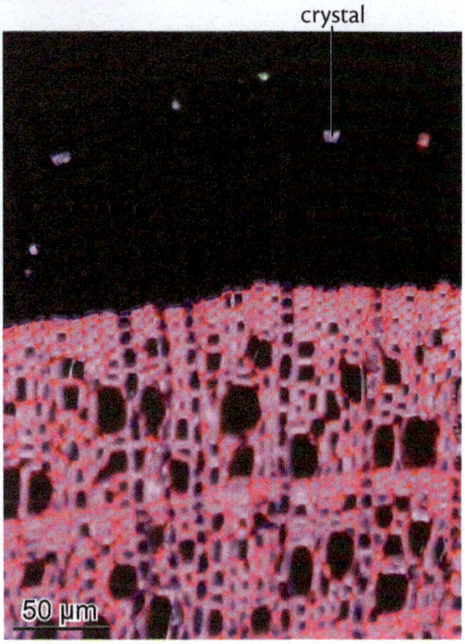

Betula nana L.

Betulaceae

dwarf birch, Zwerg-Birke, bouleau nain, betulla nana, abedul enano

Form: Low-growing, intensely branched dwarf shrub, 1 m tall.

Features: Leaves deciduous, alternate, round ovate, to 12 mm long, with sharply toothed margins. Flowers monoecious, female catkin upright, red to green, to 1 cm long, male catkin upright, yellow, to 2 cm long. Fruit a red brown, upright and nearly globose catkin.

Habitat: Subarctic Eurosiberia and North America, south to the Alps, Southern Siberia, British Columbia and Labrador.

In moist to wet, acidic bogs and subarctic tundra, to 1200 m.

Habitat: Young smoothly red gray, later glossy, reddish dark gray, with horizontal lenticels.

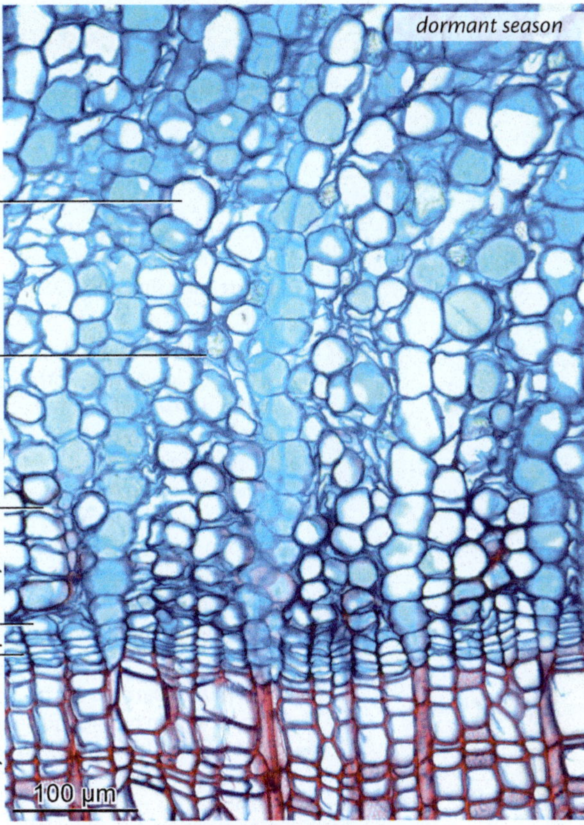

Bark anatomy: Homogeneous sieve tube-parenchyma structure with a few dense groups of sclereids.

Sieve tubes irregular shaped. Sieve tubes collapsing very soon after formation. Parenchyma cells roundish. Fibers absent. With many groups of sclereids. Ray dilatation moderate. Prismatic crystals rare, around the sclereids.

Periderm consisting of many layers of rectangular, large, thin-walled cells and small, thick-walled cells.

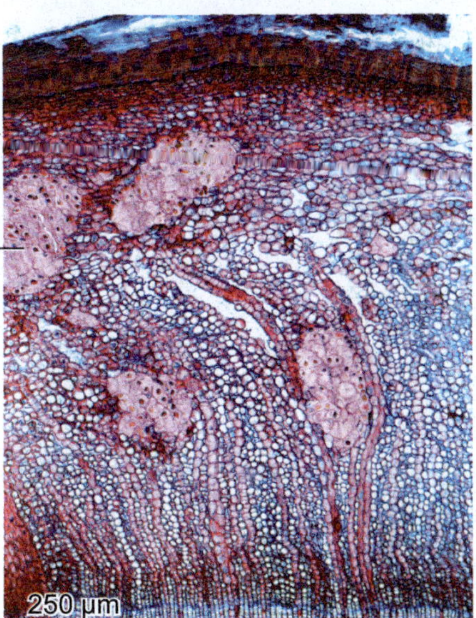

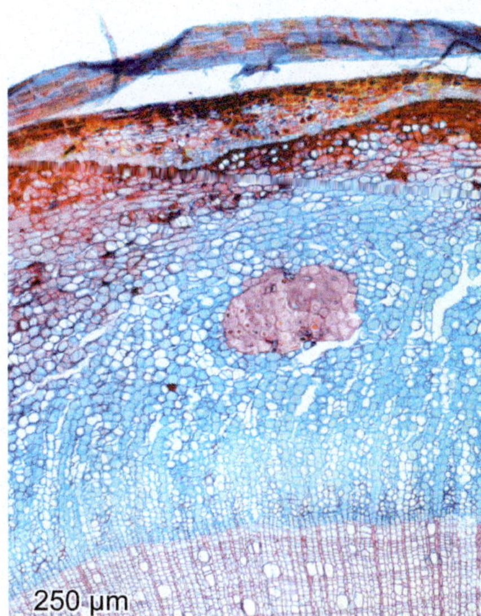

Betula papyrifera Marshall Betulaceae

paper birch, Papier-Birke, bouleau papier, betulla della carta, abedul de papel

Form: Upright tree, often multi-trunked, crown round to irregular shaped, to 25 m tall.

Features: Leaves deciduous, alternate, long ovate, with a round base and pointed tip, with sharply toothed margins, to 7 cm long. Flowers monoecious, female catkin upright, greenish, to 4 cm long, male catkin hanging, yellow brown, to 10 cm long. Fruit a brown, hanging catkin, to 10 cm long.

Habitat: Subarctic North America from Newfoundland to Western Alaska, south to Pennsylvania, Michigan and Idaho.

On dry to wet soils, acidic bogs to limestone, often gregarious after fire, to 1500 m.

Bark: Young bronze brown, later snowy white, papery, peeling in horizontal strips, curling at the ends, bark on the lower trunk can be black.

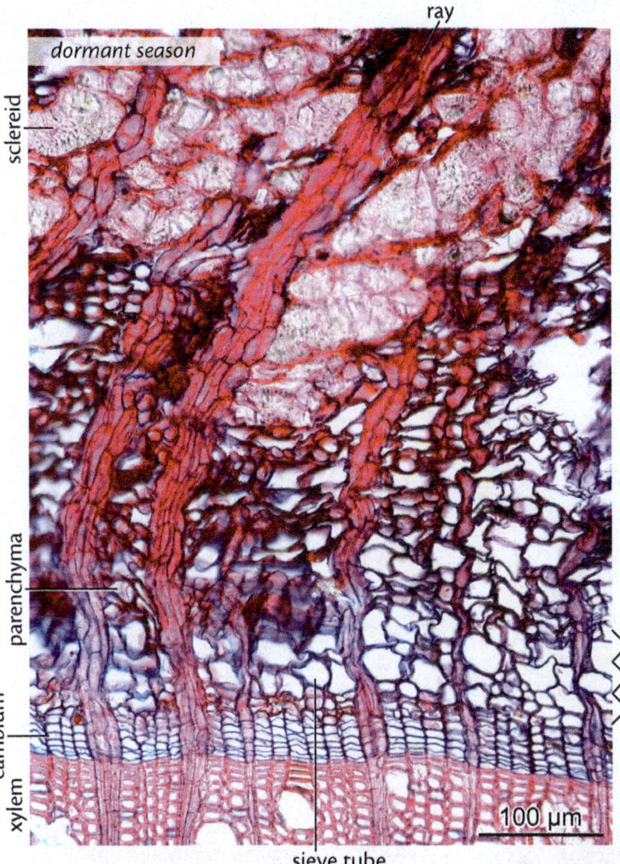

Bark anatomy: Slightly tangentially layered groups of sclereids.

Sieve tubes irregular shaped, collapsing very soon after formation. Parenchyma cells roundish. Fibers absent. Sclereids in groups, often in tangential layers. Ray cell-dilatation without cell divisions. Crystals absent.

Periderm consisting of many layers of rectangular, large, thin-walled cells and small, thick-walled cells.

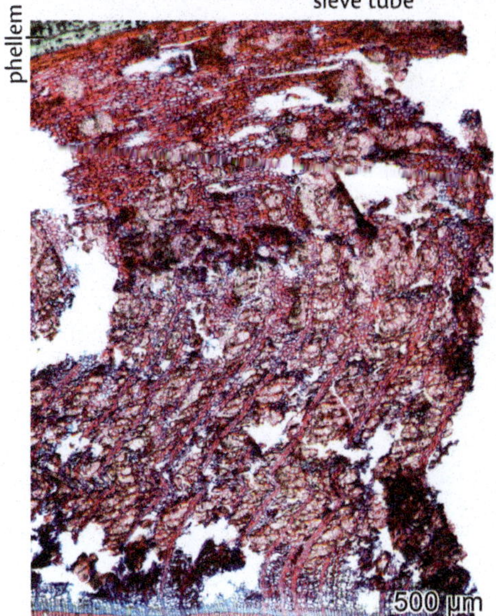

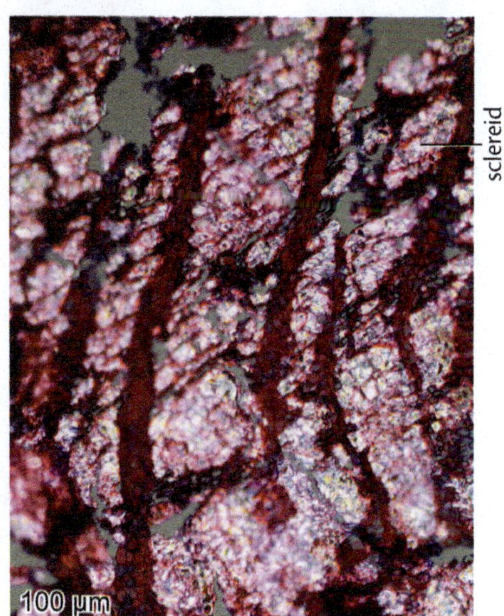

Hippophae rhamnoides L. Eleagnaceae

sea buckthorn, Sanddorn, argousier, olivella spinosa, espino amarillo

Form: Very dense, thorny shrub with rounded crown, rarely a small tree, to 6 m tall.

Features: Leaves deciduous, alternate, linear, to 6 cm long, dull gray green above, silver gray to red brown below, hairy, with an entire margin. Flowers dioecious, female and male very small and globose, yellow brown, to 5 mm wide. Fruit only on female plants, orange, fleshy, berry-like drupe, to 8 mm long.

Habitat: Eurosiberia, from France to Northern China and Siberia, south to Italy, Turkey, the Himalayas and Szechuan, north to Norway.

On dry to seasonally wet, gravelly, sandy and often instable soils in river beds, on gravelly slopes or coastal sand dunes, to 1800 m, in the Himalayas to 4200 m.

Bark: Young glossy, silvery or dark red brown, later light gray brown, with fine longitudinal fissures.

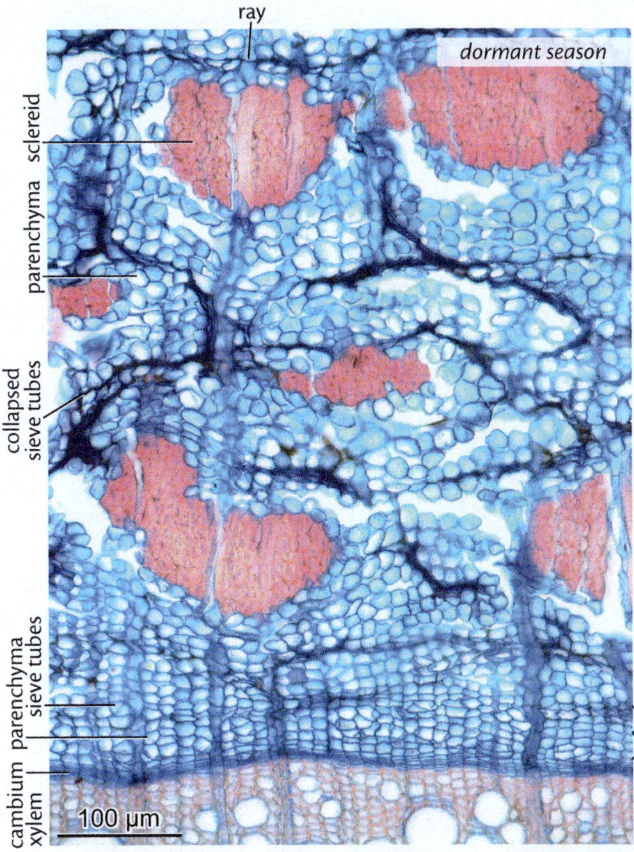

Bark anatomy: Irregular, unlignified tissue with many round to oval groups of sclereids.

Sieve tubes angular, parenchyma roundish. Sieve tubes collapsing, indicated by dark, irregular lines. Fibers absent. Sclereids in round to oval groups. Rays wavy, disappearing after a short time. Dilatations absent. Crystals absent or rare, associated with sclereids. Annual ring boundaries unrecognizable.

Phellem consisting of layers of irregular zones of large, thin-walled cells.

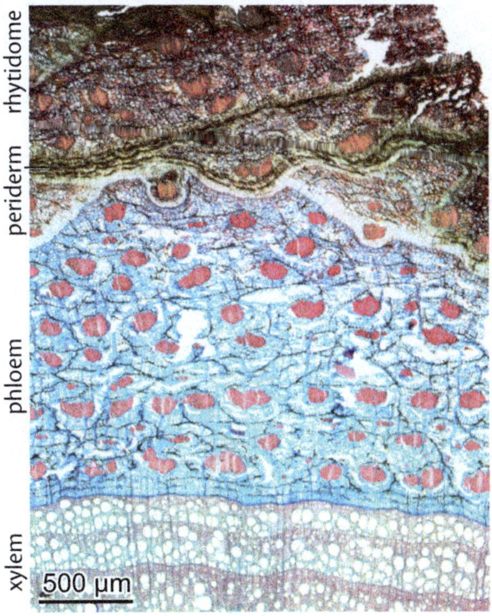

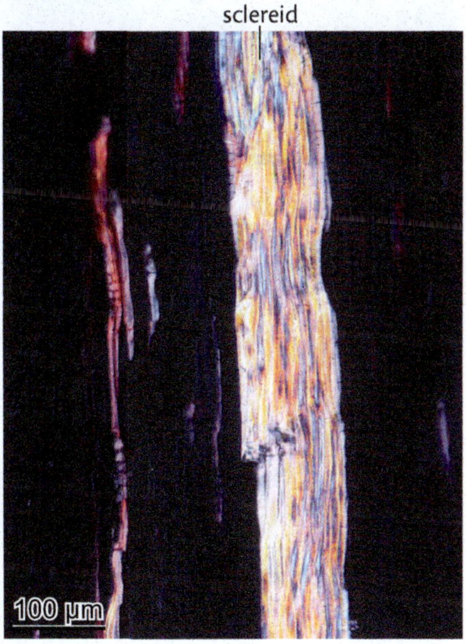

Myrica gale L. Myricaceae

bog myrtle, Gagelstrauch, galé odorante, mirto de turbera

Form: Deciduous, upright dwarf shrub, to 150 cm tall.

Features: Leaves alternate, long ovate, strongly aromatic, to 6 cm long. Flowers dioecious, female cone upright, crimson, to 6 mm long, male cone yellow brown, to 15 mm long. Fruit a brown capsule with yellow wax glands, splitting into three parts.

Habitat: Northwestern Europe and Northeastern America, from Portugal to Scandinavia and from Labrador to the Appalachians and Great Lakes.

Only in acidic bogs, to 300 m.

Bark: Light gray or red brown to dark brown, with light-colored lenticels and oil glands.

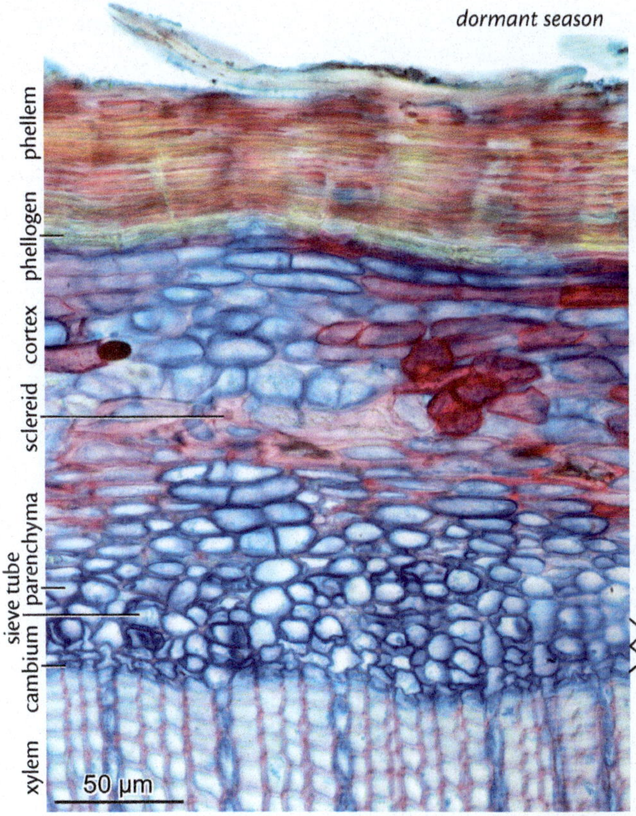

Bark anatomy: Bark narrow, consisting of a small phloem and cortex and a relatively large periderm.

Sieve tubes angular, accompanied by round to oval parenchyma cells. Sclereids irregular shaped, occuring in a continuous band. Rays dilating slightly. Small prismatic crystals and crystal druses in the sclereid belt.

Phellem cells rectangular, thick-walled, filled with dark-staining substances, arranged in radial rows.

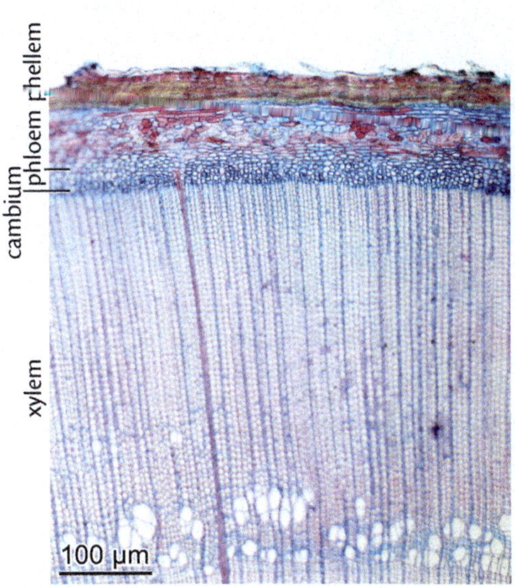

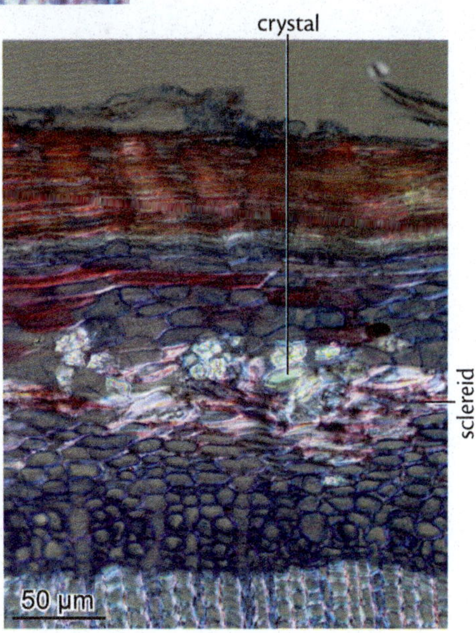

Populus tremula L. Salicaceae

aspen, Zitter-Pappel, tremble, pioppo tremulo, álamo temblón

Form: Upright, often multi-stemmed tree, crown open, pyramidal, to 20 m tall.

Features: Leaves deciduous, alternate, triangular or rhombic with an irregular, bluntly toothed margin, to 7 cm long. Flowers dioecious, female catkin hanging, green, to 4 cm long, male catkin gray red, to 10 cm long. Fruit a hanging, green catkin, to 10 cm long.

Habitat: Eurosiberia, from Morocco northeast to Iceland, North Cape and the Siberian Pacific coast, southeast to Central Asia and Northwest China, the largest distribution of any tree species, together with Scots pine.

Usually a fast-growing pioneer in full sun on rocky, sandy or bare and unstable soils, to 2300 m.

Bark: Young smooth, yellow brown, soon turning into greenish gray, with irregular, dark gray scales and short, longitudinal fissures.

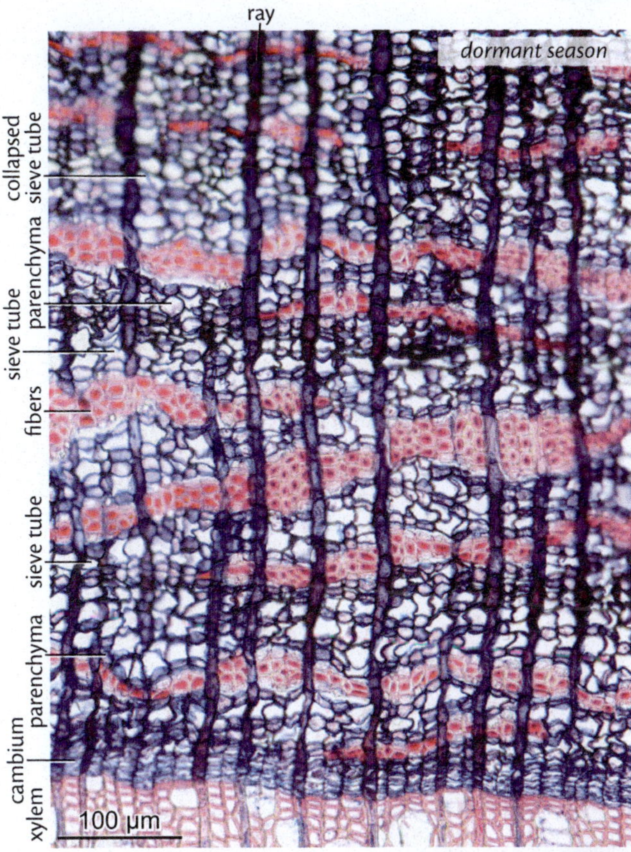

Bark anatomy: Regular tangential layers of parenchyma, sieve tubes and fibers.

Sieve tubes irregular shaped, companion cells small. Sieve tubes collapsing within a few years. Parenchyma cells roundish. Fibers in tangential layers. Sclereids absent. Prismatic crystals and crystal druses occuring along fiber bands. Tangential bark layers in fast-growing individuals not corresponding with annual growth rings.

Periderm tangentially layered, consisting of rectangular, large, thin-walled cells and small, thick-walled cells.

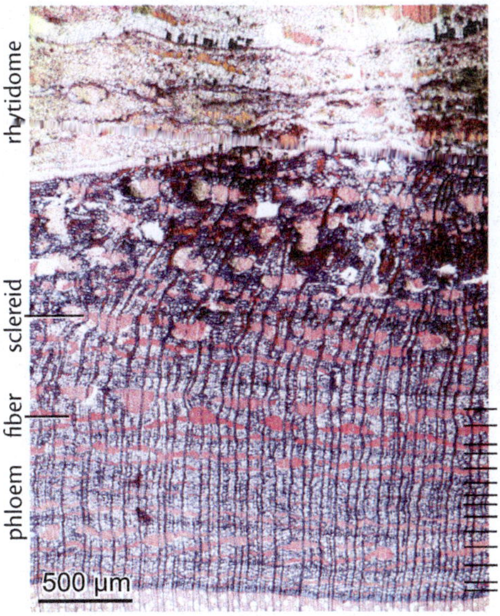

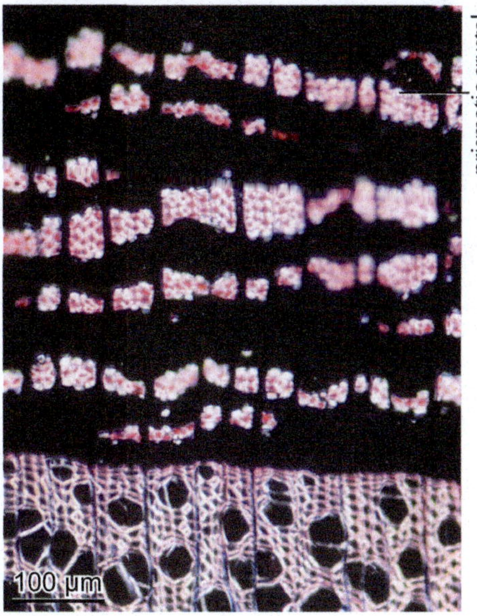

Rhododendron dauricum L. Ericaceae

Dahurian rhododendron, Dahurische Azalee, rhododendron dahuirique, rododendro daurico, rododendro daurico

Form: Semi-evergreen shrub, crown broad, lax, ramified, to 3 m tall.

Features: Leaves alternate, round-elliptic, strongly aromatic, to 4 cm long. Flowers magenta, in lax corymbs, striking with their impressive early spring display. Fruit a small brown capsule, splitting into five-parted stars.

Habitat: Northeast Siberia, south to northern Japan, Korea, Manchuria and Mongolia.

Usually as an understory to larch and birch forests on a variety of soils, to 2000 m.

Bark: Young light brown, smooth, later gray brown.

cambium active

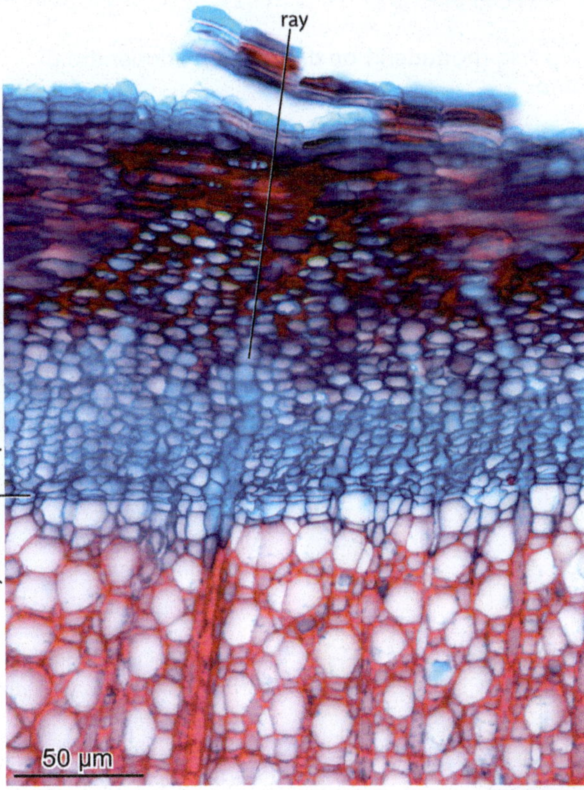

Bark anatomy: Phloem poorly structured. Sieve tubes and parenchyma cells impossible to differentiate. Fibers absent. Crystals absent.

Cortex consisting of irregular shaped parenchyma. Phellem cells thick-walled, square.

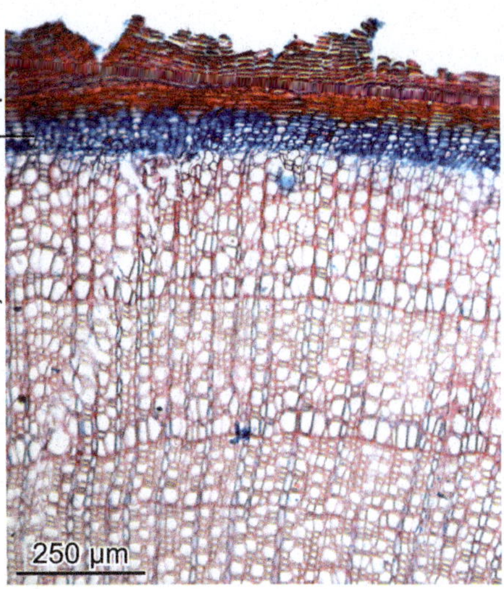

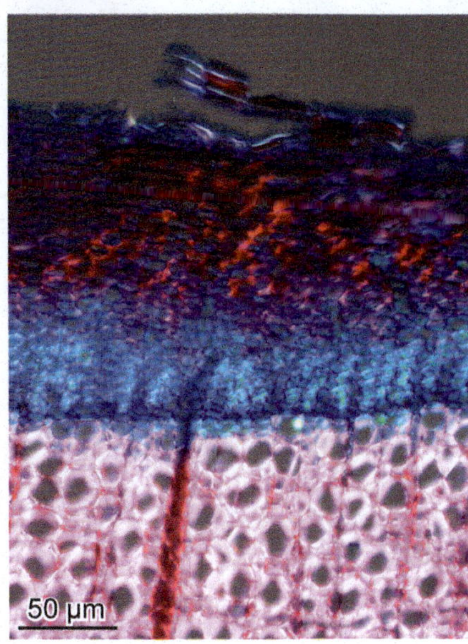

Rhododendron tomentosum Harmaja Ericaceae

European Labrador tea, Sumpfporst, rhododendron des marais, rododendro di palude, rododendro del pantano

Form: Evergreen dwarf shrub, crown broad, densely ramified, to 150 cm tall.

Features: Leaves alternate, linear, strongly aromatic, to 5 cm long, cinnamon brown indumentum below. Flowers white, fragrant, with long stamens in dense corymbs. Fruit a small brown capsule, splitting into a five-parted star.

Habitat: Northeastern Europe and Siberia, from Northern Germany eastwards to Sakhalin, Korea and Northeast China.

Only in acidic bogs, to 1500 m.

Bark: Young light brown, smooth, later gray brown.

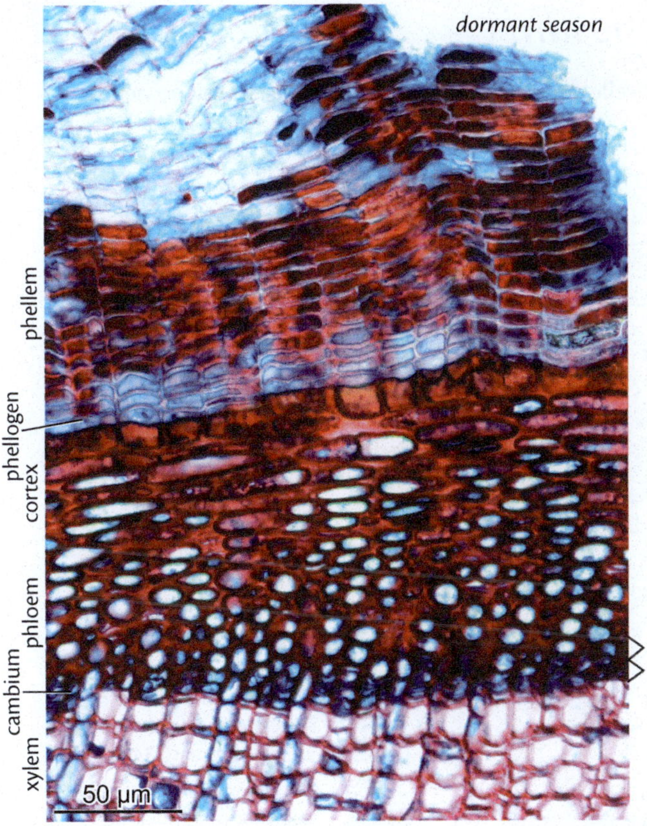

Bark anatomy: Phloem poorly structured. Sieve tubes and parenchyma difficult to distinguish. Fibers and crystals absent. Rays uniseriate, a few dilating.

Phellem cells thin-walled, rectangular, filled with dark-staining substances.

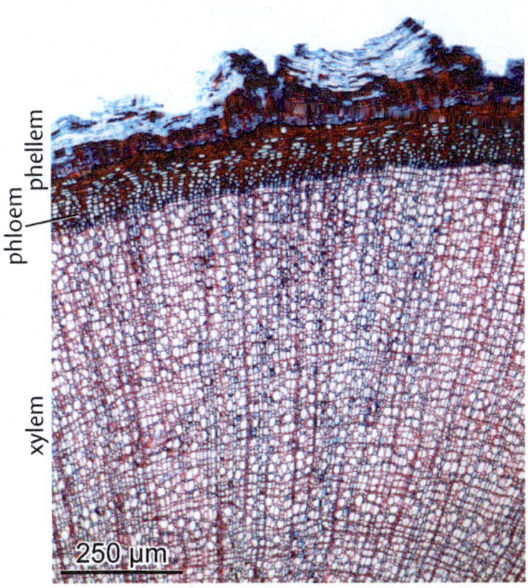

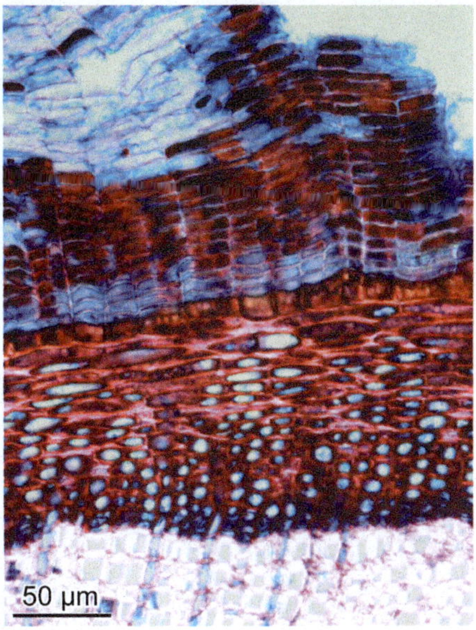

Vaccinium uliginosum L. Ericaceae

bog bilberry, Moorbeere, airelle des marais, mirtillo di palude, arándano negro

Form: Deciduous dwarf shrub, forming densely ramified mats, to 1 m tall.

Features: Leaves alternate, round ovate, blue green and gray green below, to 3 cm long. Flowers pinkish white, hanging bells. Fruit berry-like, fleshy and blue black.

Habitat: Circumpolar Eurosiberia and North America, in mountains south to Spain, the Balkans, Turkey, Korea, Japan, California and Indiana.

On moist to wet, acidic soils, mostly in subarctic and alpine tundra, to 3400 m.

Variation: Smaller subalpine and tundra populations are sometimes regarded as a separate species, *V. gaultheroides*.

Bark: Light or red brown, slightly glossy and glabrous.

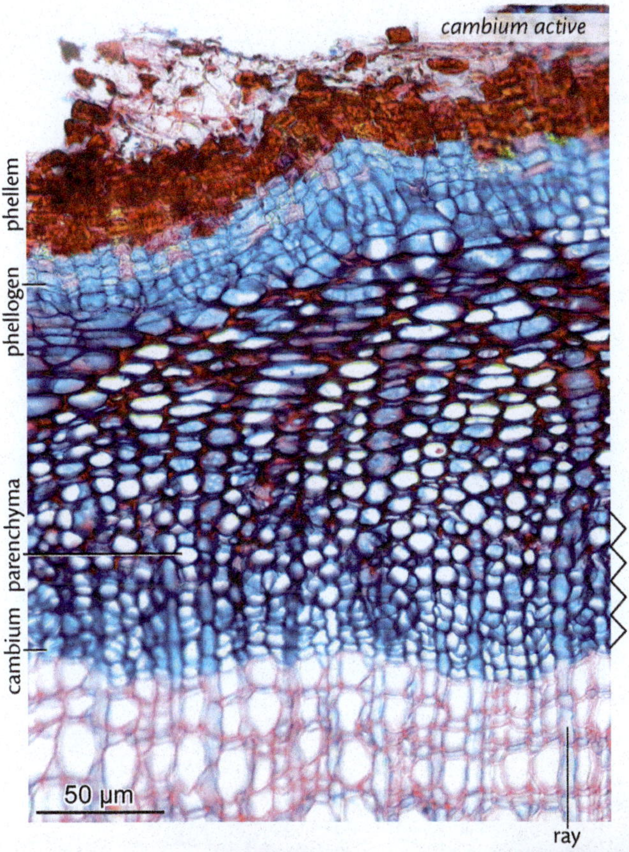

Bark anatomy: Phloem poorly structured. Sieve tubes and parenchyma difficult to distinguish. Fibers absent. Rays straight, dilating, laterally separating sieve tube and parenchyma zones. Crystals absent.

Periderm divided into a living, thin-walled phelloderm and an external zone of phellem. Phellem cells square, filled with dark-staining substances.

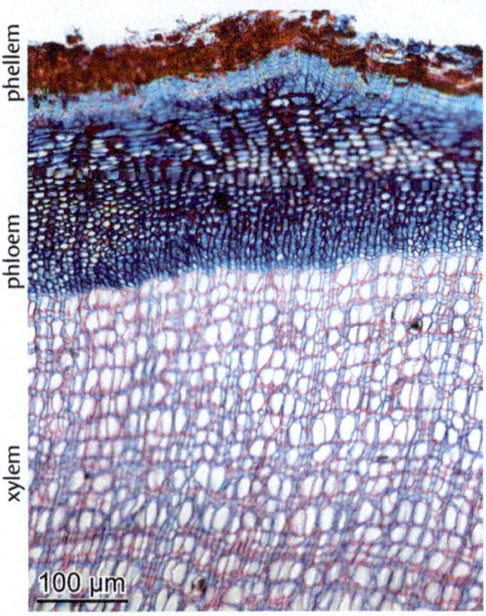

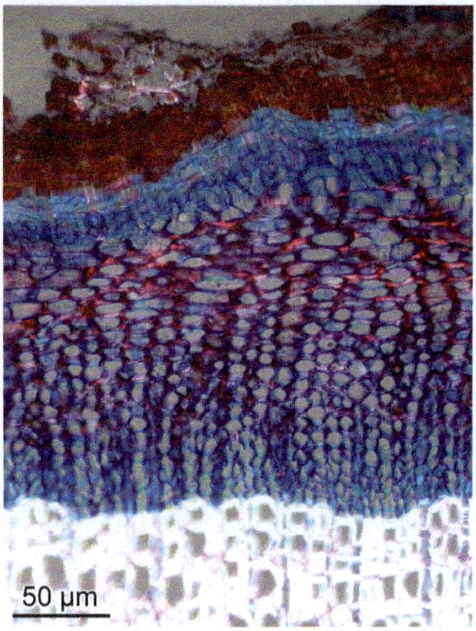

The **subarctic** biome is a circumpolar, treeless belt north of the taiga. This book includes only dwarf shrub communities, with most of the species having a circumpolar or at least Eurasian range.

Ecologically similar conditions can be found in **subalpine** dwarf shrub communities of high mountain ranges above the tree line, such as the Alps, the Caucasus, the Altai Mountains, Siberian and Alaskan mountain ranges, the Northern Rockies or mountain tops in Maine.

F. H. Schweingruber et al., *Bark Anatomy of Trees and Shrubs in the Temperate Northern Hemisphere*, https://doi.org/10.1007/978-3-030-14056-4_3

3 Subalpine and subarctic dwarf shrubs

Arctostaphylos alpina (L.) Spreng. Ericaceae

alpine bearberry, Alpen-Bärentraube, raisin d'ours des Alpes, corbezzolo alpino, manzanita alpina

Form: Low, spreading dwarf shrub, to 10 cm tall.

Features: Leaves deciduous, alternate, ovate, net-veined, leathery, to 3 cm long, deep red in autumn. Flowers small, greenish to pinkish bells, to 6 mm long. Fruit red at first, later nearly black, a berry-like drupe, to 8 mm wide.

Habitat: Subarctic Eurosiberia and North America, south to Spain, Turkey, Mongolia, California and New Hampshire.

Mostly on moist, slightly acidic and stony soils in subarctic or alpine tundra with a long-lasting snow cover, to 3000 m.

Bark: Young gray brown, later reddish, with longitudinal fissures.

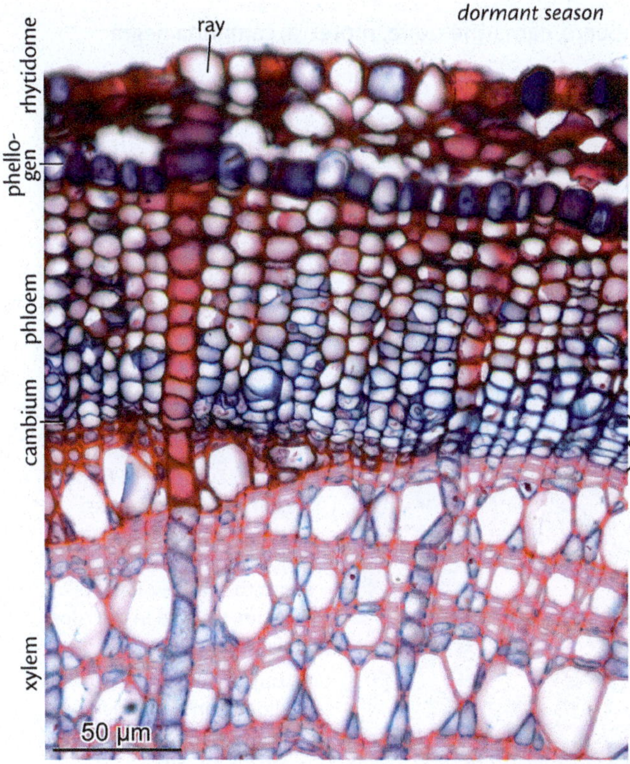

Bark anatomy: Phloem poorly structured. Sieve tubes and parenchyma difficult to distinguish. Fibers absent. Rays straight, laterally separating sieve tube and parenchyma zones. Crystals absent.

Cortex and phellem flaking.

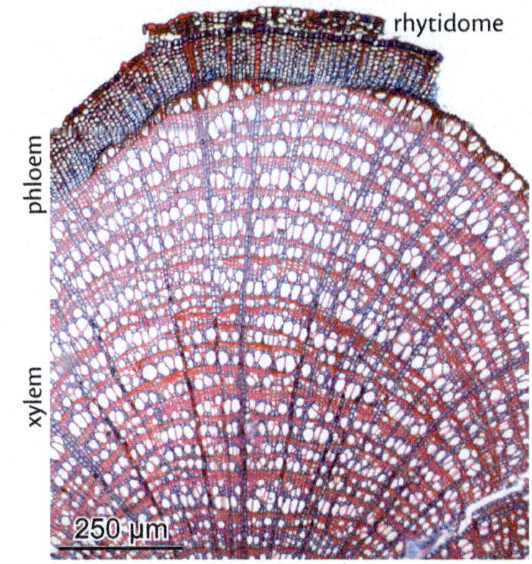

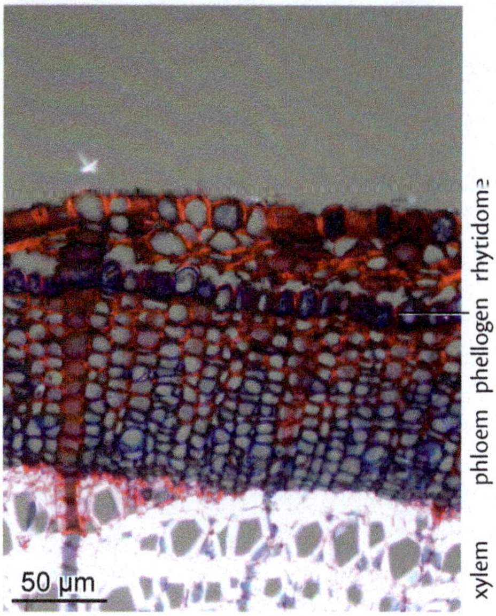

Empetrum nigrum L. Ericaceae

crowberry, Schwarze Krähenbeere, camarine noire, moretta, camarina negra

Form: Low, spreading dwarf shrub, twig tips erect, to 20 cm tall.

Features: Leaves evergreen, alternate, dark to olive green, short linear, needle-like, to 5 mm long. Flower a small, wine red star, with three sepals and petals, to 5 mm wide. Fruit a glossy, blue black, berry-like drupe, to 8 mm wide.

Habitat: Eurosiberia, from Iceland and Scotland east to the Ural Mountains and disjunct in Northern Siberia, south to Moscow, Poland, Belgium, isolated in the Alps and Carpathians.

Mostly on moist to wet, acidic soils in subarctic or alpine tundra, bogs or coastal heath, to 3100 m.

Variation: The tundra form *Empetrum hermaphroditum* is often distinguished from *E. nigrum* in lowland bogs and coastal heath.

Bark: Young green or red, later light gray.

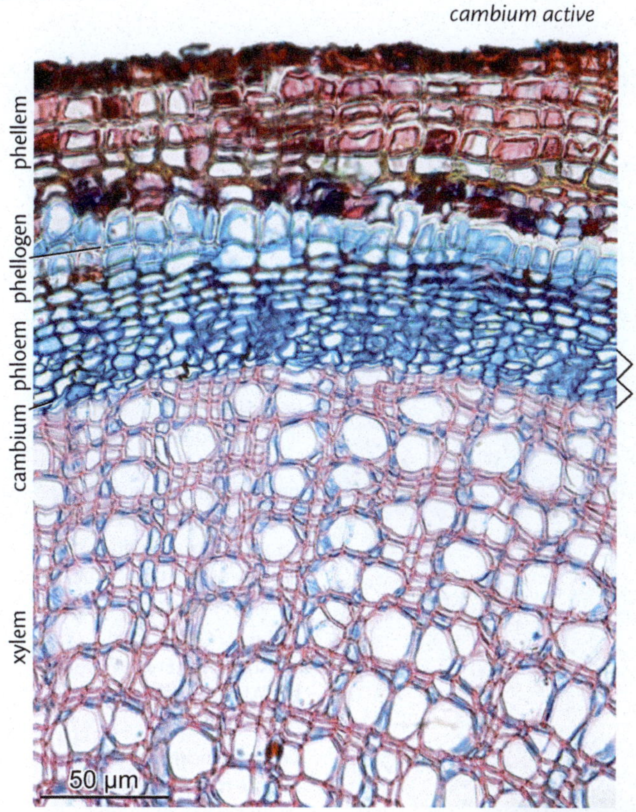

Bark anatomy: Phloem poorly structured. Sieve tubes and parenchyma difficult to distinguish. Fibers absent. Crystals absent.

Periderm divided into a living and a dead part, living part consisting of square, unlignified cells, dead part consisting of tangentially layered cells.

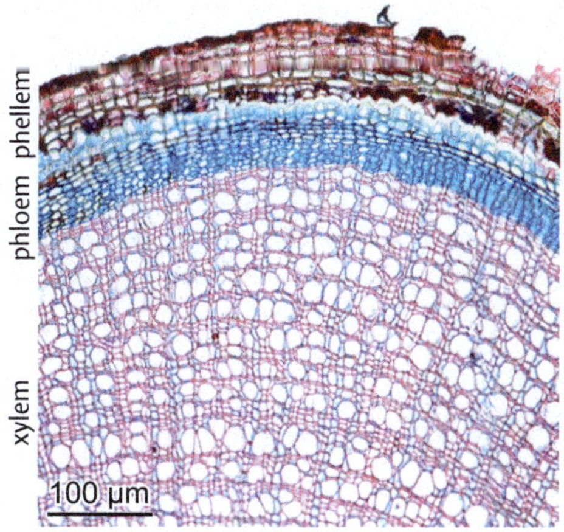

Erica carnea L. Ericaceae

winter-flowering heath, Schneeheide, bruyère carnée, scopina, brezo de invierno

Form: Low, spreading dwarf shrub, twig tips erect, to 30 cm tall.

Features: Leaves evergreen, verticillate at four, short linear, needle-like, to 1 cm long. Flowers small, carmine pink, in dense, one-sided racemes, to 7 mm wide. Fruit a small capsule, hidden in the remaining dry calyx tube, to 2 mm wide.

Habitat: Central and southeastern Europe, from the Alps south to the Balkans and Apennines.

On dry, calcareous soils in pine forests and alpine tundra, often over limestone rock or debris, to 2600 m.

Bark: Young dull, light gray, turning into red brown with age.

dormant season

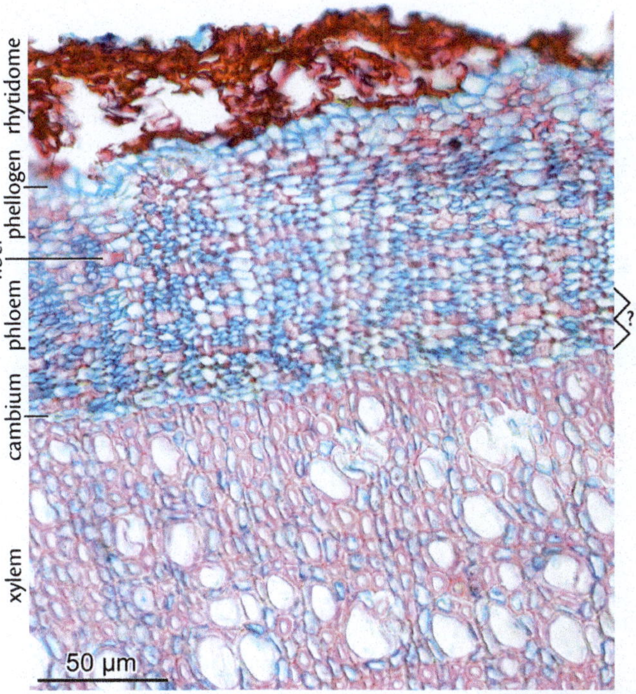

Bark anatomy: Phloem poorly structured. Sieve tubes and parenchyma difficult to distinguish. Fibers solitary.

Rhytidome consisting of irregular layered phloem and phellem cells.

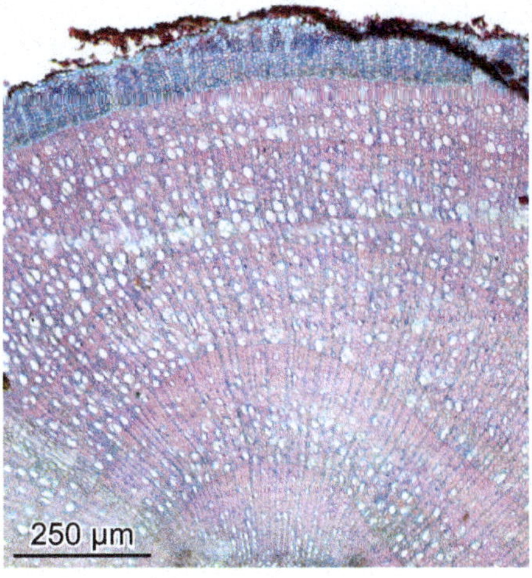

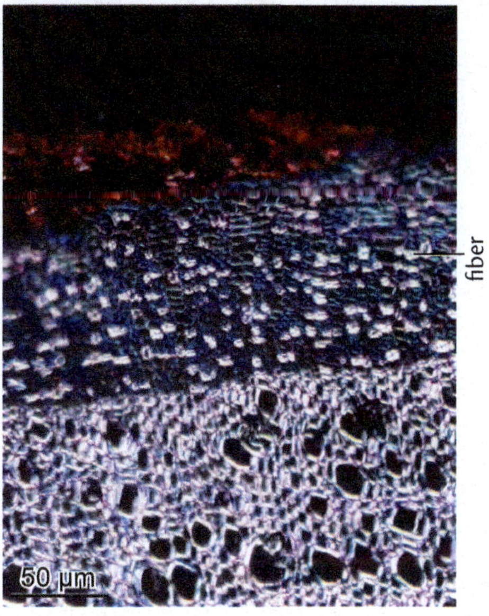

Loiseleuria procumbens (L.) Desv. — Ericaceae

trailing azalea, Alpenazalee, azalée des Alpes, azalea alpina, Loiseleuria

Form: Very low, spreading dwarf shrub, forming mats densely adpressed onto the ground, to 10 cm tall.

Features: Leaves evergreen, opposite, long ovate, to 7 mm long. Flower a small, pink star with five petals, to 4 mm wide. Fruit a brown, woody capsule, to 4 mm wide.

Habitat: Subarctic Europe and America, from Iceland and Scotland east to the Barents Sea, isolated in the Ural Mountains, Alps, Carpathians and Pyrenees; in America from Greenland to Alaska, south to British Columbia and New Hampshire.

Always on acidic, rocky or stony soils, with short or absent snow cover, in subarctic or alpine tundra, to 3100 m.

Bark: Very thin, brown gray.

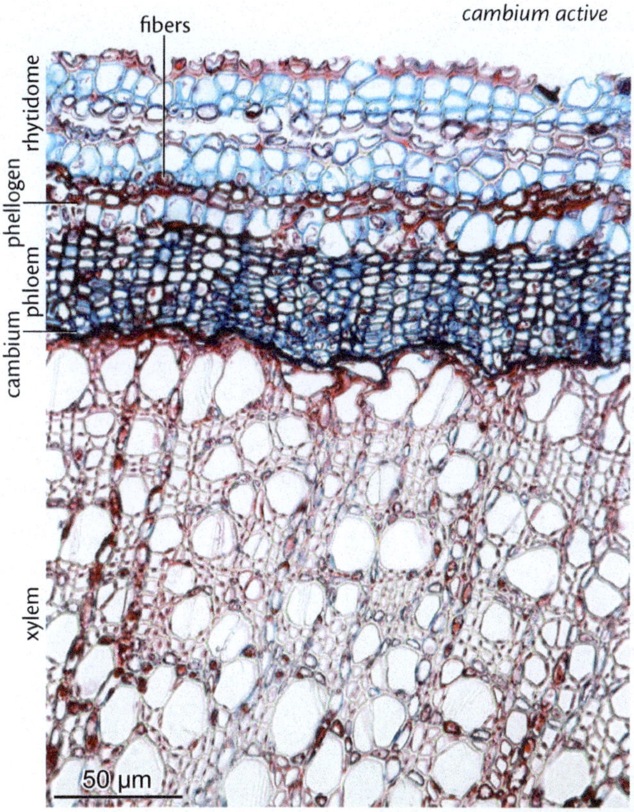

Bark anatomy: Tangentially structured phloem, consisting of sieve tubes, parenchyma cells and fibers.

Sieve tubes and parenchyma difficult to distinguish. Fibers in 1–3-seriate tangential rows. Rays and crystals absent.

Phellem not observed.

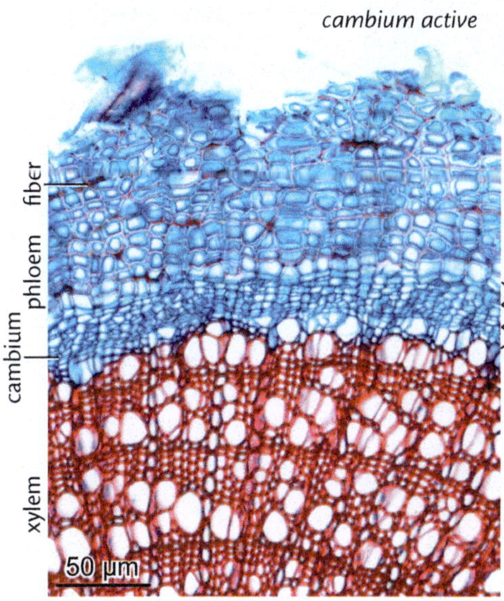

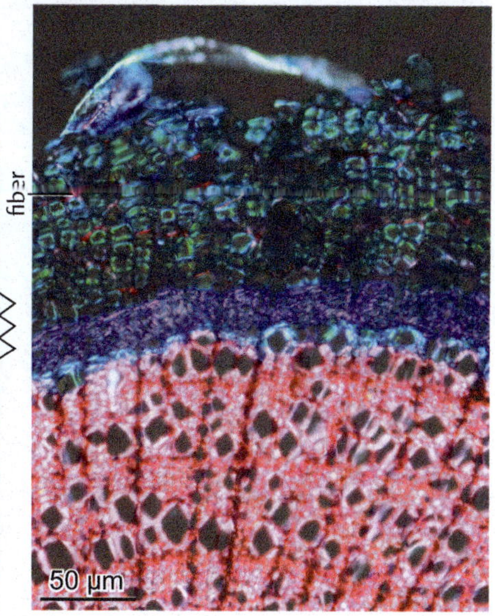

Phyllodoce caerulea (L.) Bab. Ericaceae

blue heath, Bläuliche Moosheide, andromède bleue, andromeda azurra, andromeda azul

Form: Low, spreading dwarf shrub, twig tips erect, to 30 cm tall.

Features: Leaves evergreen, alternate, linear, needle-like, to 1 cm long. Flower a hanging bell, purple, fading to pinkish-blue, to 5 mm long. Fruit a brown, woody capsule, to 4 mm wide.

Habitat: Subarctic Eurasia and America, from Iceland and Scotland east to the Barents Sea, isolated in the Ural Mountains, around Lake Baikal, Central Mongolia, northern Japan and Kamchatka; in America isolated in Alaska and the Northwest Territories, Greenland, south to Labrador and Vermont.

Always on acidic, rocky or peaty soils, in subarctic or alpine tundra, to 3300 m.

Bark: Very thin, brown gray.

dormant season

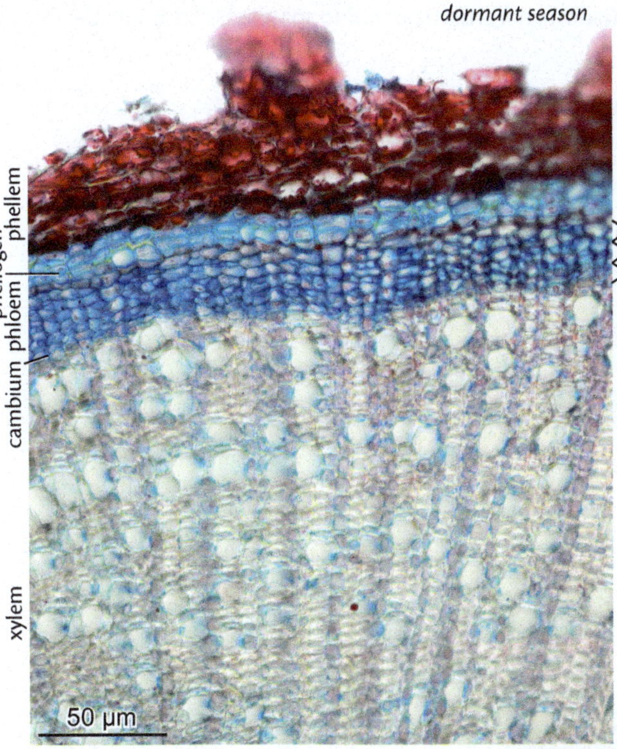

Bark anatomy: Phloem poorly structured. Sieve tubes and parenchyma difficult to distinguish. Fibers absent. Crystals absent.

Cortex consisting of irregular parenchyma. Phellem cells square, thin- to thick-walled, filled with dark-staining substances.

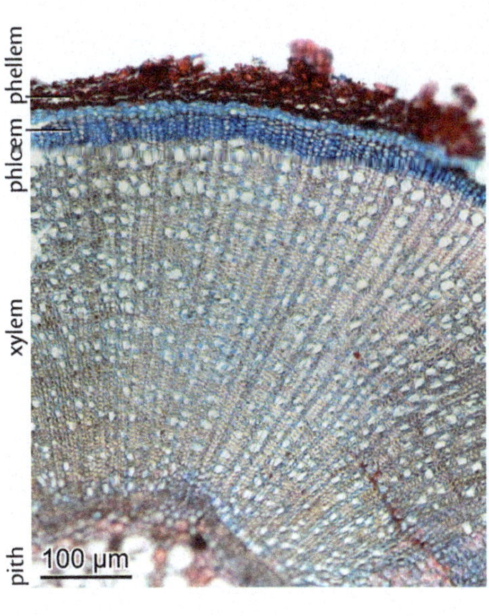

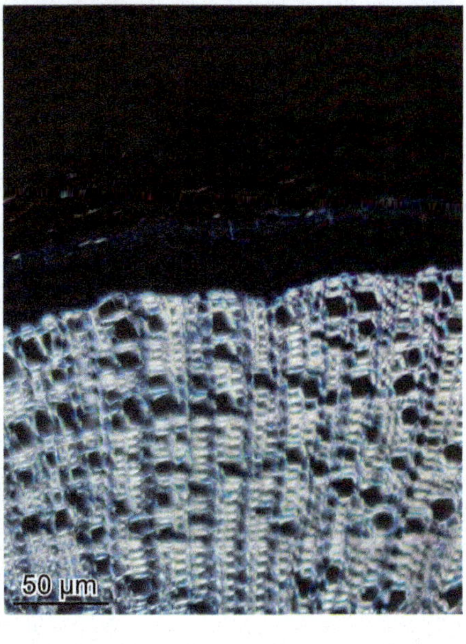

Rhamnus pumila Turra

Rhamnaceae

dwarf buckthorn, Zwerg-Kreuzdorn, nerprun nain, ranno spaccasassi, chopera de roca

Form: Low, spreading dwarf shrub, forming mats densely adpressed onto the ground, to 20 cm tall.

Features: Leaves deciduous, alternate, long ovate, finely serrated, to 4 cm long. Flower a small, yellow-green cross, to 6 mm wide. Fruit a glossy black, berry-like drupe, to 8 mm wide.

Habitat: Central and Southern Europe, from the Hoggar and Atlas mountains northeast to the Alps, dispersed through the Balkans and isolated in the Central Apennines.

Always on limestone rock in full sun, to 2800 m.

Bark: Young glossy, brown gray, turning into dull gray.

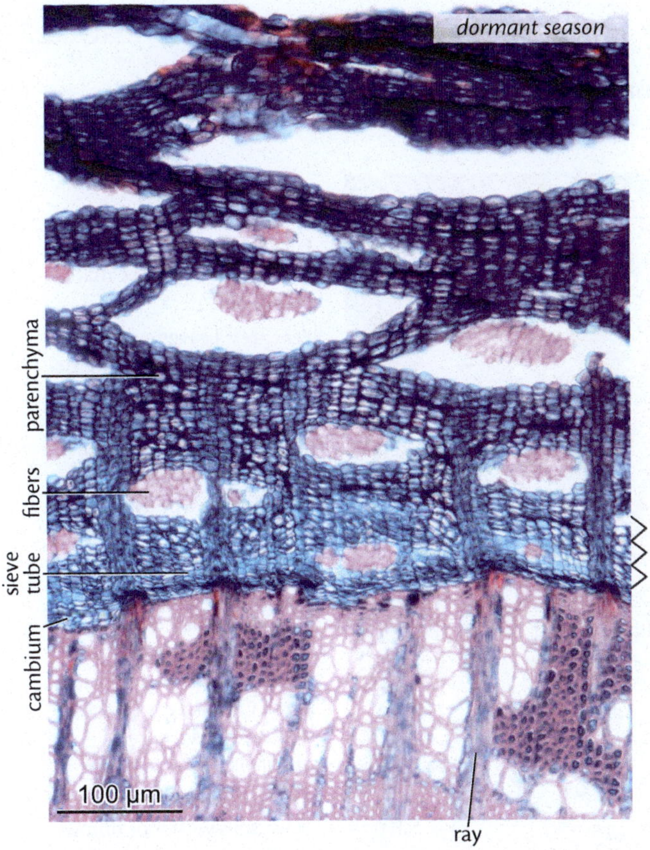

Bark anatomy: Irregular, unlignified tissue with many oval groups of fibers.

Sieve tubes angular, parenchyma roundish. Fibers in tangentially arranged oval groups, loosely connected with the sieve tube and parenchyma tissue. (Holes are preparation artifacts). Sclereids rare, occurring in groups, or absent. Rays straight, dilating. Prismatic crystals surrounding the fiber groups. Crystal druses in most axial parenchyma and ray cells.

Rhytidome cells squashed, dark-stained.

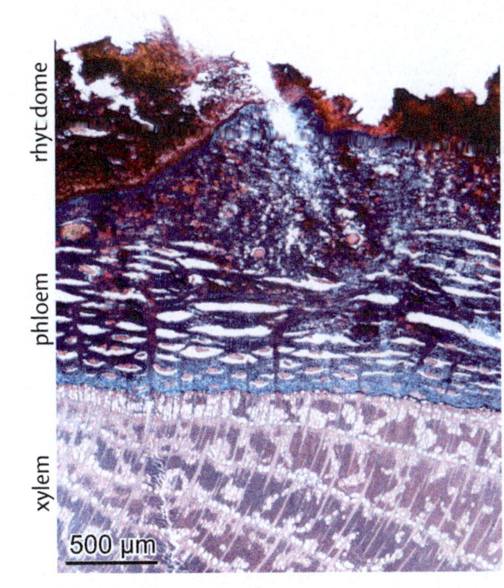

Rhododendron ferrugineum L. Ericaceae

rusty-leaved alpenrose, Rostblättrige Alpenrose, rhododendron ferrugineux, rododendro rosso, rosa de los Alpes

Form: Low, spreading dwarf shrub, forming dense mats, to 1 m tall.

Features: Leaves evergreen, alternate, leathery, aromatic, dark green above, rusty below, long ovate, to 4 cm long. Flowers pink carmine, in dense corymbs, to 1.5 cm wide. Fruit a woody capsule, splitting into a five-parted star, to 8 mm wide.

Habitat: The Alps and Pyrenees, isolated in the southern Jura, North Apennines, Croatia, replaced in the Carpathians by the similar *Rhododendron myrtifolium*.

Always on acidic, deep and moist soils with mor humus, in subalpine conifer forest or tundra, often forming large stands, to 2800 m.

Bark: Young brown gray, turning into light gray.

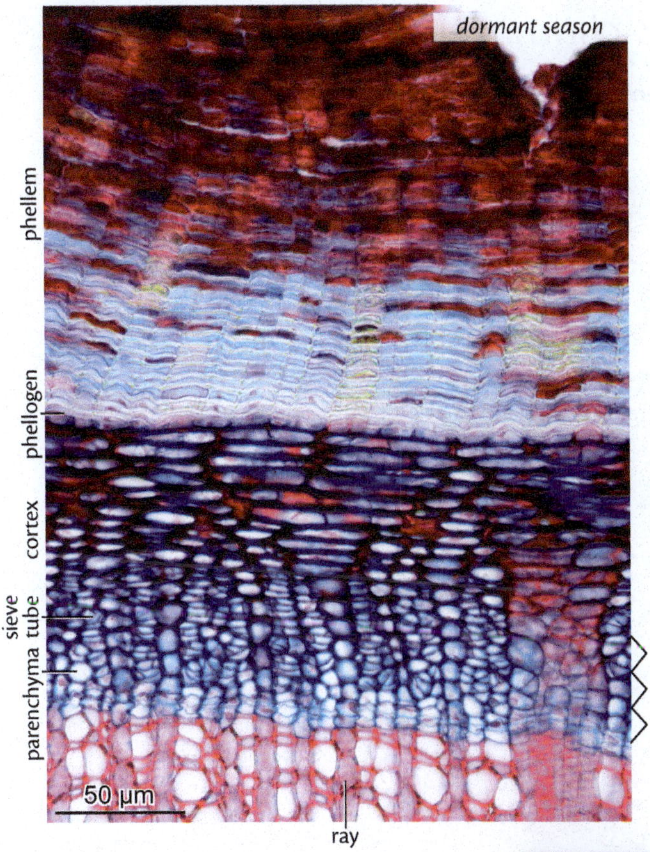

Bark anatomy: Bark thin, consisting of a phloem, a small cortex and a large periderm.

Sieve tubes numerous, accompanied by relatively large companion cells and round to oval parenchyma cells. Fibers and sclereids absent. Rays large, slightly dilating. Crystals absent.

Phellem cells rectangular, thick-walled, partially filled with dark-staining substances, arranged in radial rows.

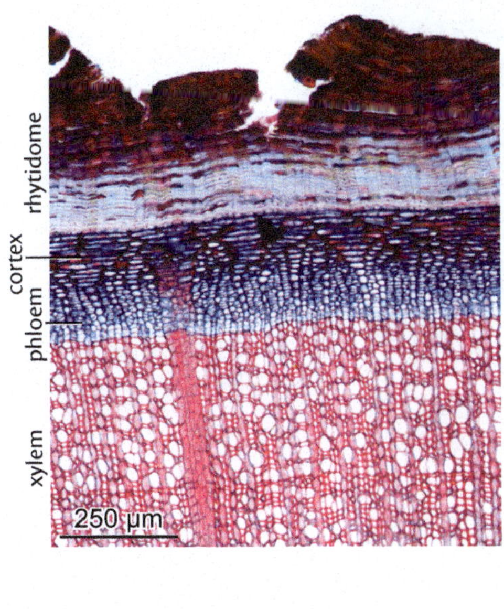

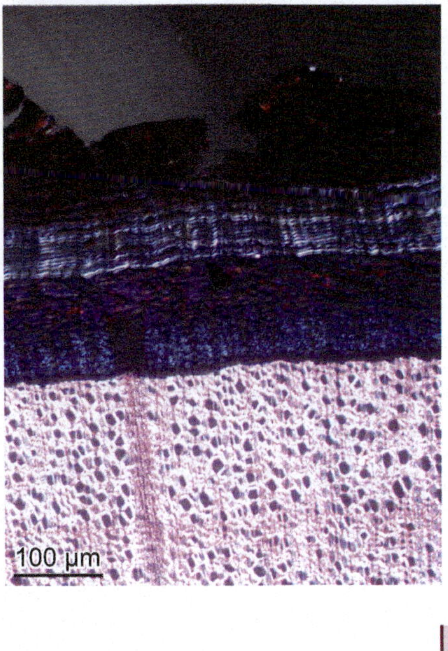

Rhodothamnus chamaecistus (L.) Rchb. Ericaceae

ground cistus, Zwerg-Alpenrose, rhodothamne ciste nain, rododendro cistino, rosa de los Alpes enana

Form: Low, spreading dwarf shrub, forming dense mats, to 40 cm tall.

Features: Leaves evergreen, alternate, leathery, long ovate, margin ciliated, to 1.5 cm long. Flower a light pink star with five petals, to 3 cm wide. Fruit a small, woody capsule, splitting into a five-parted star, to 5 mm wide.

Habitat: The Eastern Alps from western Bavaria to Slovenia and isolated in the Croatian Dinarides.

Always on limestone rock in full sun, in subalpine conifer forest or tundra, to 2400 m.

Bark: Young light brown, later turning into light gray brown, peeling in long, fissured flakes.

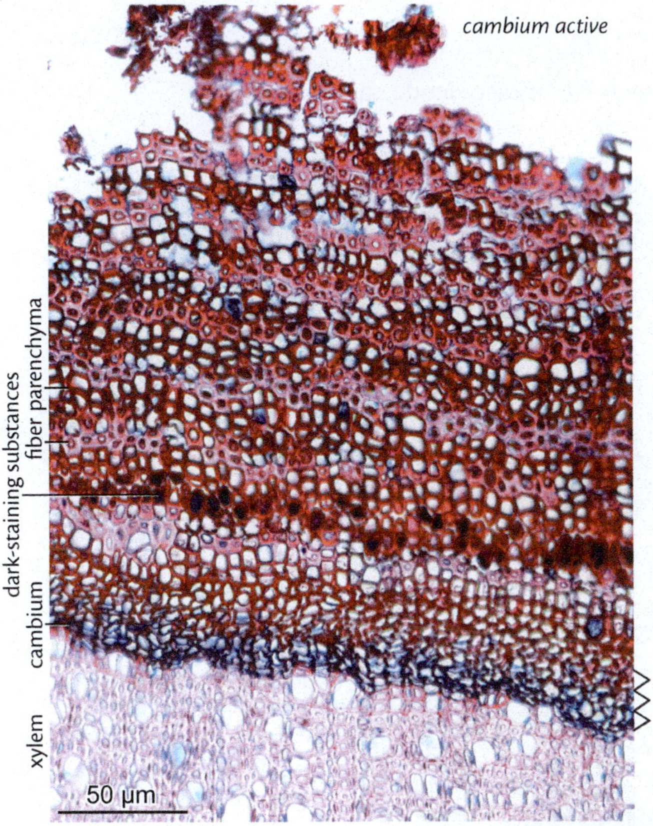

Bark anatomy: Bark thin, consisting of a layered phloem, a small cortex and a large periderm.

Alternating layers of parenchyma/sieve tubes and fibers. Rays and crystals absent.

Phellem cells rectangular, thick-walled, partially filled with dark-staining substances.

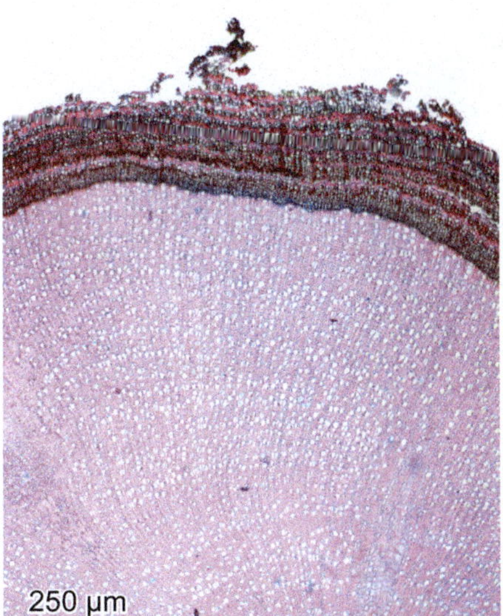

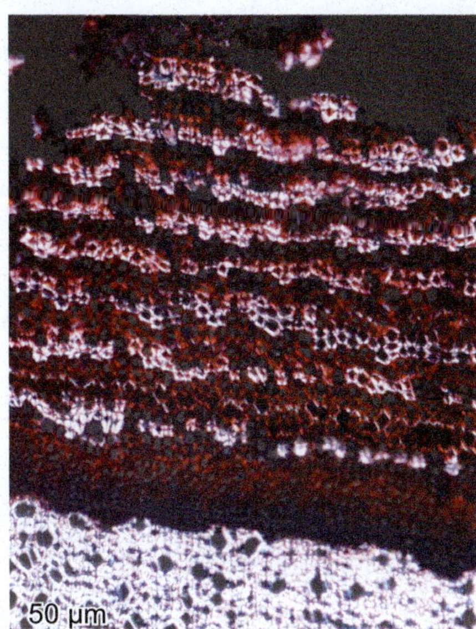

Salix foetida Schleich. Salicaceae

fishy willow, Stink-Weide, saule fétide, salice fetido, sauce hediondo

Form: Low, spreading shrub, twigs bent upwards in arcs, to 2 m tall.

Features: Leaves deciduous, alternate, long ovate, margin finely serrated, to 4 cm long. Flowers dioecious, female catkin upright, green, male catkin upright, yellow, to 3 cm long. Fruit catkin upright, green, with seed capsules, to 3 cm long.

Habitat: The Pyrenees and Western Alps and from the Maritime Alps east to Tyrol.

Always on acidic, constantly moist scree with a long snow cover, to 2600 m.

Bark: Young glossy, red to brown gray, turning into dull gray.

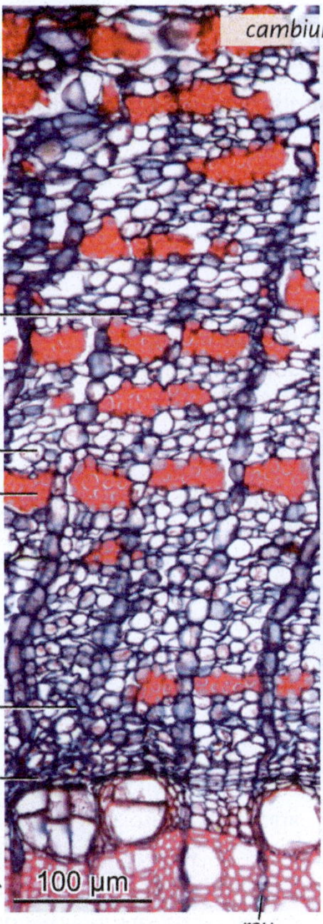

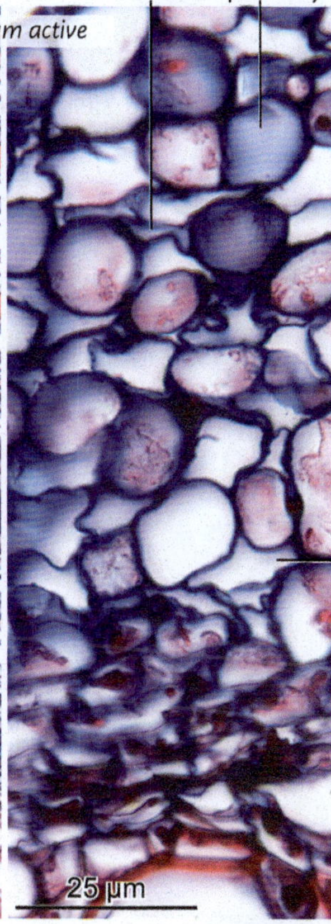

Bark anatomy: Regular tangential layers of parenchyma, sieve tubes and fibers.

Sieve tubes irregular shaped, companion cells small. Sieve tubes collapsing very soon after formation. Parenchyma cells roundish. Fibers in tangential layers. Sclereid groups in older parts of the phloem. Outer parts of the phloem with a few triangular dilations. Prismatic crystals and a few crystal druses occurring along fiber bands. Tangential layers of fibers not corresponding with annual rings.

Periderm consisting of one layer of rectangular, large, thin-walled cells and small, thick-walled cells.

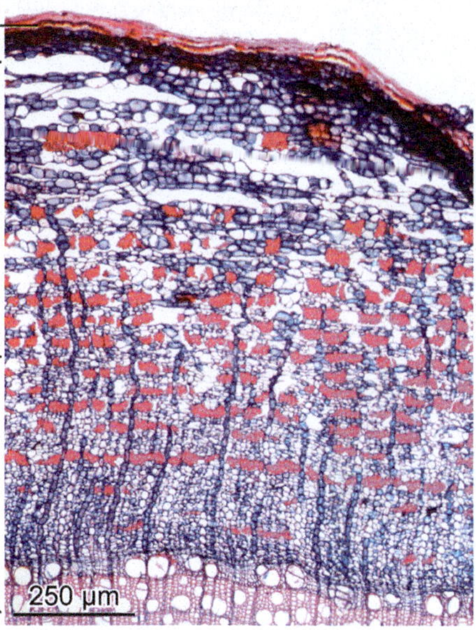

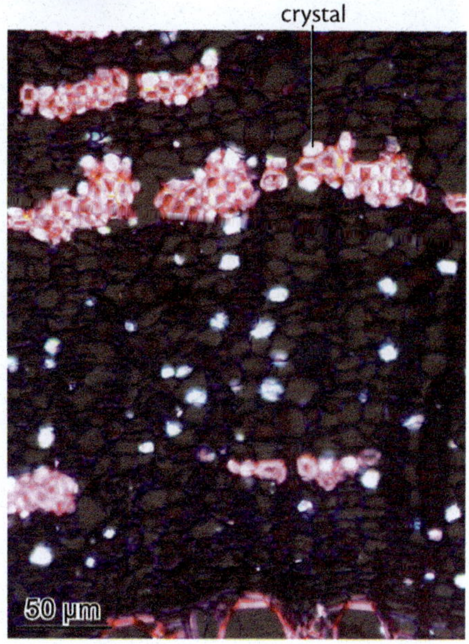

Salix reticulata L. — Salicaceae

net-leaved willow, Netz-Weide, saule à réseau, salice reticulato, sauce reticulado

Form: Very low, spreading shrub, forming mats, adpressed onto the ground, to 10 cm tall.

Features: Leaves deciduous, alternate, orbicular, margin entire, to 4 cm long. Flowers dioecious, female catkin upright, reddish, to 2 cm long, male catkin upright, yellow-red, to 4 cm long. Fruit catkin upright, reddish, with seed capsules, to 3 cm long.

Habitat: Arctic-alpine, circumpolar without Iceland and Svalbard, isolated in Scotland, the Alps, Jura and Pyrenees.

Always on calcareous scree or rock with a long snow cover, to 3000 m.

Bark: Young glossy, red to brown gray, turning into dull gray.

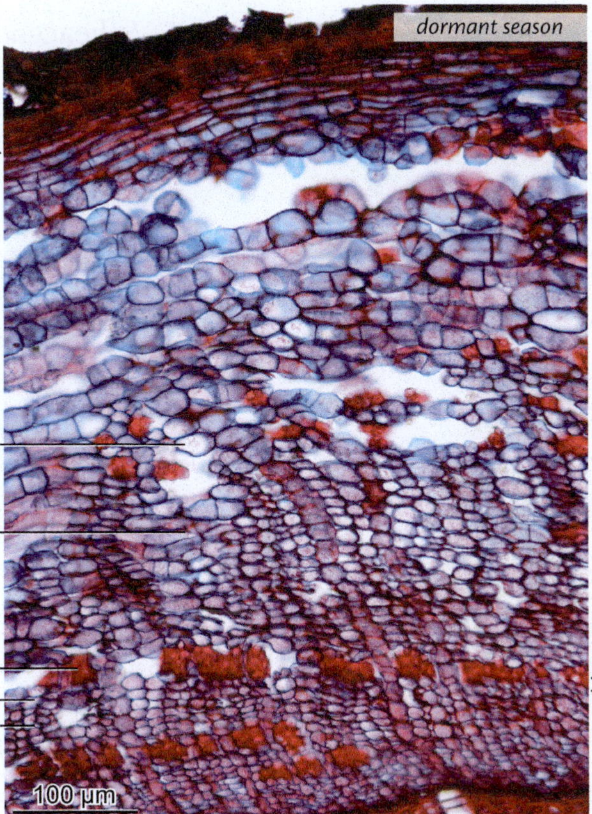

Bark anatomy: Few regular tangential layers of parenchyma, sieve tubes and fibers.

Sieve tubes irregular shaped, companion cells small. Sieve tubes collapsing very soon after formation. Parenchyma cells roundish. Fibers in tangential layers. Sclereid groups in older parts of the phloem. Outer parts of the phloem with a few triangular dilations. Prismatic crystals and a few crystal druses occurring along fiber bands. Tangential layers of fibers not corresponding with annual rings.

Periderm consisting of one layer of rectangular, large, thin-walled cells and small, thick-walled cells.

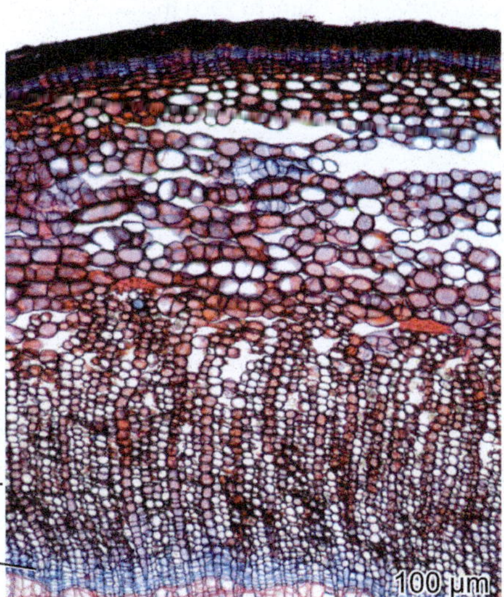

Salix retusa L. Salicaceae

stunted willow, Stumpfblättrige Weide, saule à feuilles émoussées, salice retuso, sauce de hoja truncada

Form: Very low, spreading shrub, forming mats, adpressed onto the ground, to 10 cm tall.

Features: Leaves deciduous, alternate, long ovate, glossy, tips stunted, margin entire, to 2 cm long. Flowers dioecious, female and male catkin similar, upright, yellow, to 2 cm long. Fruit catkin upright, brown green, with seed capsules, to 1 cm long.

Habitat: Mountains of Central and Southern Europe with the Alps, Jura, Pyrenees and Apennines. In the Carpathians and Tatras replaced by the similar *Salix kitaibeliana*.

On calcareous or slightly acidic scree or rock, in full sun, to 2900 m.

Bark: Young glossy, brown, turning into dull gray.

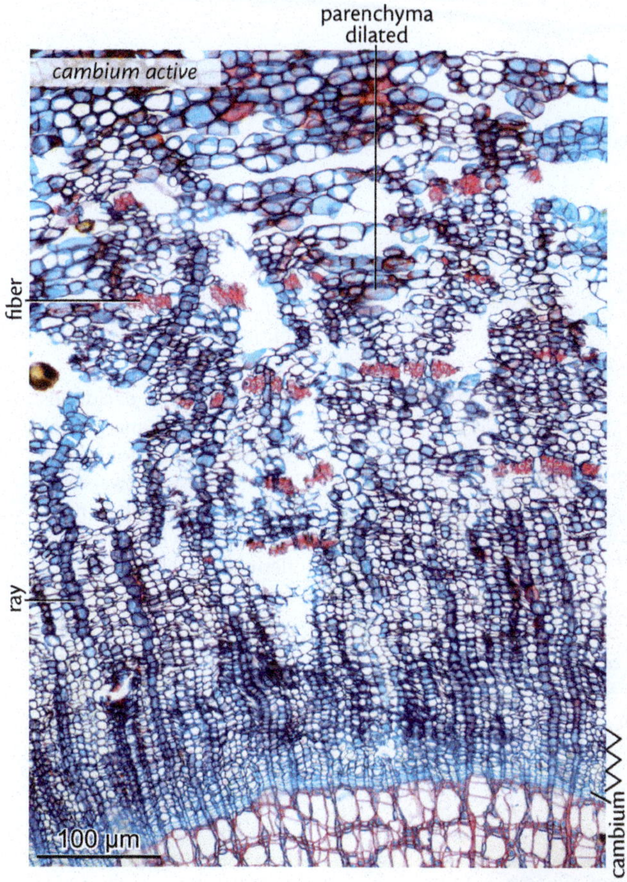

Bark anatomy: Few irregular tangential layers of parenchyma, sieve tubes and fibers.

Sieve tubes irregular shaped, companion cells small. Sieve tubes collapsing very soon after formation. Parenchyma cells roundish. Fibers in tangential layers. Sclereid groups in older parts of the phloem. Outer parts of the phloem with a few triangular dilations. Prismatic crystals and a few crystal druses occurring along fiber bands. Tangential layers of fibers not corresponding with annual rings.

Periderm consisting of one layer of rectangular, large, thin-walled cells and small, thick-walled cells.

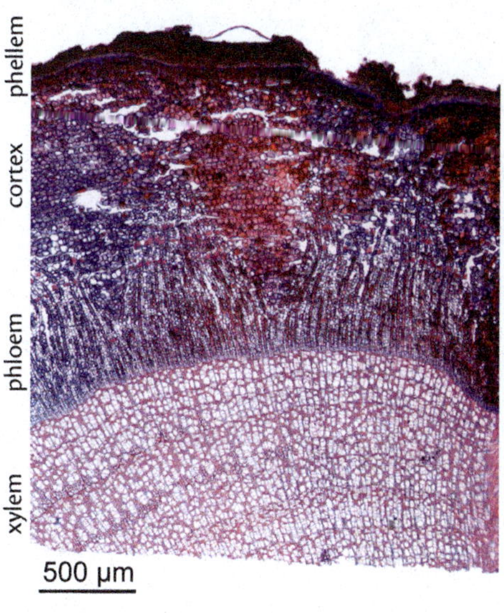

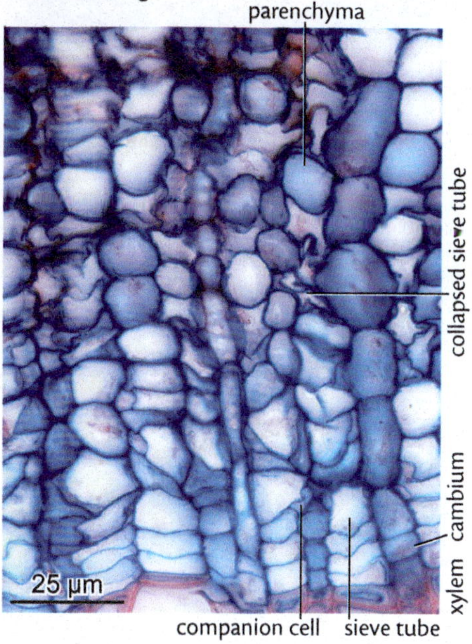

Vaccinium myrtillus L. — Ericaceae

bilberry, Heidelbeere, myrtille, mirtillo nero, arándano

Form: Low, spreading dwarf shrub, forming dense mats, to 50 cm tall.

Features: Leaves deciduous, alternate, ovate, margin finely serrated, to 4 cm long, deep red in autumn. Flowers a reddish to greenish, hanging bell, to 5 mm long. Fruit fleshy, berry-like, black-blue, to 8 mm wide.

Habitat: Eurosiberia, from Northern Spain northeast to the North Cape, Upper Yenisei River and Altai Mountains, south to the Caucasus, Balkans, Apennines and Southern France.

On acidic, deep and moist soils with mor humus, in forest or alpine and subarctic tundra, often forming large stands, to 2900 m.

Bark: Smooth, green, on sharply winged stems.

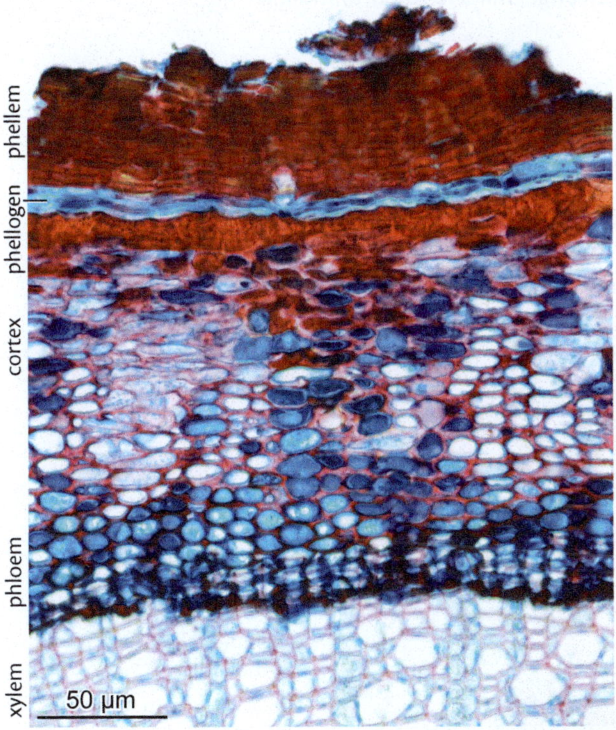

Bark anatomy: Phloem poorly structured. Sieve tubes and parenchyma difficult to distinguish. Fibers absent. Rays straight, dilating, laterally separating sieve tube and parenchyma zones. Crystals absent.

Periderm divided into a living, thin-walled phelloderm and an external zone of phellem. Phellem cells square, filled with dark-staining substances.

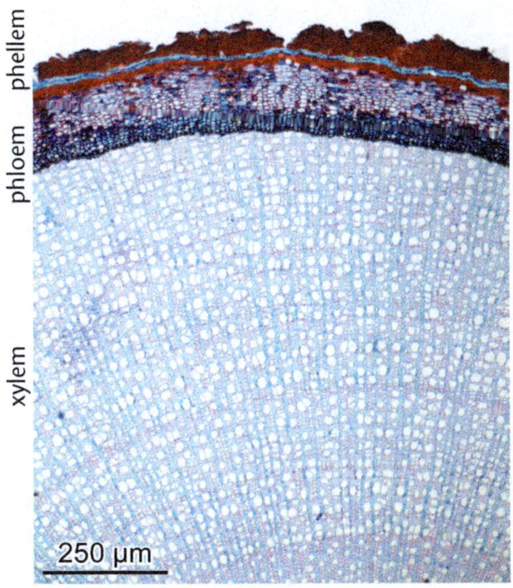

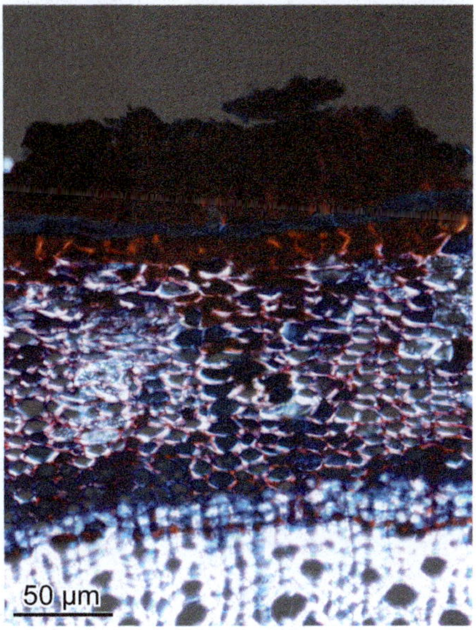

Vaccinium vitis-idaea L. Ericaceae

cowberry, Preiselbeere, airelle rouge, mirtillo rosso, arándano rojo

Form: Low, spreading dwarf shrub, forming dense mats, to 30 cm tall.

Features: Leaves evergreen, alternate, ovate, leathery, margin entire, to 3 cm long. Flowers a pinkish white, hanging bell, to 8 mm long. Fruit fleshy, berry-like, glossy red, to 8 mm wide.

Habitat: Eurosiberia and America, from Northern Spain northeast to the North Cape, circumpolar to Alaska, Greenland and Iceland, south to the Apennines, Balkans, Altai, Sakhalin, California, New Mexico and Newfoundland.

On acidic, dry to moist soils with mor humus, in heath, forest or alpine and subarctic tundra, often forming large stands, to 3500 m.

Bark: Light brown or dark red.

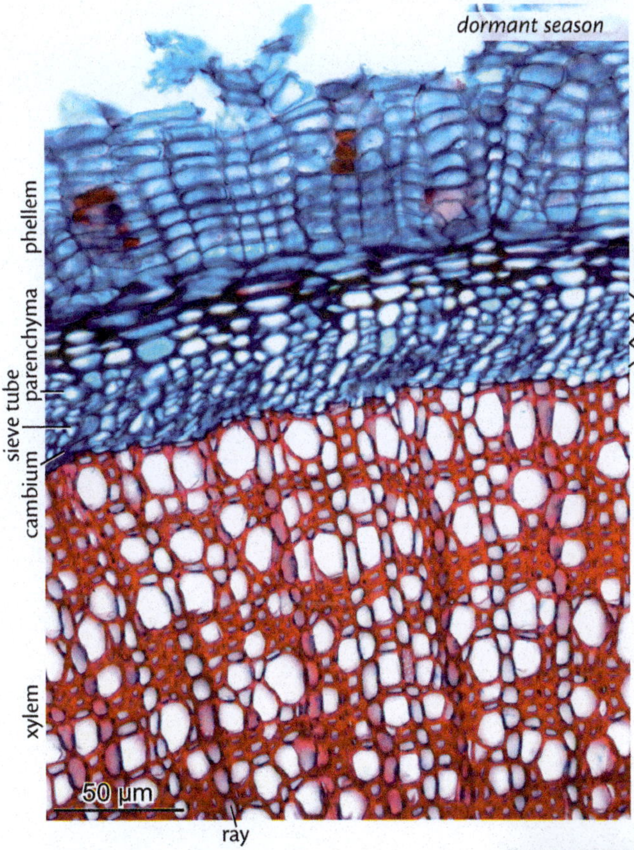

Bark anatomy: Phloem poorly structured. Sieve tubes and parenchyma difficult to distinguish. Fibers absent. Rays straight, dilating, laterally separating sieve tube and parenchyma zones. Crystals absent.

Periderm divided into a living, thin-walled phelloderm and an external zone of phellem. Phellem cells square, filled with dark-staining substances.

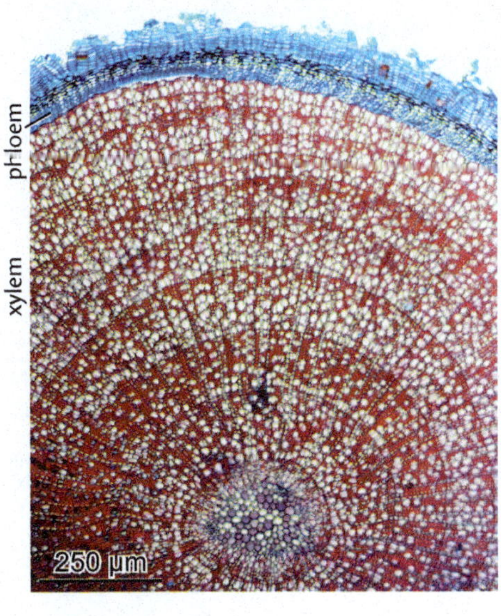

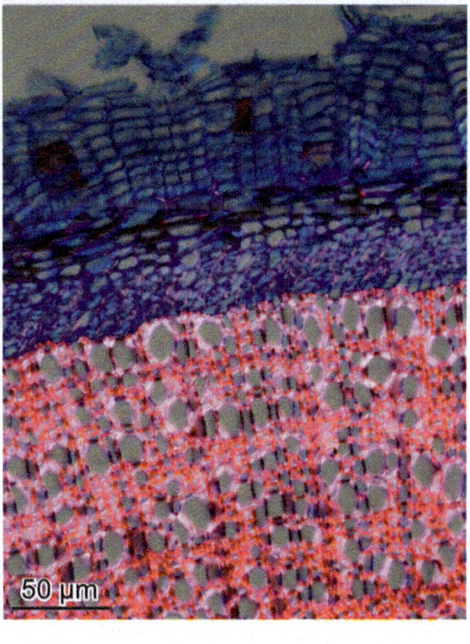

The **mountain ranges of Eurasia** are represented in this book only by European species, with the exception of *Cedrus atlantica* from the Mediterranean mountains. The dominant conifers of the pine family in this biome are accompanied by various deciduous shrubs.

Central Asian and Himalayan mountains are not included in this book, simply because of a lack of available bark material.

F. H. Schweingruber et al., *Bark Anatomy of Trees and Shrubs in the Temperate Northern Hemisphere*, https://doi.org/10.1007/978-3-030-14056-4_4

4 | Mountain ranges in Eurasia

Abies alba Mill. Pincaceae

European silver fir, Tanne, sapin blanc, abete bianco, abete blanco

Form: Upright conifer, crown narrow to broad, to 60 m tall.

Features: Needles evergreen, soft, flat, aromatic, with two parallel white stripes on the underside, to 3 cm long. Flowers monoecious, female cone upright, pale green, short, male cone hanging, yellow, to 2 cm long. Seed cones brown, upright, soft-scaled, to 12 cm long, disintegrating at the tree.

Habitat: Mountains of Central and Southern Europe from the Pyrenees to the Vosges, Thuringia, the Carpathians, south to the Balkans, the Apennines, Corsica.

On deep, moist soils, in montane forest, to 1900 m.

Bark: Young smooth, silver gray, later tessellated-flaky, light gray or reddish brown.

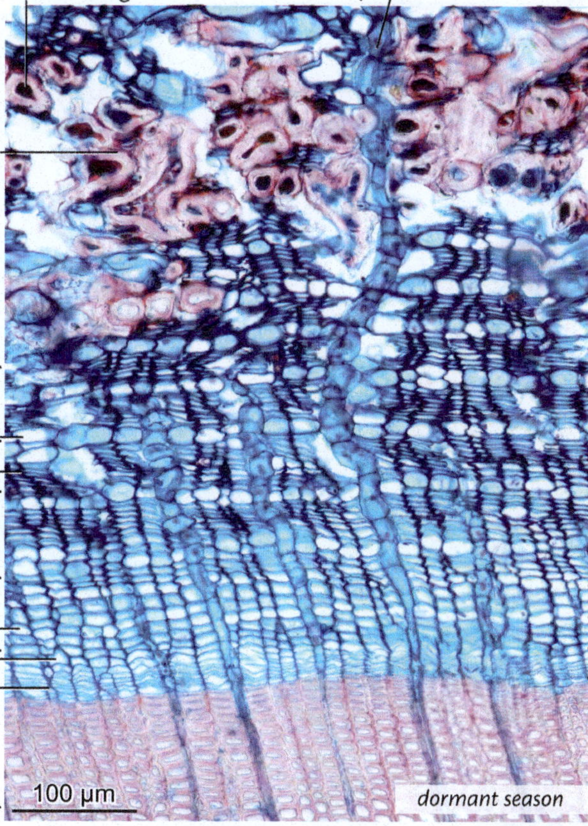

Bark anatomy: Regular tangential layers in the young phloem, adult phloem with many irregular shaped groups of sclereids.

Sieve cells rectangular, collapsing within a few years. Parenchyma cells small and oval in the juvenile phloem, enlarged in the adult phloem. Fibers absent. Sclereids very frequent, often filled with dark-staining substances in the adult phloem. Rays radially wavy, in the cortex often with slime-producing, large cells. Rays frequently dilating. Prismatic crystals occurring outside of sclereids. Resin ducts absent.

Annual rings distinct for a few years near the cambium. Rings indistinct on the sclerotic part of the bark. Periderm small. Phellem cells rectangular, thin-walled.

Cedrus atlantica (Endl.) Manetti ex Carrière Pinaceae

Atlantic cedar, Atlas-Zeder, cèdre de l'Atlas, cedro dell'Atlante, cedro del Atlas

Form: Upright conifer, crown conical at first, turning with age into a very distinct broad form, to 40 m tall.

Features: Needles evergreen, soft, flat, dark or blue green, in star-shaped bundles of 10–30 on short shoots, to 3 cm long. Flowers monoecious, female cone upright, reddish green, short, male cone upright, yellow, to 5 cm long. Seed cones brown, upright, barrel-shaped, soft-scaled, to 8 cm long.

Habitat: Atlas and Rif mountains in Morocco.

On dry, rarely moist, stony soils, often over rocks, in montane and subalpine forest, only between 1000 to 1900 m.

Bark: Young smooth, silver gray, later tessellated-flaky, deeply fissured, dark to black gray.

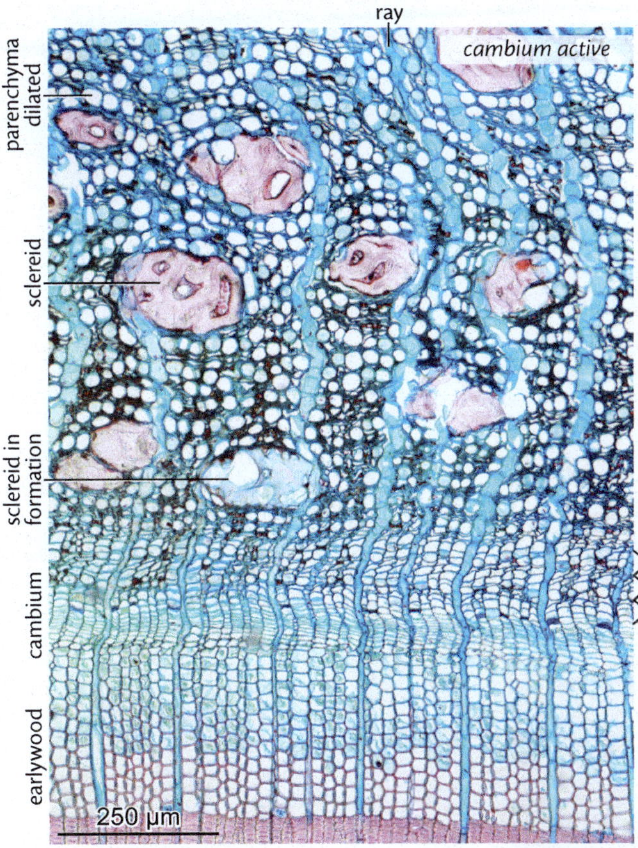

Bark anatomy: Tangential layers absent. Round groups of sclereids are characteristic.

Sieve cells rectangular, collapsing within a few years. Parenchyma cells small and oval in the juvenile phloem, enlarged in the adult phloem. Fibers absent. Sclereids arranged in round groups. Rays irregular shaped, wavy. Prismatic crystals numerous, occurring in long, tube-like cells. Large resin ducts in the cortex. Annual rings indistinct.

Periderm small. Phellem cells rectangular, thin- to thick-walled.

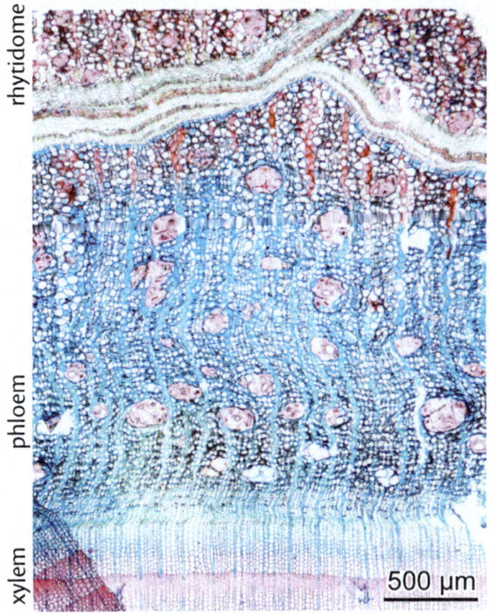

Juniperus sabina L. Cupressaceae

savin juniper, Sevistrauch, genévrier sabine, ginepro sabino, sabina rastrera

Form: Coniferous shrub, low spreading, forming large and dense stands, to 150 cm tall.

Features: Needles evergreen, young needle-like, later scaly, cross-opposite, olive or blue green, strongly aromatic, to 2 mm long. Flowers monoecious or dioecious, female and male flowers in tiny yellowish globes, to 5 mm wide. Cones with a berry-like, light bluish structure of coalescing scales, to 7 mm long.

Habitat: Eurasia from the Sierra Nevada east to the Alps, the Balkans, the Caucasus, Central Asia and Mongolia.

On very dry soils, mostly over rocks of different types and acidity, always in full sun, can cover whole slopes with extensive stands, to 3300 m.

Bark: Young red brown or gray, with papery scales, later gray, with distinct longitudinal fissures.

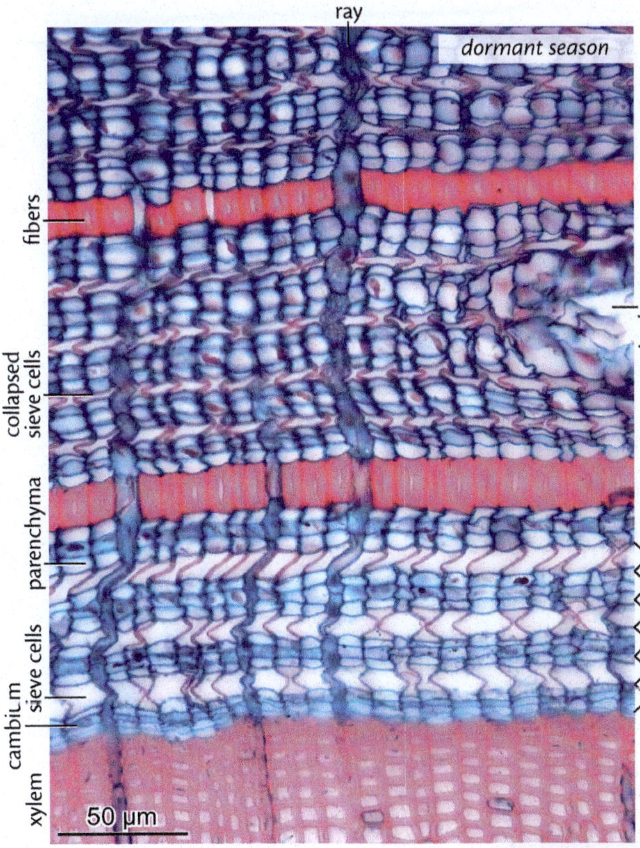

Bark anatomy: Regular tangential layers of sieve cells, fibers and parenchyma cells.

Sieve cells thin-walled, rectangular, parenchyma cells round to oval, fibers thick-walled, square and rectangular. Fiber bands periodically missing. Sieve cells tending to collapse. Rays uniseriate, slightly wavy, dilating in later stages. Rarely with large resin ducts. Crystal sand in radial intracellular layers.

Annual rings mostly indistinct. Phellem cells thin-walled, irregularly shaped.

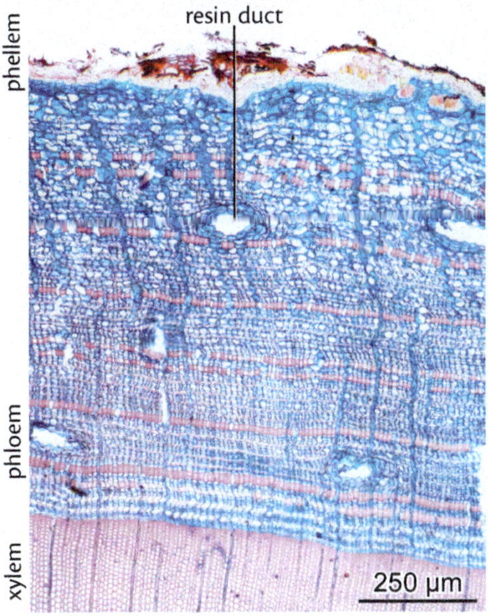

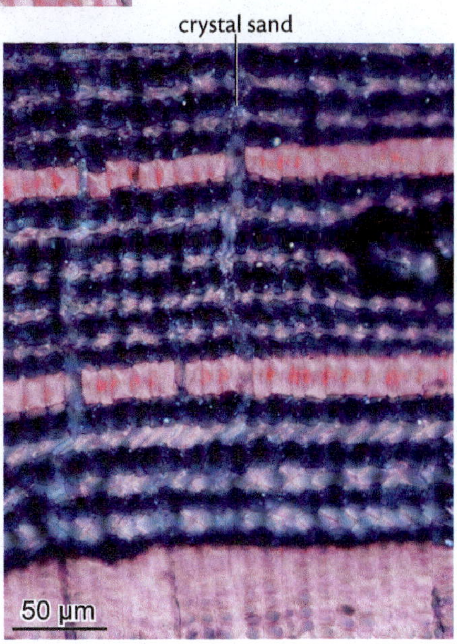

Larix decidua Mill. Pinaceae

European larch, Lärche, mélèze, larice, alerce europeo

Form: Conifer, crown broadly columnar, later rounded, with horizontal branches and hanging twigs, to 55 m tall.

Features: Needles deciduous, soft, to 3 cm long, in star-shaped bundles of 20–40 on short shoots. Flowers monoecious, female cone upright, red, 1–2 cm long, male cone hanging, yellow, globular, to 1 cm long. Seed cones soft, globular, gray brown, to 4 cm long.

Habitat: The Alps and Carpathians, from the Maritime Alps east to Lower Austria and Croatia, one subspecies in the hills of southern Poland, to 2600 m.

Bark: Young smooth, gray brown, later red brown, thick scaly-corky with deep longitudinal furrows, exposing the wine red inner bark.

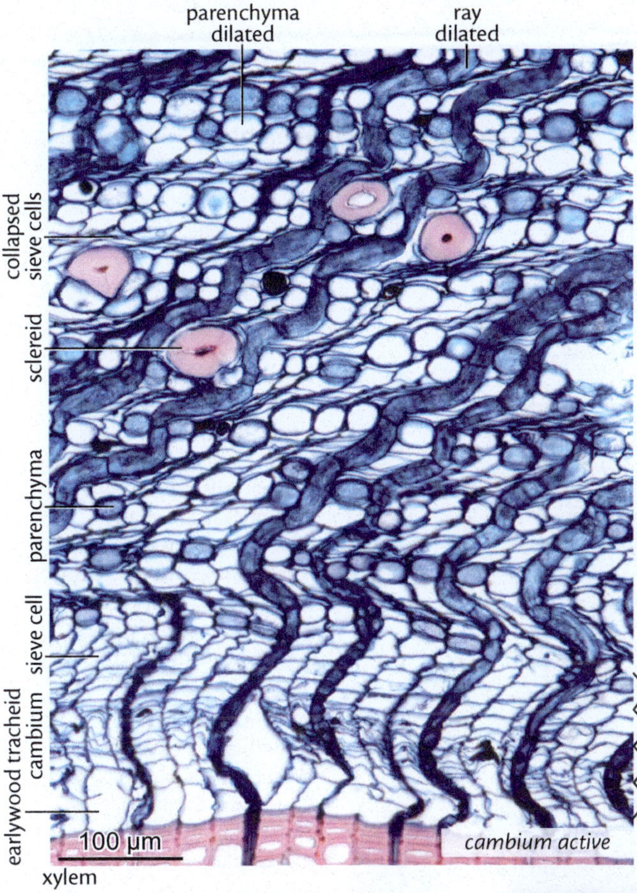

Bark anatomy: Regular tangential layers in the young phloem. Isolated sclereids are characteristic.

Sieve cells rectangular, arranged in 3–5-seriate tangential rows, collapsing within a few years. Parenchyma cells small and oval in the juvenile phloem, enlarged in the adult phloem. Fibers absent. Sclereids in form of isolated, round, very thick-walled, lignified cells. Rays radially wavy. Small prismatic crystals outside of sclereids.

Annual rings distinct for a few years near the cambium. Rings indistinct in the adult phloem due to sieve-cell collapse.

Periderm active for several years. Phellem cells rectangular, thin-walled.

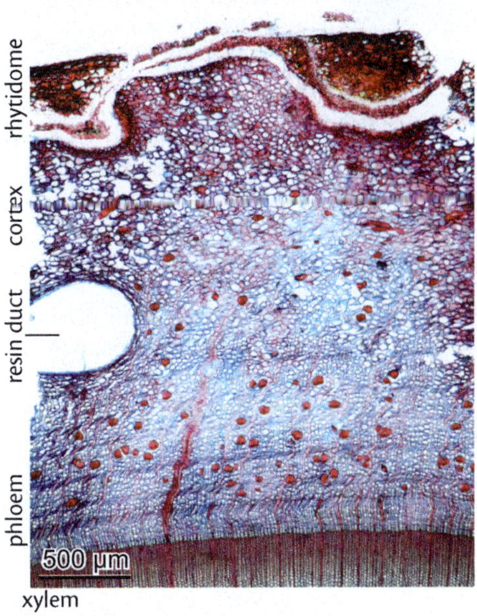

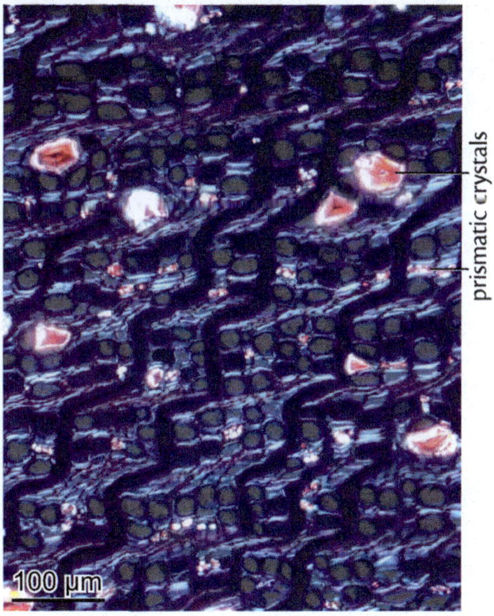

Pinus cembra L.

Pinaceae

cembran pine, Arve/Zirbe, pin cembro, pino cembro, pino cembra

Form: Conifer, crown columnar at first, later rounded, with age developing gnarled individual forms, to 25 m tall.

Features: Needles evergreen, in bundles of five, soft, dark green above, blue green below, to 11 cm long. Flowers monoecious, female cone upright, lilac red, male cones hanging, yellow violet, crowded together at the top of twigs. Seed cones soft, ovate, young green violet, mature brown violet, to 9 cm long.

Habitat: The Alps and Carpathians, from the Maritime Alps east to Lower Austria and Carinthia, isolated in the High Tatras and Carpathians.

Always on mature soils with acidic mor humus, also over limestone rock, only in subalpine forest, often forming the tree line in the Central Alps, mostly together with European larch as a pioneer, to 2500 m.

Bark: Young smooth, silver gray, with resinous pockets, later red brown to brown, thick scaly with deeply tessellated flakes.

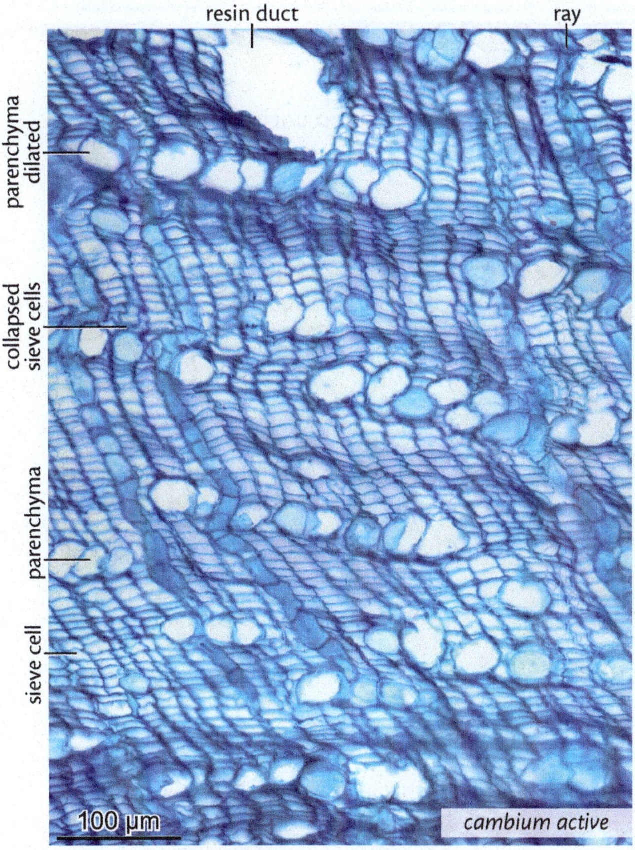

Bark anatomy: Irregular tangential layers in the young phloem. Absence of fibers and isolated sclereids is characteristic.

Sieve cells rectangular, arranged in 3–5-seriate tangential rows, collapsing within a few years. Parenchyma cells small and oval in the juvenile phloem, enlarged in the adult phloem. Sclereids and fibers absent. Rays radially wavy. With large resin ducts in the older phloem. Crystals absent in phloem.

Annual rings mostly indistinct due to the irregular occurrence of parenchyma cells.

Phellem irregularly layered, consisting of rectangular, thick-walled (sclereids) and thin-walled cells. Many small prismatic and acicular crystals in the periderm.

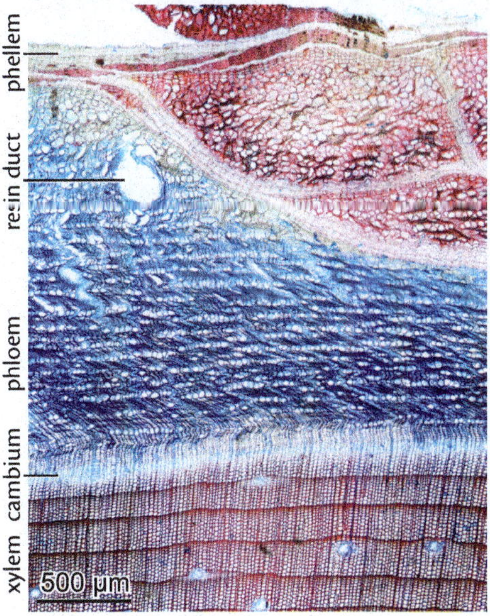

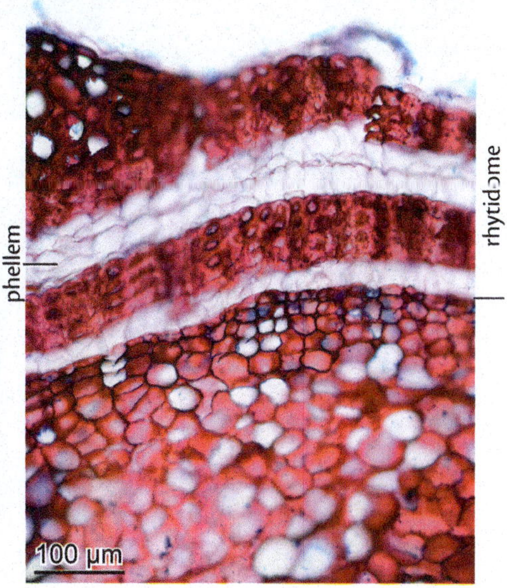

Pinus mugo subsp. *uncinata*
(Ramond ex DC.) Domin

Pinaceae

mountain pine, Aufrechte Bergföhre, pin à crochets, pino uncinato, pino negro

Form: Conifer, crown upright and columnar, to 25 m tall.

Features: Needles evergreen, in bundles of two, stiff, with a pointy tip, dark to olive green, to 7 cm long. Flowers monoecious, female cone globular, red, male cones straw yellow, crowded together at the top of twigs. Seed cones broad ovate, young green, mature gray to yellow brown, to 7 cm long.

Habitat: Mountains of Central and Southern Europe, from the Pyrenees east to the Massif Central, Jura, Western Alps east to Bavaria and Tyrol.

Always on very dry and rocky sites (limestone and acidic rocks), or very wet and acidic boggy soils, only in montane and subalpine forest, to 2600 m (in the Central Pyrenees reaching the uppermost tree line in Europe).

Variation: The shrubby form subsp. *mugo* is much more widespread, also in the Eastern Alps and the Balkans.

Bark: Fine flaky-scaly to tessellated, always uniformly gray to gray brown.

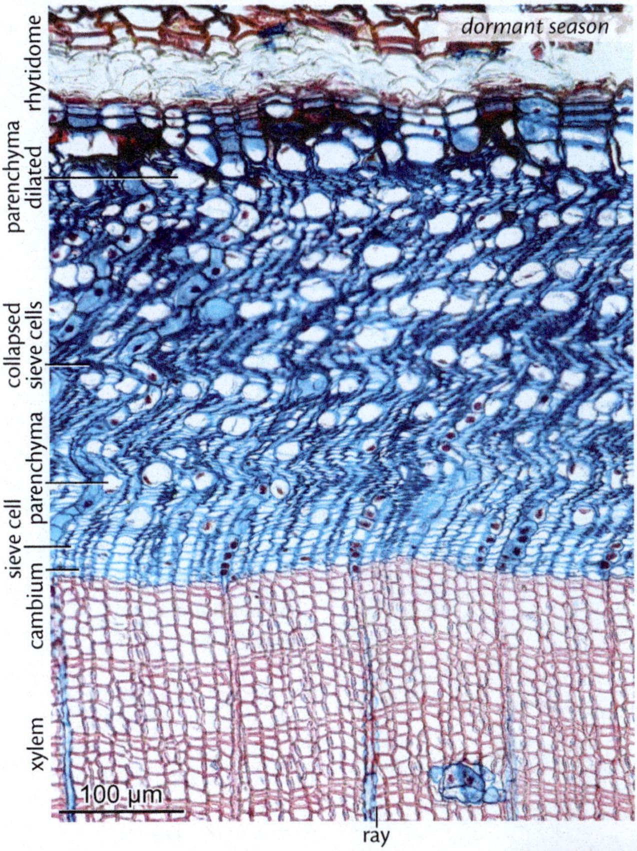

Bark anatomy: Anatomy very similar to that of *Pinus sylvestris*. Differs from *P. sylvestris* in the less compact phellem.

Sieve cells rectangular, arranged in 3–5-seriate, tangential rows, collapsing within a few years. Parenchyma cells small and oval in the juvenile phloem, enlarged in the adult phloem. Sclereids and fibers absent. Rays radially wavy. With elongated prismatic crystals outside of rays, mostly in axially elongated cells (crystal tubes).

Annual rings distinct for a few years near the cambium. Rings indistinct in the adult phloem due to sieve-cell collapse and dilatation of parenchyma cells.

Phellem irregularly layered, consisting of rectangular thick-walled (sclereids) and thin-walled cells.

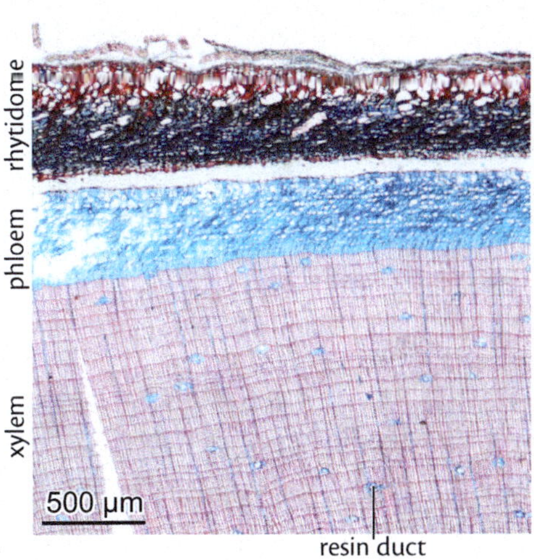

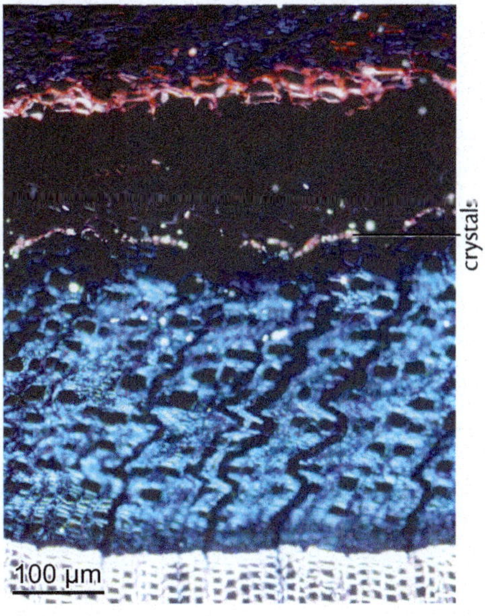

Pinus nigra subsp. *nigra* J.F.Arnold

Pinaceae

Austrian pine, Schwarz-Kiefer, pin noir d'Autriche, pino nero, pino salgareño

Form: Conifer, crown upright and columnar, later turning into a broad umbrella shape, to 30 m tall.

Features: Needles evergreen, in bundles of two, stiff, dark green, to 15 cm long. Flowers monoecious, female cone globular, red, male cone straw yellow, crowded together at the top of twigs. Seed cones round ovate, young green, mature gray to yellow brown, to 8 cm long.

Habitat: Mountains of Southern Europe, with five different subspecies from Morocco to Spain, Corsica, Italy, Austria, the Balkans, Crimea and Turkey. The subsp. *nigra* from the southeastern alpine rim in Austria to the Balkans.

Always on very dry and rocky (limestone and serpentine) soils, in montane forest, to 1300 m.

Bark: Fine flaky-scaly to coarsely tessellated, always uniformly gray to gray brown.

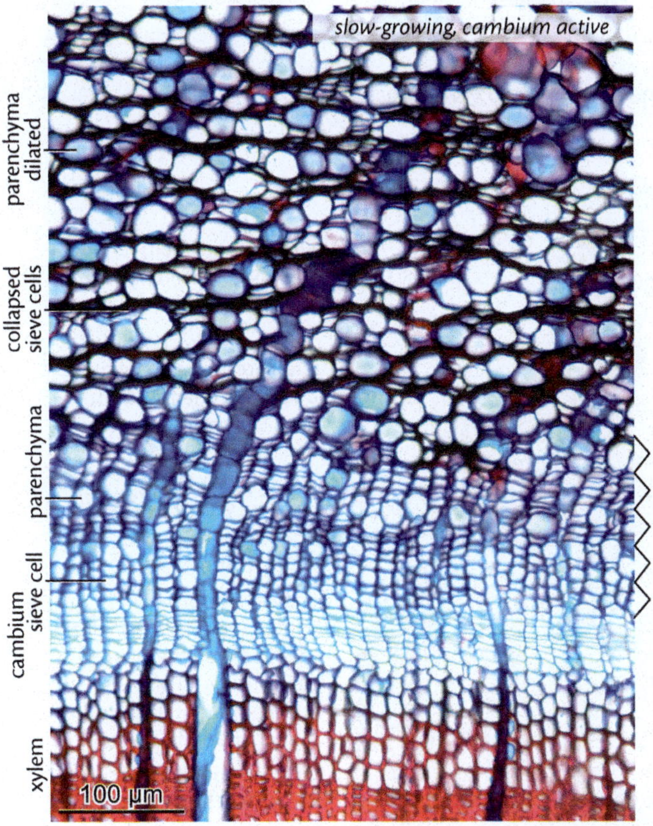

Bark anatomy: Anatomy very similar to that of *Pinus sylvestris*. Differs from *P. sylvestris* in the less compact phellem.

Sieve cells rectangular, arranged in 3–5-seriate, tangential rows, collapsing within a few years. Parenchyma cells small and oval in the juvenile phloem, enlarged in the adult phloem. Sclereids and fibers absent. Rays radially wavy. With elongated prismatic crystals outside of rays, mostly in axially elongated cells (crystal tubes).

Annual rings distinct for a few years near the cambium. Rings indistinct in the adult phloem due to sieve-cell collapse and dilatation of parenchyma cells.

Phellem irregularly layered, consisting of rectangular thick-walled (sclereids) and thin-walled cells.

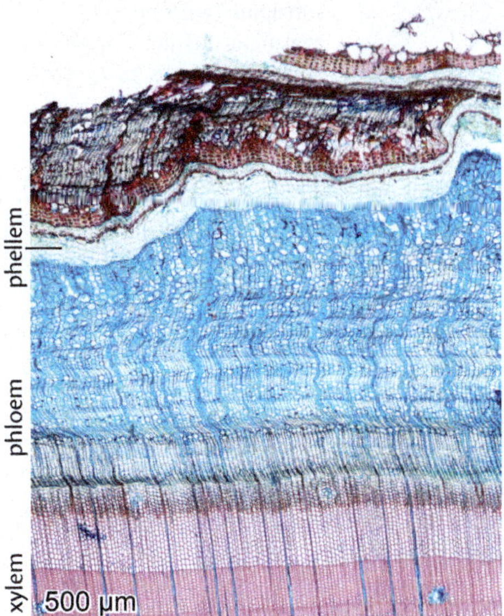

Alnus viridis (Chaix) DC. — Betulaceae

green alder, Grün-Erle, aulne vert, ontano verde, aliso verde

Form: Shrub, multi-stemmed, ascending in arcs, to 4 m tall.

Features: Leaves deciduous, alternate, ovate, margin double-toothed, to 6 cm long. Flowers monoecious, female cone upright, dark red, short, male cone hanging, yellow, to 4 cm long. Seed cones red brown, ovate, to 2 cm long.

Habitat: Mountains of Central and Southern Europe, the Alps from the Dauphiné east, the Black Forest, Sudetes, Carpathians to the mountains of the Balkans.

On permanently moist to wet soils in humid, shadowy locations in montane and subalpine forest or open avalanche channels, often in gregarious stands, to 2300 m.

Bark: Young green to gray brown, smooth, silver gray, later dark gray brown, often with dots or longitudinal strips.

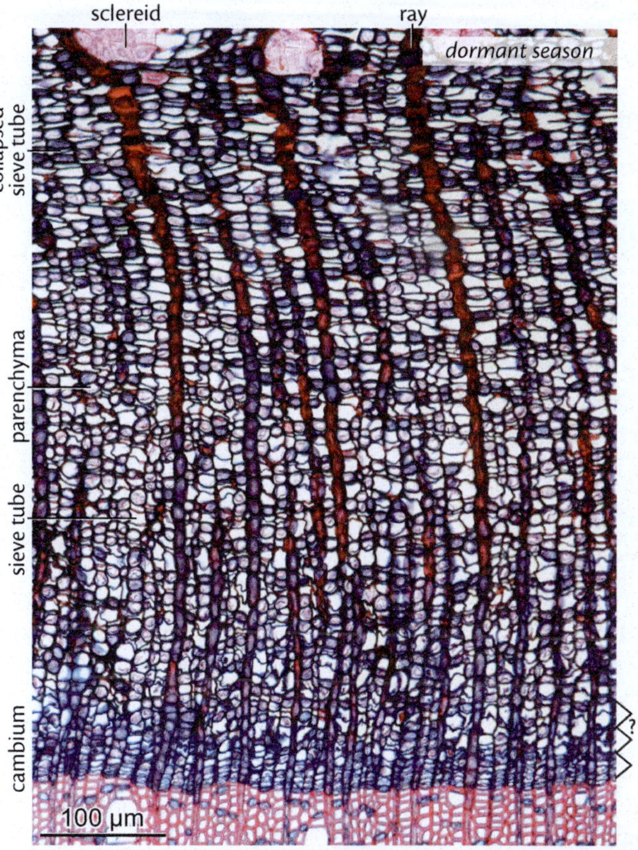

Bark anatomy: Structure uniform or tangentially layered. Sieve tube and parenchyma tissue with many dense groups of sclereids. Tangential layers of sieve tubes at the beginning of the early bast. Parenchyma cells roundish. Fibers absent or in small tangential layers. Sclereids in groups in the older phloem. Dilatations in the shape of a few large, triangular zones. Many crystal druses and prismatic crystals present.

Periderm consisting of layers of rectangular, large, thin-walled cells and small, thick-walled cells.

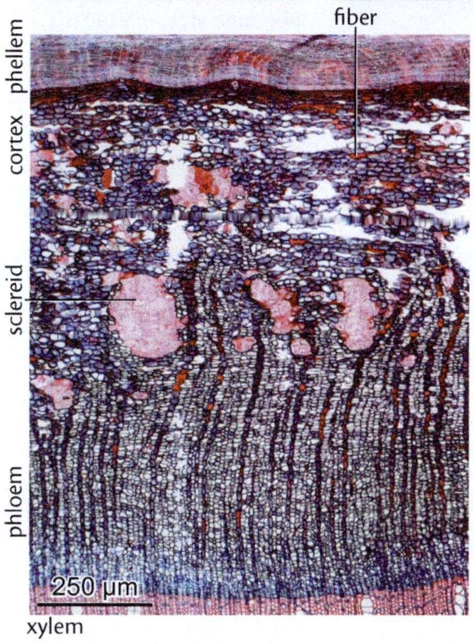

Amelanchier ovalis Medik. Rosaceae

snowy mespilus, Felsenbirne, néflier des roches, pero corvino, guillomo

Form: Shrub, dense and upright, to 3 m tall.

Features: Leaves deciduous, alternate, ovate, margin finely serrated, to 4 cm long. Flowers white, with five large petals, to 5 cm wide. Fruit a blue black, blueberry-like drupe, to 1 cm wide.

Habitat: Mountains of Europe and Southwest Asia, from the mountains of the Maghreb northeast to Luxembourg, Poland, the Caucasus and Turkey.

On shallow, dry, often calcareous soils, in full sun, often on escarpments and limestone rock, to 2100 m.

Bark: Smooth, gray brown.

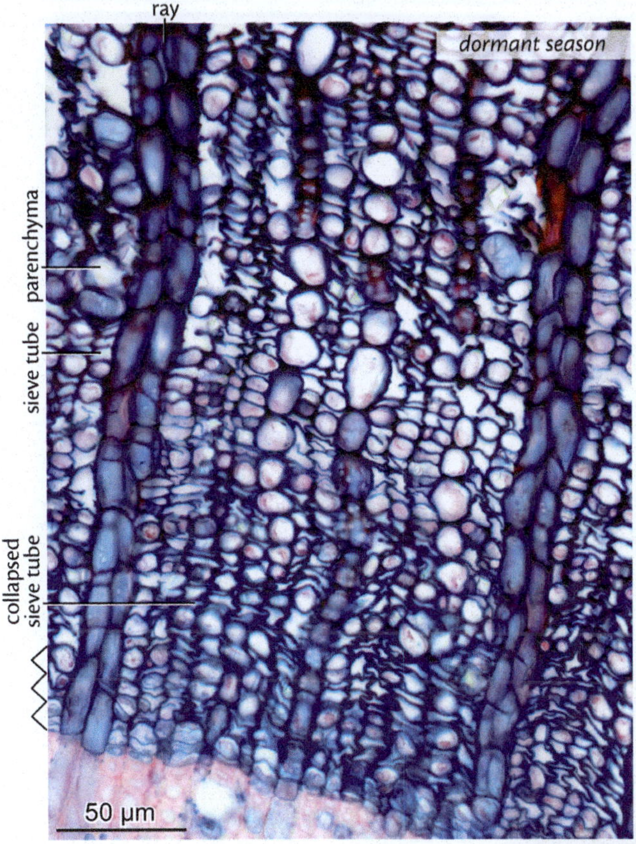

Bark anatomy: Structure with few groups of sclereids. Sieve tubes and parenchyma cells radially and tangentially arranged. Sieve tubes collapsing soon after formation. Fibers absent. Sclereids in small, tangentially arranged groups, occurring mainly in older tissues. Rays slightly wavy, with a few dilatations. Many prismatic crystals in axial parenchyma cells.

Phellem cells rectangular, thick-walled, filled with dark-staining substances.

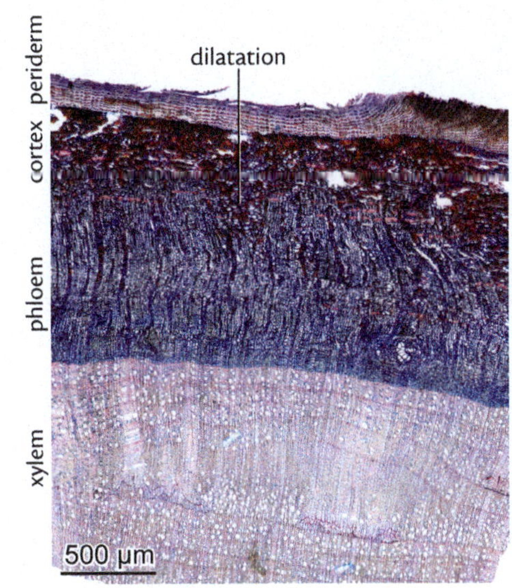

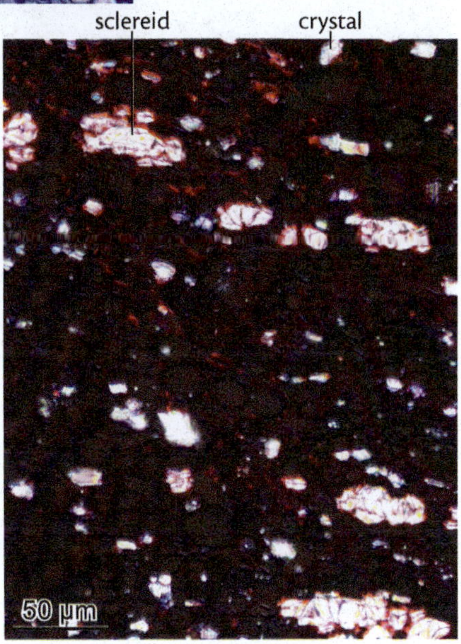

Myricaria germanica (L.) Desv. Tamaricaceae

German tamarisk, Deutsche Tamariske, tamarin d'Allemagne, tamerice alpina, taray europeo

Form: Shrub, multi-stemmed, erect, to 2 m tall.

Features: Leaves deciduous, alternate, in densely tiled scales, blue gray green, to 5 mm long. Flower a pink star, with five petals, fragrant, to 1 cm wide. Fruit a gray green capsule, to 12 mm long, releasing seeds with white flying hairs.

Habitat: Mountains of Eurasia, from the Pyrenees northeast to the Alps, Mid-Scandinavia, the Balkans, Turkey, Central Asia and the Himalayas.

Only riverine, on moist mud to fine sand in full sun, to 4400 m.

Bark: Young glossy, red brown, later dull, gray brown, with fine longitudinal fissures.

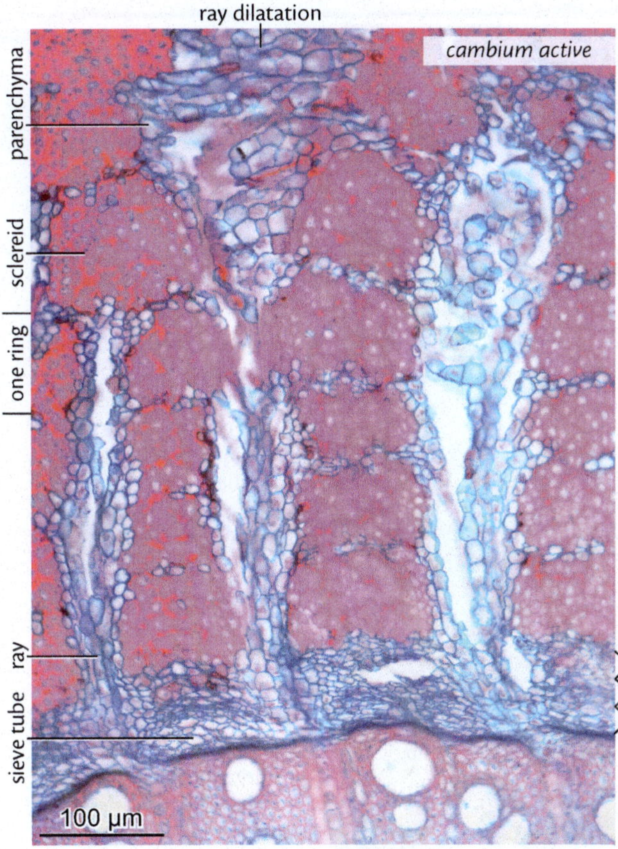

Bark anatomy: Radially structured tissue with many blocks of fiber groups.

Sieve tubes angular, parenchyma cells roundish. Sieve tubes collapsing soon after formation. Fibers absent. Sclereids in more or less square blocks, arranged in radial strips. Groups tangentially separated by unlignified parenchyma cells. Rays large, unlignified, slightly dilating. Small crystal druses occupying almost all axial parenchyma, arranged in regular tangential lines.

Phellem cells large, rectangular, in tangential lines, filled with dark-staining substances.

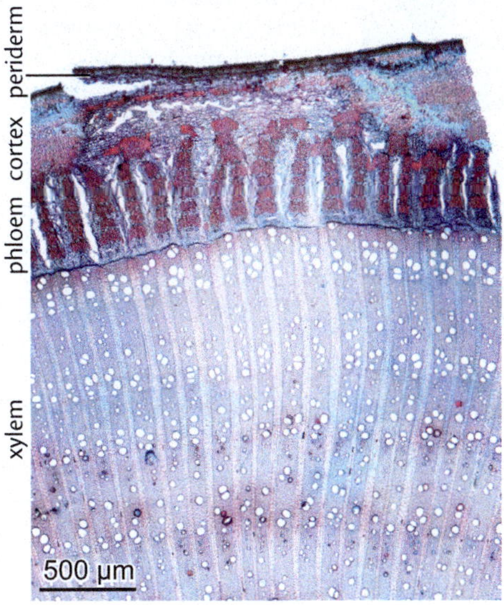

Rhamnus alpina L. Rhamnaceae

alpine buckthorn, Alpen-Kreuzdorn, nerprun des Alpes, ranno alpino, pudio bastardo

Form: Shrub, erect and often multi-stemmed, spineless, to 3 m tall.

Features: Leaves deciduous, alternate, ovate, margin finely serrated, conspicuous side veins parallel, to 6 cm long. Flowers yellow green, cross-shaped, to 12 mm wide. Fruit a glossy black, berry-like drupe, to 1 cm wide.

Habitat: Mountains of Central and Southern Europe, from the Sierra Nevada northeast to the Jura and Calcareous Alps of Central Switzerland, Southern Alps and the Apennines.

On dry, stony and calcareous soils, mostly over limestone rock or debris, to 1500 m.

Variation: From the Karawanks south to the Balkans replaced by the very similar *Rhamnus fallax*.

Bark: Young reddish gray, glossy, with light-colored lenticels, later dull, gray.

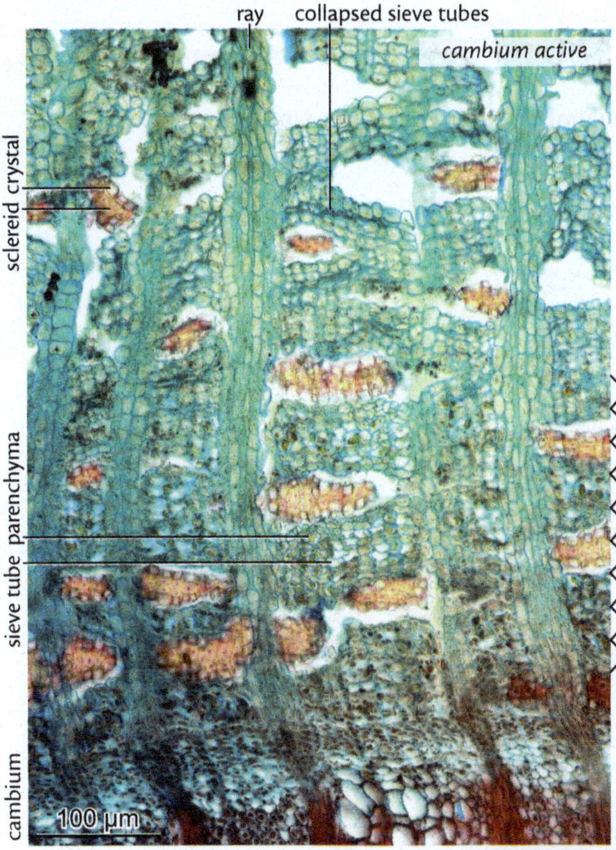

Bark anatomy: Irregular, unlignified tissue with many oval groups of fibers.

Sieve tubes angular, companion cells numerous and parenchyma cells roundish. Sieve tubes collapsing, indicated by dark, irregular lines. Fibers in tangentially arranged, oval groups, loosely connected with the sieve tube and parenchyma tissue. Sclereids rare, in groups towards the periderm. Rays straight, dilatations absent. Prismatic crystals surrounding fiber groups. Crystal druses occupying most axial parenchyma and ray cells. Each layer consisting of large, thick-walled cells.

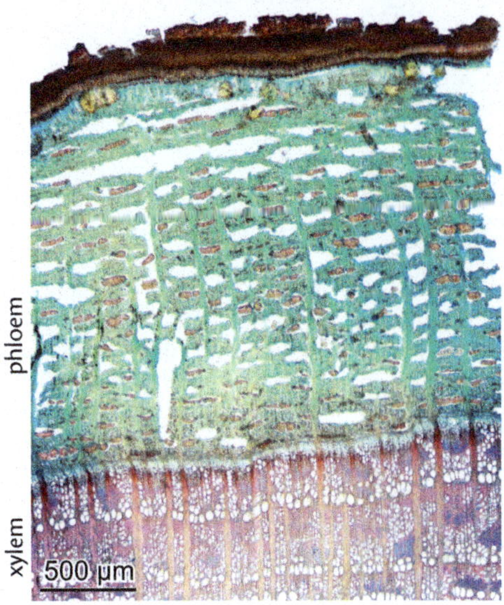

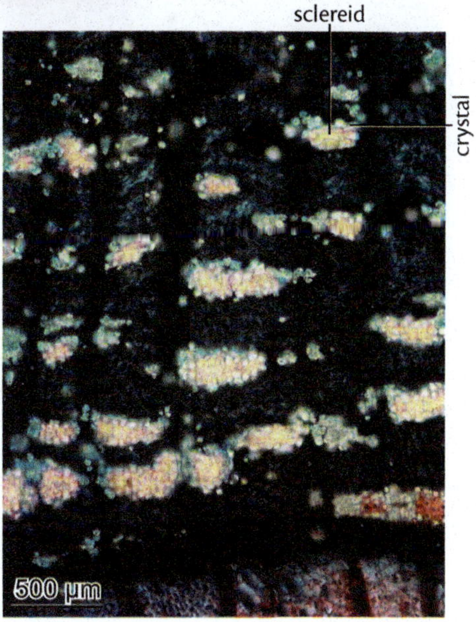

Ribes alpinum L. Grossulariaceae

mountain currant, Alpen-Johannisbeere, groseillier des Alpes, ribes alpino, grosellero de montaña

Form: Shrub, erect and mostly multi-stemmed, to 2 m tall.

Features: Leaves deciduous, alternate, three- to five-lobed, margin toothed, to 4 cm long. Flowers yellow green, with five petals, in an upright grape, to 6 mm wide. Fruits red berries in a hanging panicle, to 7 mm wide.

Habitat: Europe, from the Sierra de Cantabria northeast to Scandinavia, Lithuania, Ukraine, south to Bulgaria, Central Italy and Southern France.

On moist, stony and calcareous soils in shadowy locations, mostly forest and scree, to 2000 m.

Bark: Reddish or silvery gray brown, with age flaky and with fine longitudinal fissures.

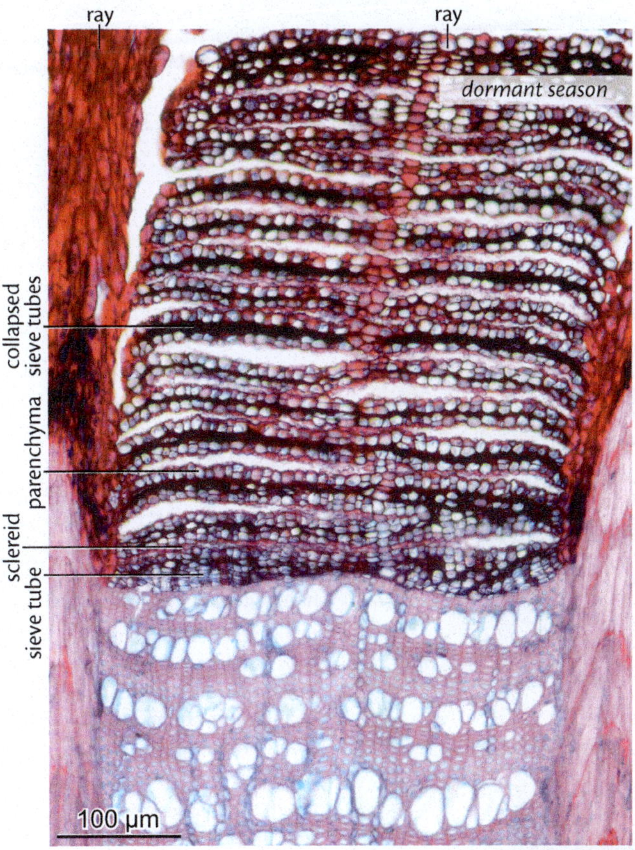

Bark anatomy: Regular, thin, tangential layers of parenchyma, sieve tubes and crystals.

Sieve tubes irregular shaped, parenchyma cells round. Sieve tubes collapsing soon after formation, forming dark lines. Fibers and sclereids absent. Crystal druses small, strictly arranged in tangential lines. Rays large, straight or slightly bent.

Periderm tangentially layered, consisting of rectangular, small, thin- to thick-walled cells.

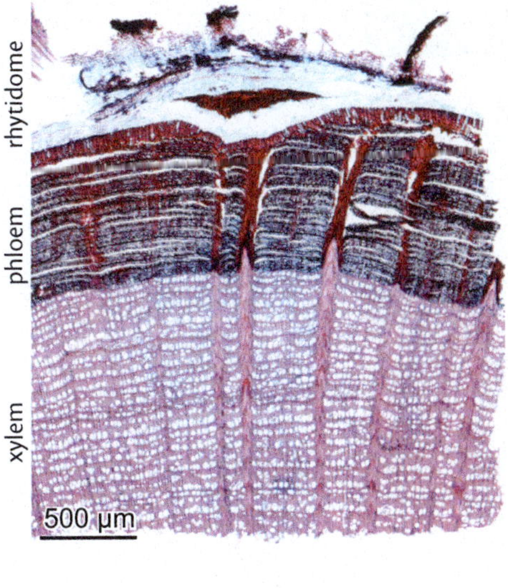

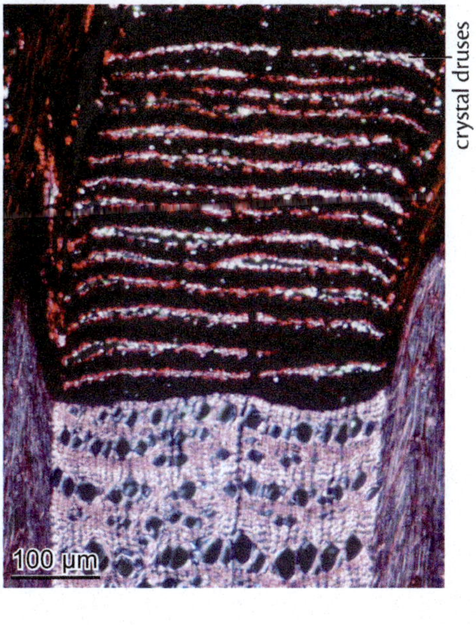

Ribes petraeum Wulfen Grossulariaceae

rock currant, Felsen-Johannisbeere, groseillier des rochers, ribes dei sassi, grosellero de las rocas

Form: Shrub, erect and mostly multi-stemmed, to 3 m tall.

Features: Leaves deciduous, alternate, three- to five-lobed, margin toothed, to 10 cm long. Flowers brown to wine red, with five petals, in a hanging grape, to 8 mm wide. Fruits dark red berries in a hanging grape, to 7 mm wide.

Habitat: Eurasia, from the Alps and Jura northeast to the Sudetes, Carpathians, Ural Mountains and east in Siberian mountains to the Amur River.

On moist, acidic, stony soils in shadowy locations, montane to subalpine scree and forest, to 2400 m.

Bark: Reddish or silvery gray brown, with horizontal lenticels, with age flaky and with fine longitudinal fissures.

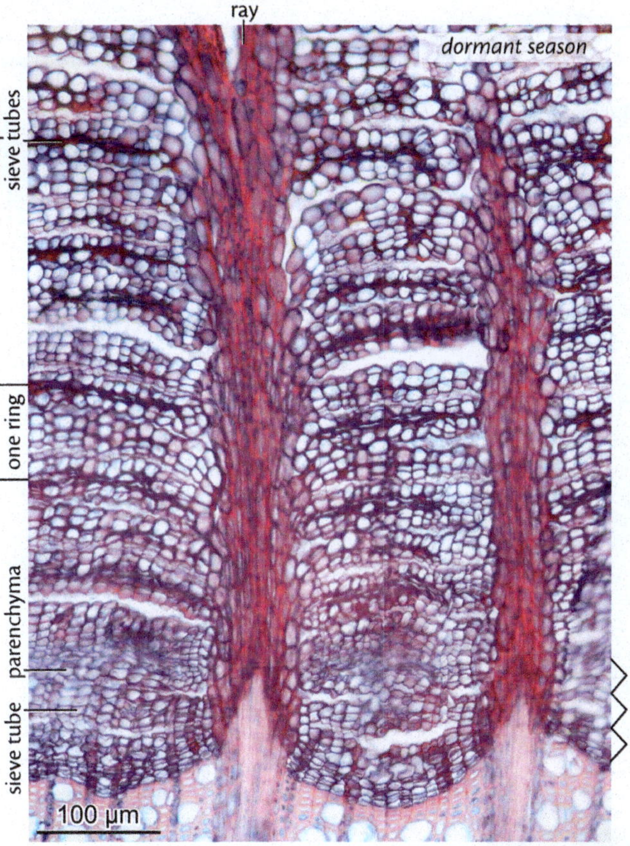

Bark anatomy: Regular, thin, tangential layers of parenchyma sieve tubes and crystals.

Sieve tubes irregular shaped, parenchyma cells round. Sieve tubes collapsing soon after formation, forming dark lines. Fibers and sclereids absent. Crystal druses small, strictly arranged in tangential lines. Rays large, straight or slightly bent.

Periderm tangentially layered, consistings of rectangular, small, thin- to thick-walled cells.

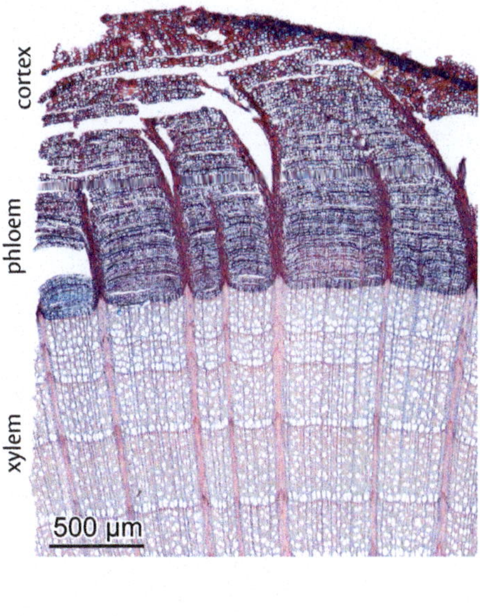

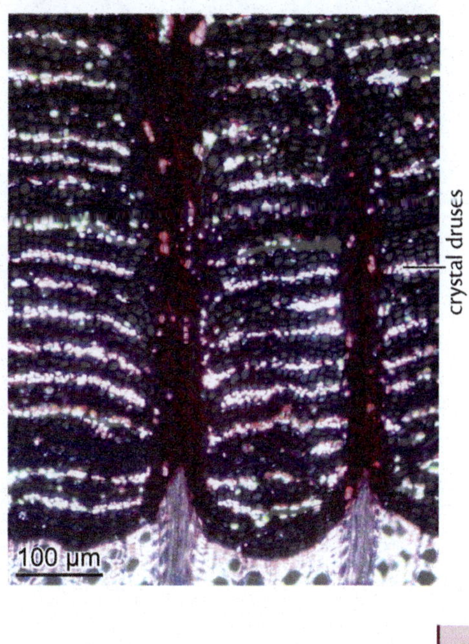

Salix appendiculata Vill.

Salicaceae

large-leaved willow, Großblättrige Weide, saule appendiculé, salice stipolato, sauce apendiculado

Form: Shrub, erect and mostly multi-stemmed, to 6 m tall.

Features: Leaves deciduous, alternate, long ovate, margin irregular toothed or entire, to 12 cm long. Flowers dioecious, female catkin erect, gray green, short, male catkin erect, yellow, to 3 cm long. Fruit a hanging, green catkin, to 5 cm long.

Habitat: The Alps, Jura, Black Forest and Bavarian Forest, Croatia.

On moist, calcareous and loamy soils, mostly in humid and shadowy locations, montane to subalpine ravines and forest, to 2200 m.

Bark: Young smooth, yellow or red brown, later smooth, light gray.

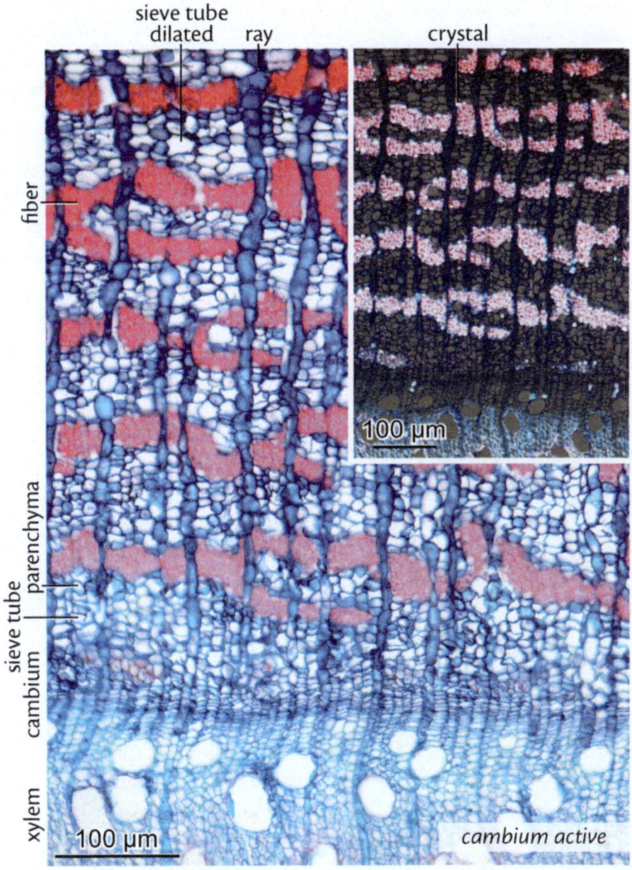

Bark anatomy: Regular tangential layers of parenchyma, sieve tubes and fibers.

Sieve tubes irregular shaped, companion cells small. Sieve tubes rarely collapsing. Parenchyma cells roundish. Fibers in tangential layers. Sclereids in groups in older parts of the phloem. Outer parts of the phloem with a few triangular dilations. Prismatic crystals and a few crystal druses along fiber bands.

Tangential layers of fibers not corresponding with annual rings. Periderm of old stems irregularly tangentially layered, consisting of rectangular, large, thin-walled cells and small, thick-walled cells.

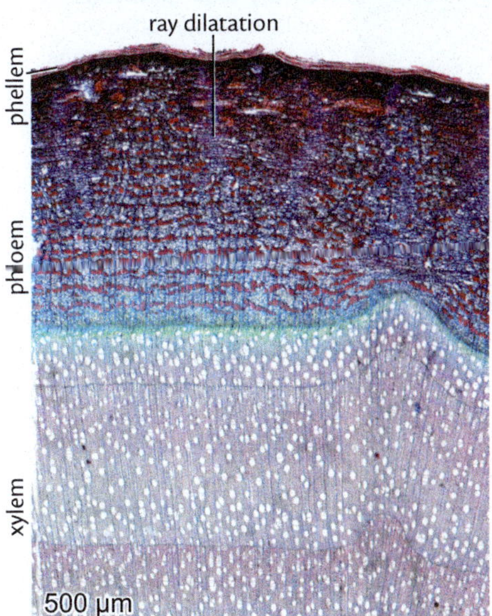

Sorbus aria (L.) Crantz Rosaceae

whitebeam, Echte Mehlbeere, alisier blanc, farinaccio, serbal blanco

Form: Upright tree, often multi-stemmed, crown round, to 15 m tall.

Features: Leaves deciduous, alternate, broad ovate, lower side densely white hairy, margin regularly double-teethed, to 15 cm long. Flowers white, in umbels. Fruits red orange, berry-like pomes in dense umbels, individual fruit to 15 mm wide.

Habitat: Mountains of Europe, from the Atlas Mountains northeast to Southern England, Germany, the Carpathians, south to Bulgaria, Northern Greece and Sicily.

On dry, often calcareous soils, in full sun, often on escarpments and limestone rock, to 1700 m.

Bark: Smooth, reddish gray to gray, often with conspicuous diamonds.

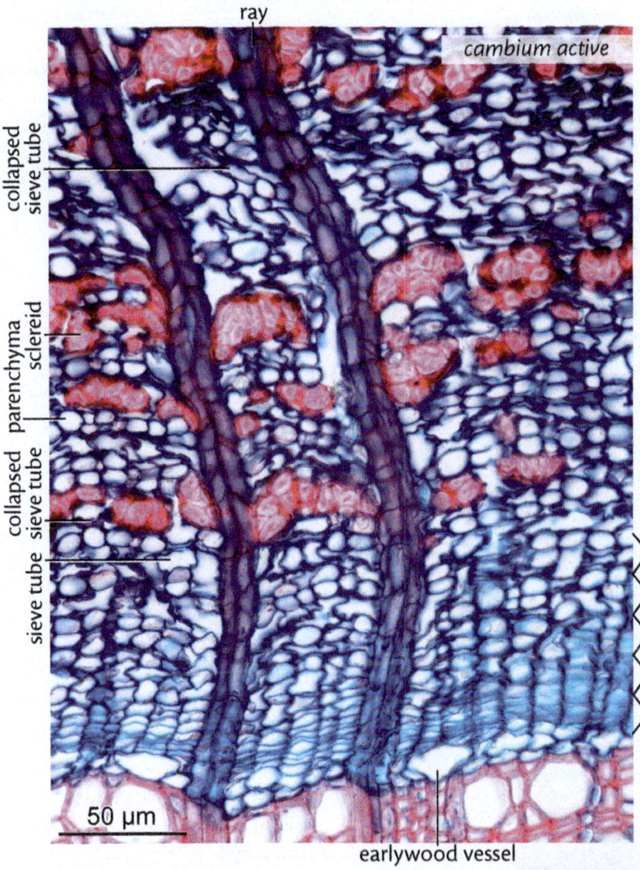

Bark anatomy: Tangentially layered structure. Sieve tubes collapsing before sclereid formation begins. Fibers absent. Tangential elongated groups of sclereids in small tangential layers. Rays straight at the beginning, later in more or less large waves. Few dilatations initiated outside of large rays. Many prismatic crystals, mainly in axial parenchyma cells. Very large prismatic crystals parallel the large rays.

Phellem cells in layers, rectangular, thick-walled, filled with dark-staining substances.

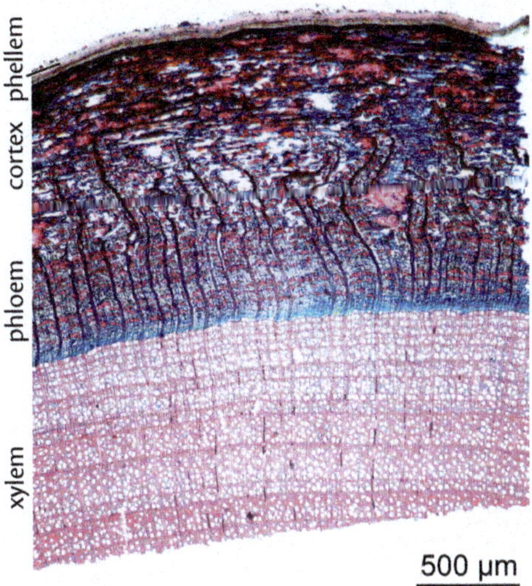

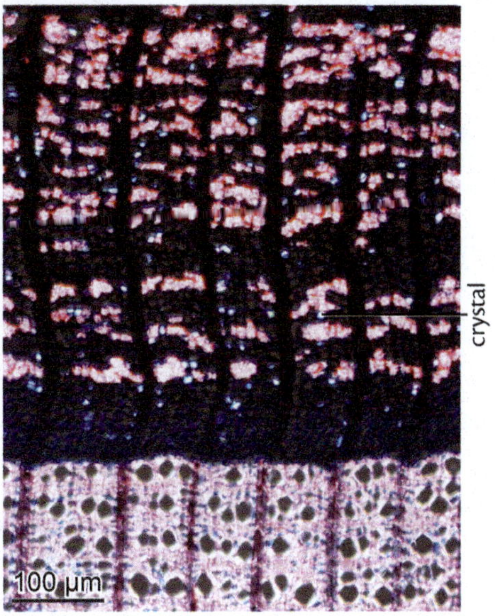

Sorbus chamaemespilus (L.) Crantz

Rosaceae

dwarf medlar, Zwerg-Mehlbeere, sorbier petit néflier, sorbo alpino, serbal enano

Form: Shrub, low and broad-crowned, to 2 m tall.

Features: Leaves deciduous, alternate, broad ovate, glossy green above, margin regularly teethed, to 7 cm long. Flowers pink, in umbels. Fruits red, berry-like pomes in umbels, individual fruit to 15 mm wide.

Habitat: Mountains of Central and Southern Europe, the Pyrenees, Jura, Alps, Carpathians, Dinarides, Balkans and North Apennines.

On stony, often calcareous soils, in montane and subalpine conifer forest and heath, to 2300 m.

Bark: Smooth, reddish gray to gray, often with conspicuous diamonds.

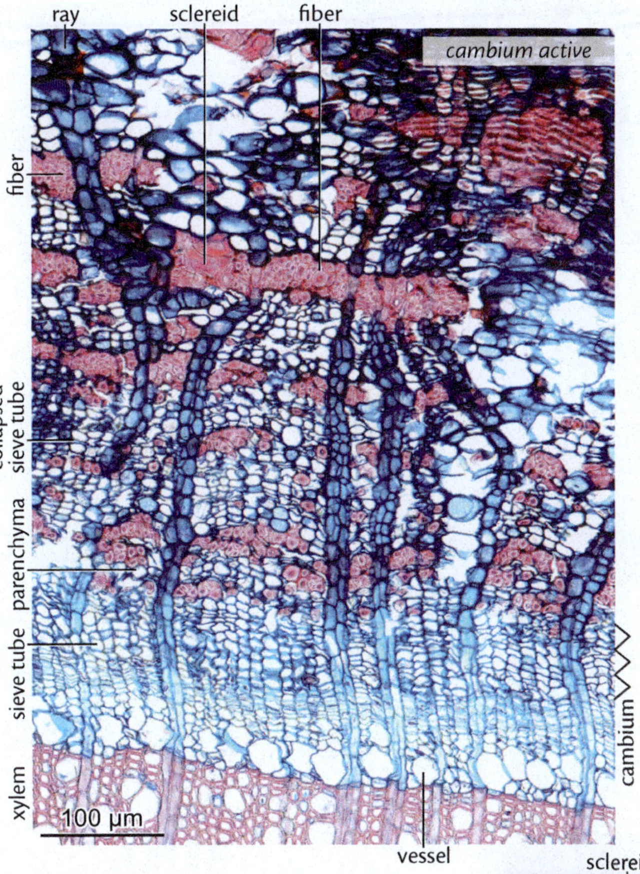

Bark anatomy: Tangentially layered structure. Sieve tubes collapsing just before sclereid formation begins. Fibers absent. Sclereids in tangential layers, solitary and in smaller or larger groups. Rays straight at the beginning, later in more or less large waves. Few dilatations initiated outside of large rays. Many small and large prismatic crystals in axial parenchyma cells.

Phellem cells in layers, rectangular, thick-walled.

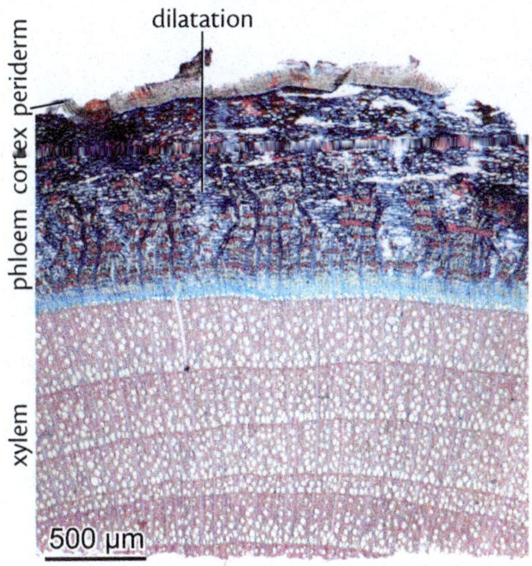

The main focus of this book, due to the huge variety of sampled bark material, is the **deciduous forest of temperate Europe**, represented by a variety of deciduous trees and shrubs.

In many areas of Central Europe, *Fagus sylvatica* is able to form vast, pure stands, whereas in Atlantic or more continental areas, mixed oak forests dominate.

F. H. Schweingruber et al., *Bark Anatomy of Trees and Shrubs in the Temperate Northern Hemisphere*, https://doi.org/10.1007/978-3-030-14056-4_5

5 Deciduous forest of temperate Europe

Taxus baccata L.

Taxaceae

yew, Eibe, if commun, tasso, tejo común

Form: Upright conifer, crown broad, conical, to 20 m tall.

Features: Needles evergreen, soft, flat, aromatic, dull green below, to 3 cm long. Flowers monoecious, female flower tiny, greenish globular, male flower tiny, yellowish globular. Brown black seed embedded in a fleshy, red, berry-like arillus.

Habitat: Europe and Western Asia, from North Africa northeast to Southern Scandinavia, Latvia, east to northwestern Iran, south to the Crimean and Mediterranean mountains.

On moist, deeply loamy, calcareous soils, often in dense shadow, to 1800 m.

Bark: Smooth and red brown, with age peeling in papery, deep red to gray brown flakes, always red brown underneath.

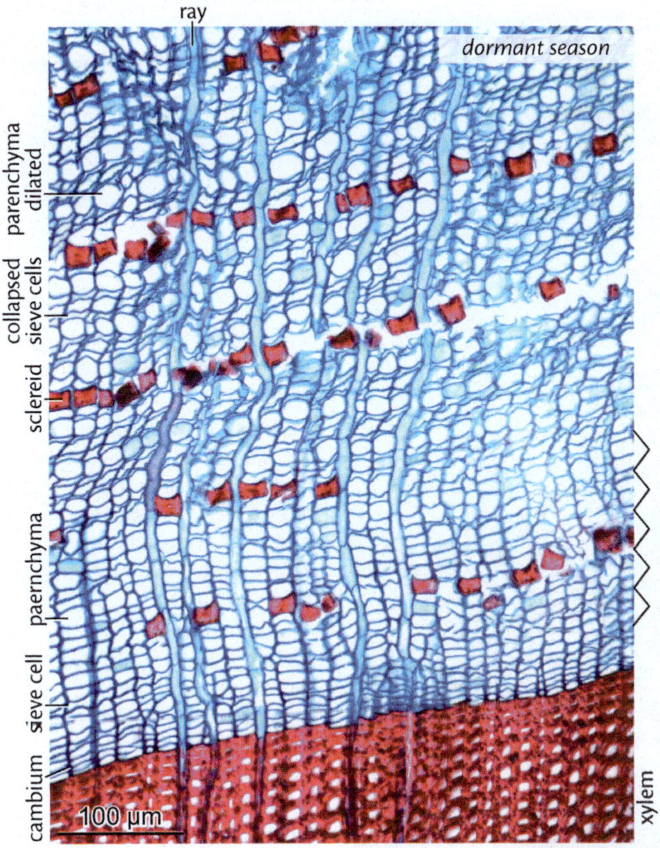

Bark anatomy: Regular tangential layers, consisting of crystal fibers, fiber-like sclereids, parenchyma cells and sieve cells.

Crystal fibers thin-walled, unlignified. Fiber-like sclereids lignified, thick-walled, square and rectangular. Parenchyma cells round to oval. Sieve cells rectangular, tending to collapse in later stages. Rays uniseriate, wavy at later stages. Crystal sand in tangentially oriented crystal fibers and in fiber-like sclereids. Crystals absent in parenchyma cells and rays. Dilatation in the form of larger parenchyma cells.

Annual rings distinct in zones with regular sclereid formation. Rings indistinct in zones without sclereids.

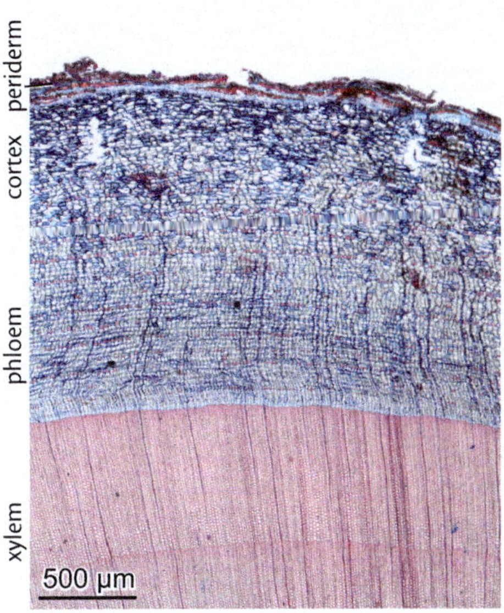

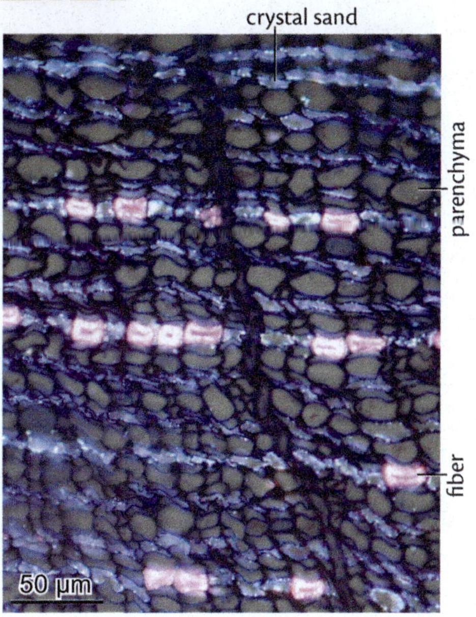

Acer campestre L. Sapindaceae

field maple, Feldahorn, érable champêtre, acero oppio, arce común

Form: Upright, often multi-stemmed tree, crown round, to 15 m tall.

Features: Leaves deciduous, opposite, three- to five-lobed, margin rounded, to 10 cm long. Flowers yellow green, in panicles. Fruits splitting, with two wings at a nearly horizontal angle.

Habitat: Europe and Western Asia, from Algeria northeast to Northern England, Southern Sweden and Poland, southeast to the Volga River, Caspian Sea, Turkey and Sicily.

On moist or dry, often calcareous soils, to 1400 m.

Bark: Twigs dull, light gray brown or markedly corky, later with longitudinal or net-like fissures, often with pinkish spots.

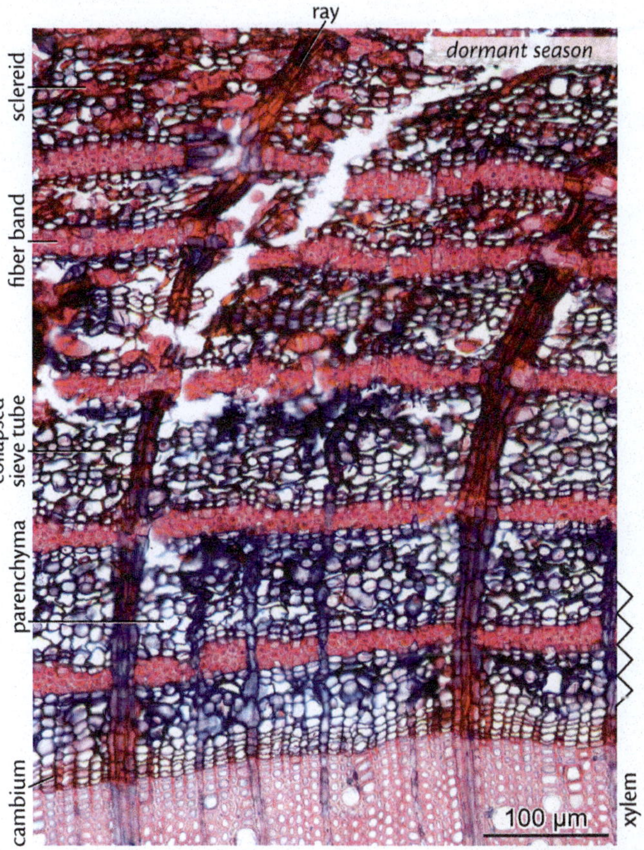

Bark anatomy: Regular tangential layers consisting of bands of fibers, and sieve tubes and parenchyma zones.

Early phloem formed by irregular shaped sieve tubes and roundish parenchyma zones. Sieve tubes collapsing within a few years. Fibers in 2–4-layered, tangential bands. Sclereids in large, irregular shapes. Ray structure lasting for many years. Prismatic crystals mainly within fibers and around fiber zones. Crystals absent in rays.

Annual ring boundaries generally corresponding with fiber bands.

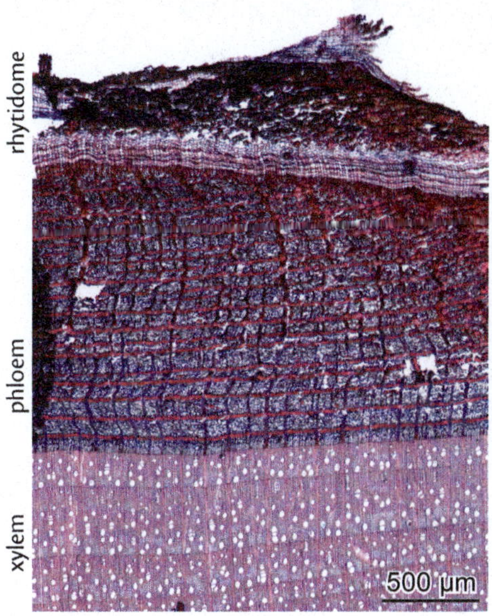

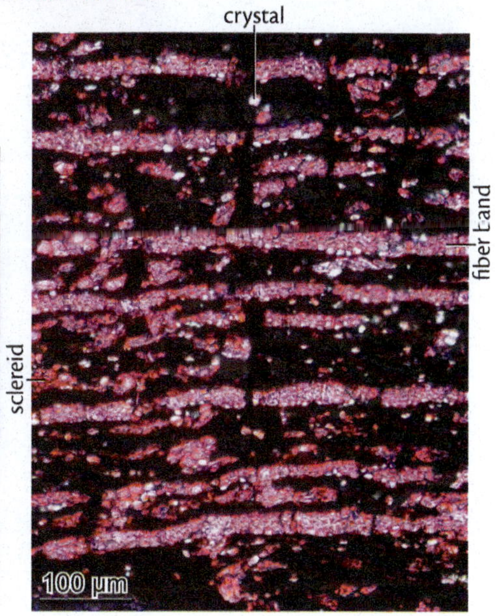

Acer platanoides L.

Sapindaceae

Norway maple, Spitzahorn, érable plane, acero riccio, arce real

Form: Upright tree, crown round to pyramidal, to 30 m tall.

Features: Leaves deciduous, opposite, three- to five-lobed, lobes ending in sharp tips, to 20 cm long. Flowers yellow green, in upright pseudo-umbels, emerging before the foliage. Fruits splitting, with two wings at an obtuse angle.

Habitat: Europe and Western Asia, from the Pyrenees northeast to Southern Sweden, Finland and the Ural Mountains, southeast to Northern Iran, Turkey and Greece.

On moist, often calcareous soils, to 1600 m.

Bark: Young smooth, light gray brown, later with characteristic longitudinal fissures.

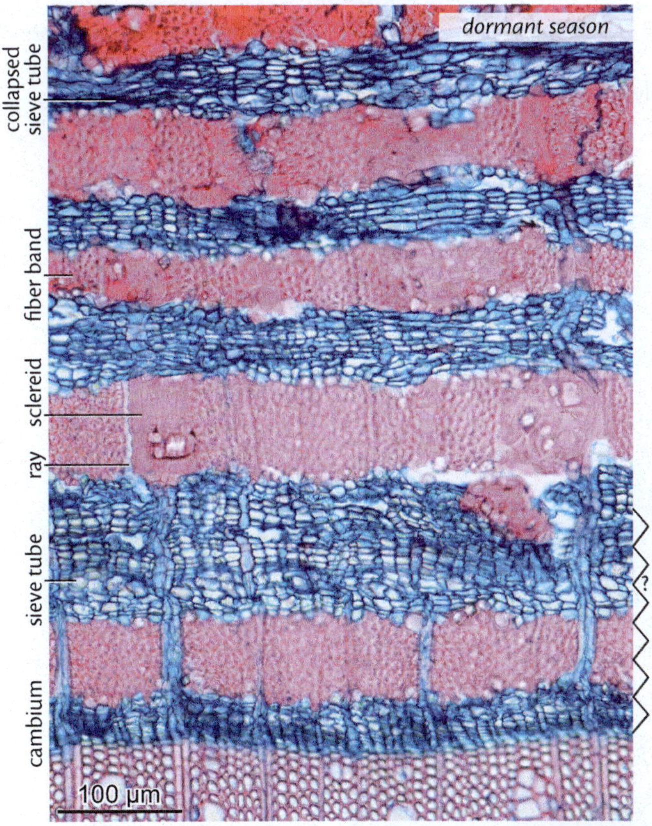

Bark anatomy: Regular tangential layers of fiber bands, sieve tubes and parenchyma zones.

Early phloem formed by irregular shaped sieve tubes and roundish parenchyma zones. Fibers in 5–10-layered tangential bands. Sclereids in irregular shaped groups. Ray structure lasting for a few years. Prismatic crystals around and in between the fiber bands as well as in sclereids within the in rays.

Annual ring boundaries generally corresponding with fiber bands.

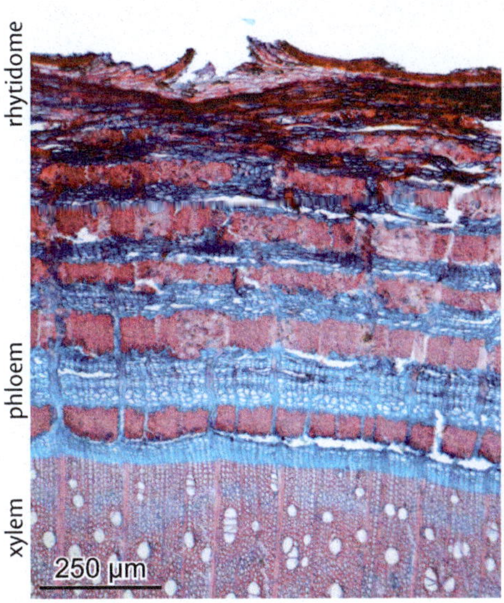

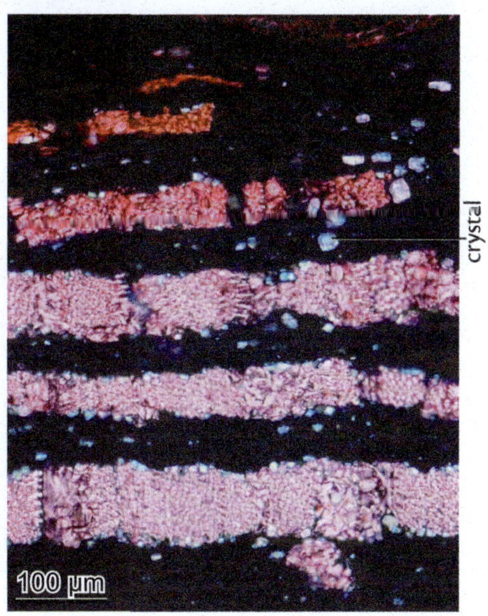

Acer pseudoplatanus L.

Sapindaceae

sycamore, Bergahorn, érable sycomore, acero di monte, arce blanco

Form: Upright tree, crown round to pyramidal, to 35 m tall.

Features: Leaves deciduous, opposite, three- to five-lobed, lobes ending in rounded tips, to 20 cm long. Flowers pale yellow green, in hanging panicles, emerging with the foliage. Fruits splitting, with two wings at right angles.

Habitat: Europe and Western Asia, from the Pyrenees northeast to southern Poland, Ukraine and the Caucasus, southeast to northern Turkey, Greece and Sicily.

On deep, moist, often calcareous soils, in shadowy or humid locations, to 1900 m.

Bark: Young smooth, light gray brown, later with characteristic irregular shaped plates, peeling to reveal pink or reddish colors underneath. Old trees often with lush lichen or moss cover.

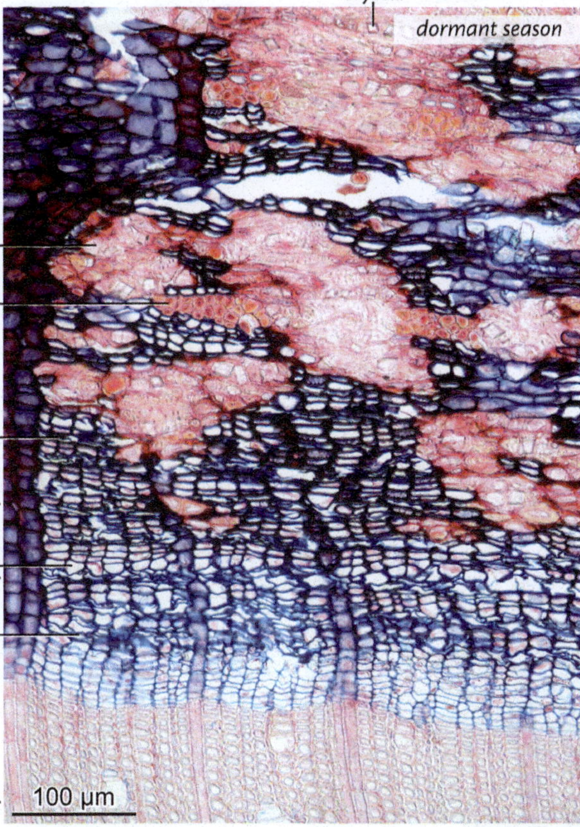

Bark anatomy: Differs from *Acer campestre* and *A. platanoides* in the absence of tangential layering.

Sieve tubes rectangular. Parenchyma cells irregular shaped to roundish, in tangential bands. Sieve tubes collapsing very early. Fibers almost absent. Sclereids very frequent in irregular shaped groups. Rays disappearing after some years due to dilation. Prismatic crystals mainly within sclereids. Crystals absent in rays.

Annual ring boundaries almost absent.

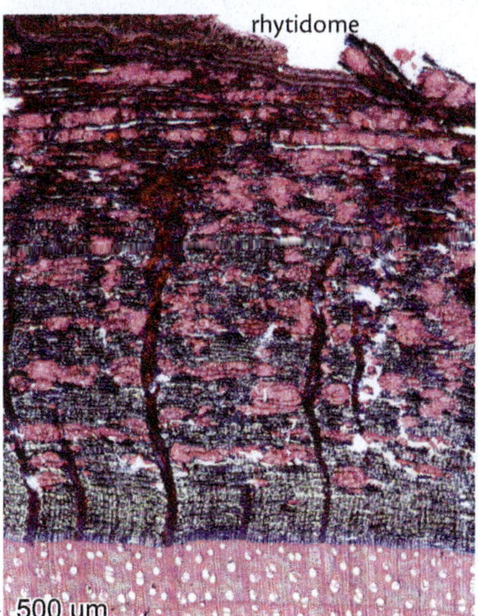

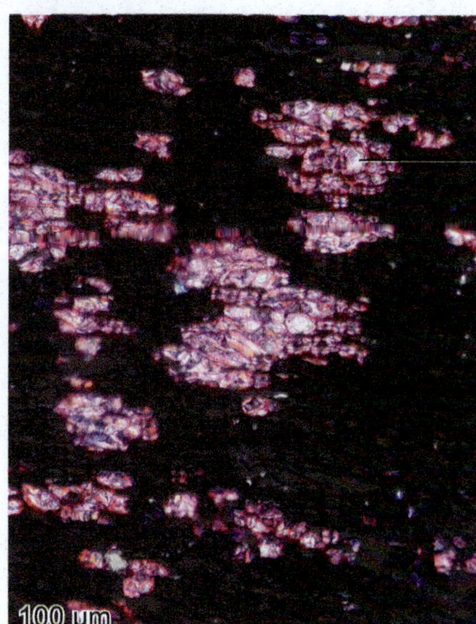

Alnus glutinosa (L.) Gaertn. Betulaceae

alder, Schwarzerle, aulne glutineux, ontano comune, aliso negro

Form: Upright, often multi-stemmed tree, to 20 m tall.

Features: Leaves deciduous, alternate, ovate, margin double-toothed, tip pointed, to 10 cm long. Flowers monoecious, female catkin dark red, short, male catkin hanging, yellow, to 4 cm long. Seed cones dark brown, ovate, to 2 cm long.

Habitat: Northeastern Europe, from the Rhône River northeast to the whole of Scandinavia, northern Russia to the Ural Mountains, south to the Balkans and Northern Italy.

On moist to wet, sandy to gravelly soils, mostly riverine, to 1000 m.

Bark: Young smooth, green brown, later characteristically dark brown to nearly black, with longitudinal fissures cracking into angular plates.

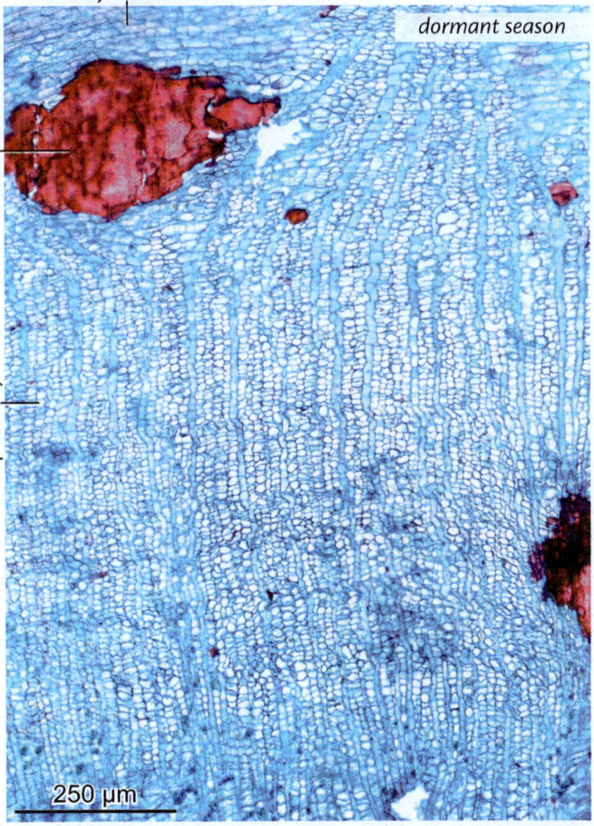

Bark anatomy: Structure variable, likely impossible to distinguish from *Alnus incana*.

Homogeneous structure of sieve tubes and parenchyma with a few dense groups of sclereids. Sieve tubes irregular shaped, collapsing very soon after formation. Parenchyma cells roundish. Fibers absent. Sclereids in groups, often in continuation of aggregate rays. Ray cell dilatation without cell divisions. Prismatic crystals, if present, located around fiber zones.

Periderm consisting of many layers of rectangular, large, thin-walled and small, thick-walled cells.

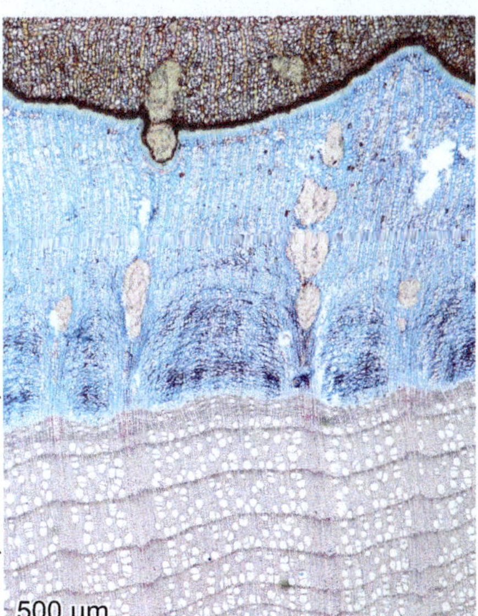

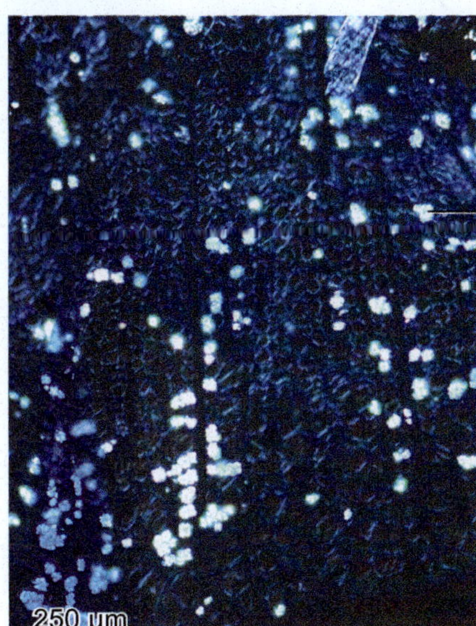

Alnus incana (L.) Moench

Betulaceae

gray alder, Grauerle, aulne blanchâtre, ontano bianco, aliso gris

Form: Upright tree, trunk straight, branches horizontal, to 30 m tall.

Features: Leaves deciduous, alternate, ovate, margin double-toothed, tip pointed, to 10 cm long. Flowers monoecious, female catkin dark red, short, male catkin hanging, yellow, to 4 cm long. Seed cones dark brown, ovate, to 2 cm long.

Habitat: Europe, from northern Morocco northeast to Southern Scandinavia, the Ural Mountains and the Volga Delta, around the Black Sea, south to Greece and Tunisia.

On constantly moist to wet, deep soils on swampy ground, to 1800 m.

Bark: Light gray, remaining smooth with age, often dotted or banded with white lichens.

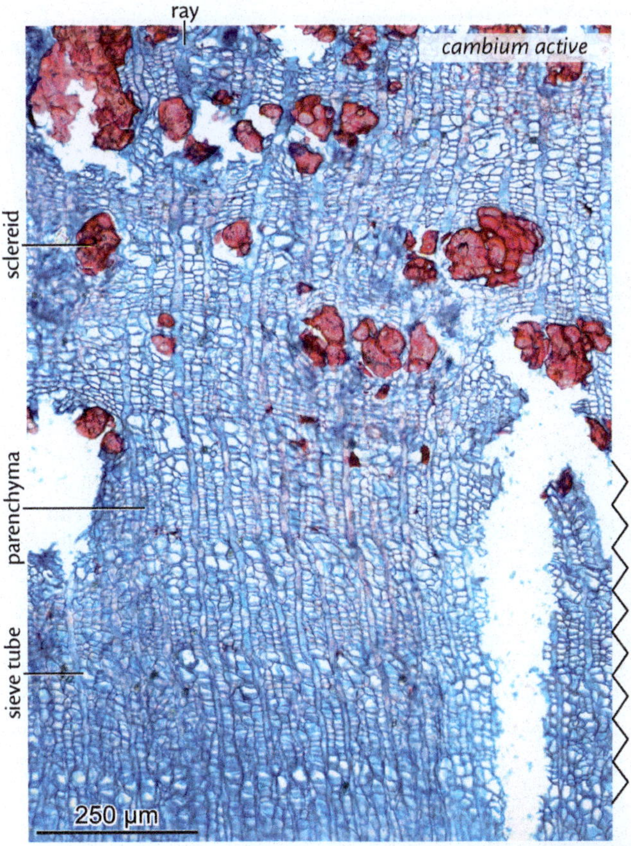

Bark anatomy: Structure variable, likely impossible to distinguish from *Alnus glutinosa*. Tangentially layered structure of sieve tubes and parenchyma with many dense groups of sclereids.

Tangential layers of sieve tubes at the beginning of the early bast. Parenchyma cells roundish. Fibers absent. Sclereids in groups. Ray cell dilatation without cell divisions. Many crystal druses and prismatic crystals.

Periderm consisting of layers of rectangular, large, thin-walled cells and small, thick-walled cells.

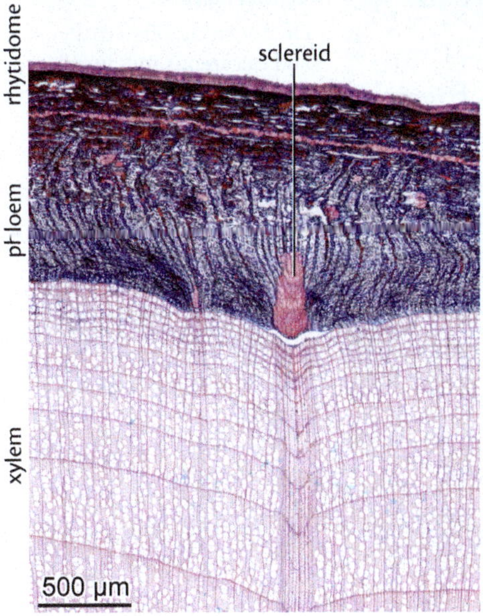

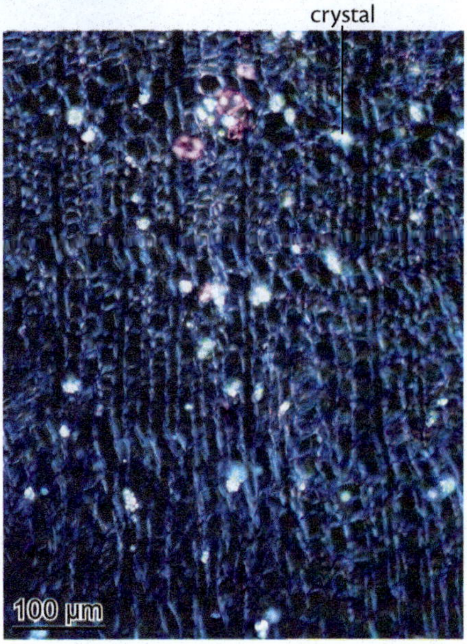

Berberis vulgaris L. Berberidaceae

barberry, Berberitze, épine-vinette, crespino comune, agracejo

Form: Multi-stemmed, upright and thorny shrub, to 3 m tall.

Features: Leaves deciduous, alternate in clusters, ovate, margin sharply toothed, to 6 cm long, accompanied by three-branched, sharply pointed, light brown thorns. Flowers lemon yellow, with six petals and six sepals, in multi-flowered, hanging panicles. Fruit an elongated, scarlet berry, edible.

Habitat: Southern Europe and Western Asia, from the Sierra Nevada northeast to Southern Scandinavia, the Lower Volga Region, southeast to Northern Iran and Turkey, Greece and Central Italy.

On dry, stony, often rocky and calcareous soils in full sun, to 2600 m.

Bark: Light gray, cracking with age.

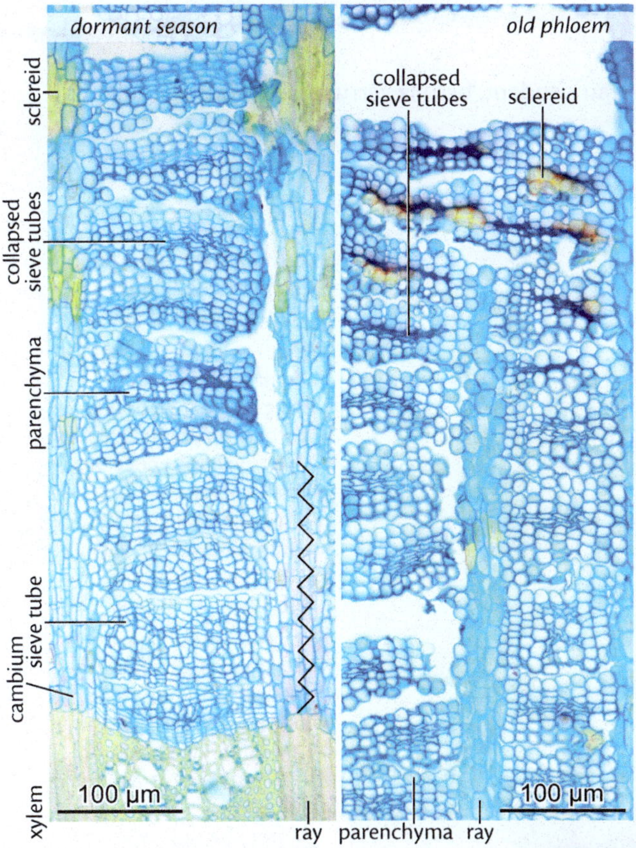

Bark anatomy: Regular tangential layers of parenchyma and sieve tubes and occasionally sclereids.

Sieve tubes irregular shaped, companion cells small, in groups. Sieve tubes collapsing within a few years. Parenchyma cells roundish. Fibers absent. Sclereids in groups in older parts of the phloem and in irregular groups within the large rays. Rays large, dilating after some years. Prismatic crystals mainly in rays.

Tangential layers not corresponding with annual rings. Periderm tangentially layered, consisting of rectangular, large, thin-walled cells and small, thick-walled cells.

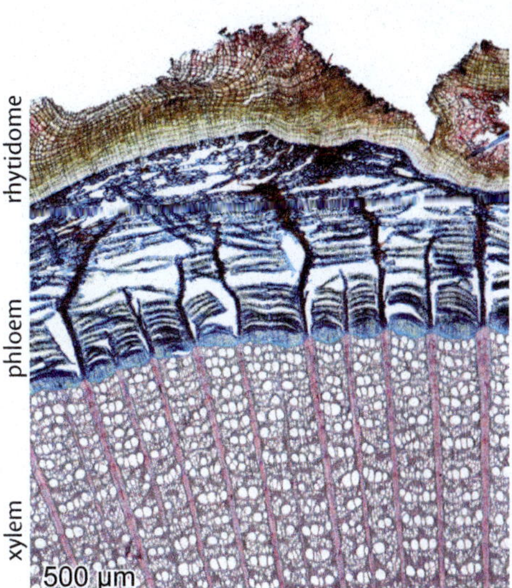

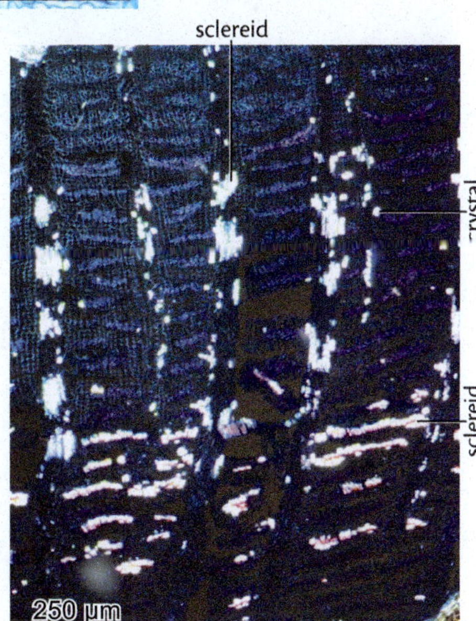

Betula pendula Roth

Betulaceae

silver birch, Hängebirke, bouleau pendant, betulla verrucosa, abedul péndulo

Form: Upright, often multi-stemmed tree, branches drooping, to 25 m tall.

Features: Leaves deciduous, alternate, rhomboid to triangular, margin double-toothed, tip pointed, to 6 cm long. Flowers monoecious, female catkin small, green, male catkin hanging, yellow, to 10 cm long. Seed catkins dark brown, cylindrical, hanging, to 3 cm long.

Habitat: Eurosiberia, from Catalonia northeast to the whole of Scandinavia, the northern Yenisei River to the Altai Mountains, south to Northern Iran, Macedonia and Sicily.

On dry to moist, often sandy or peaty soils, acting as a pioneer, to 1800 m.

Bark: Twigs glossy gray brown, young stem bark pure white and smooth, peeling off in papery flakes, with age spotted black and roughly borky.

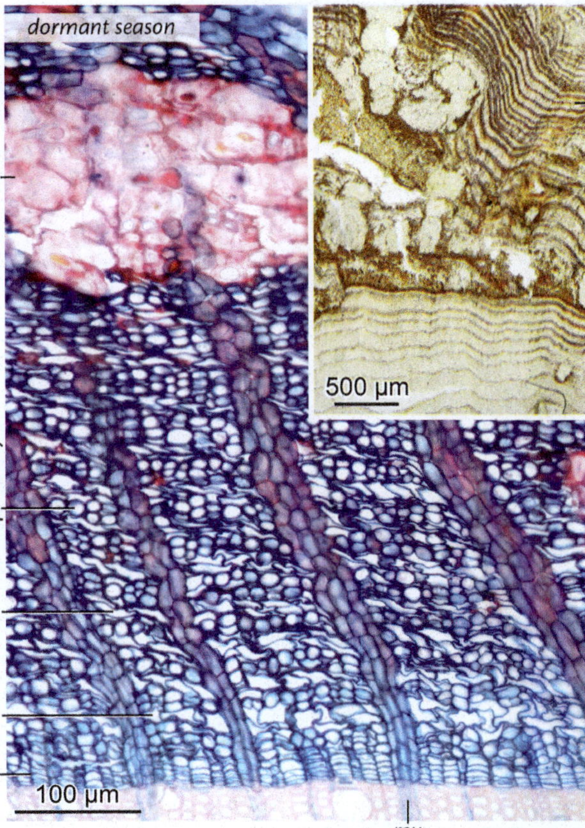

Bark anatomy: Slightly tangentially layered groups of sclereids. Sieve tubes irregular shaped, collapsing very soon after formation. Parenchyma cells roundish. Fibers absent. Sclereids in groups, often in tangential layers. Ray cell dilatation without cell divisions. Crystals absent.

Periderm consisting of many layers of rectangular, large, thin-walled cells and small, thick-walled cells.

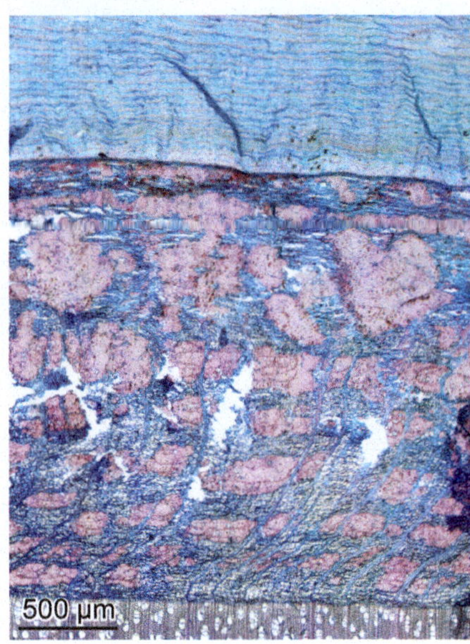

Carpinus betulus L.

Betulaceae

hornbeam, Hainbuche, charme, carpino comune, carpe

Form: Upright tree, crown irregular shaped, to 25 m tall.

Features: Leaves deciduous, alternate, ovate, margin double-toothed, tip pointed, to 10 cm long. Flowers monoecious, female catkin small, yellow green, hanging, male catkin yellow, hanging, to 6 cm long. Fruit an ovate nut, embedded in three-parted wings to 5 cm long, fruits bundled in seed catkins.

Habitat: Europe, from the Pyrenees northeast to South Sweden, the Baltic states and Ukraine, south to the Caucasus, the Balkans and the Apennines.

On moist soils, often drying out in summer, preferring warm locations, to 1000 m.

Bark: Young gray and smooth, stems later conspicuously fluted.

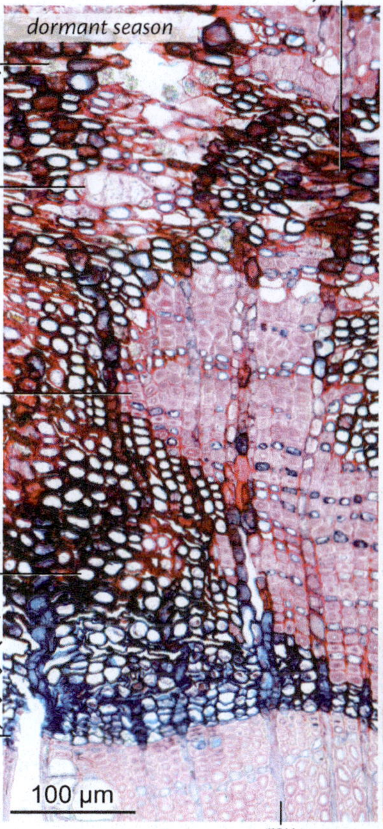

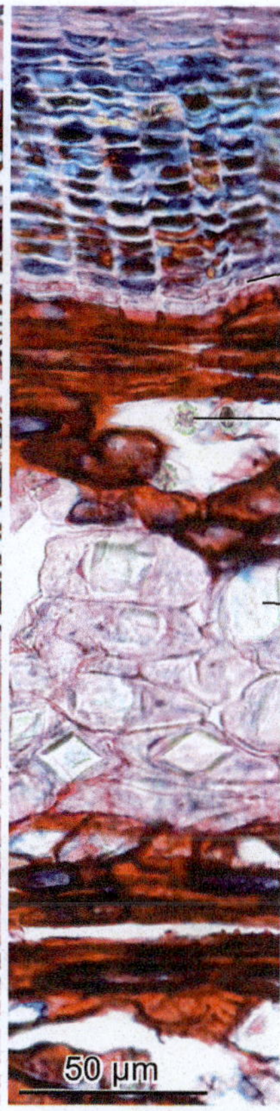

Bark anatomy: Dense, irregular structure with large groups of sclerenchyma.

Sieve tubes irregular shaped, collapsing very soon after formation. Parenchyma cells roundish. Fibers in large, often flame-like groups in continuation of aggregate rays. Sclereids in irregular groups. With a few ray dilatations. Prismatic crystals and crystal druses in all cell types.

Periderm consisting of many layers of rectangular, large, thin- to thick-walled cells.

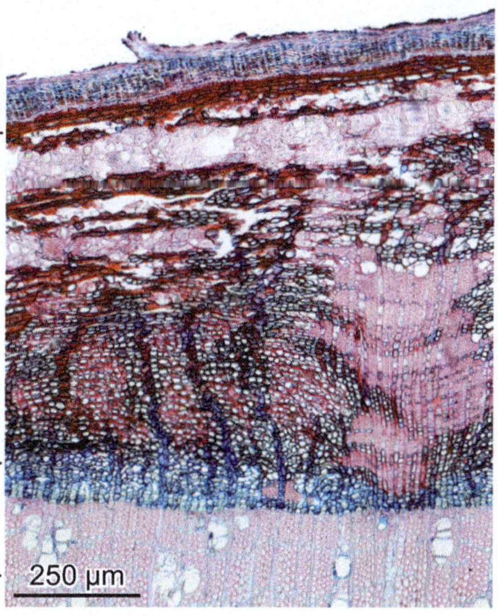

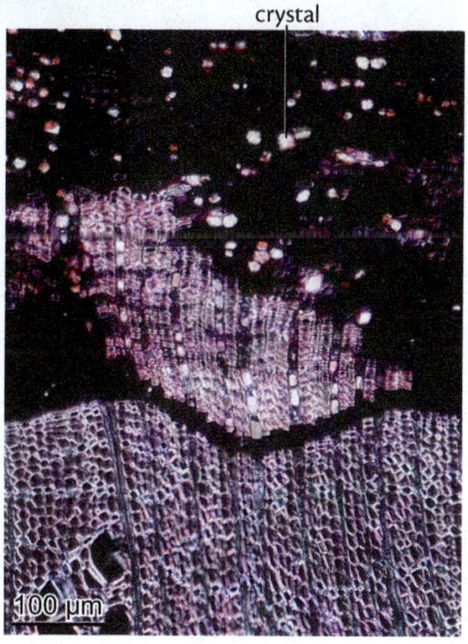

Cornus sanguinea L. Cornaceae

dogwood, Roter Hartriegel, cornouiller sanguin, corniolo sanguinello, cornejo sanguino

Form: Upright shrub, twigs dark red on the upper side, green on the underside, to 4 m tall.

Features: Leaves deciduous, opposite, ovate, with prominent veins in an arc-like bend, margin entire, to 8 cm long. Flowers white, four-parted, in pseudo-umbels. Fruit a blue black, berry-like drupe, in pseudo-umbels.

Habitat: Europe, from Northern Portugal northeast to Southern Scandinavia, Moscow, southeast to Greece, Sicily and Southern France.

On dry to wet soils, to 1500 m.

Bark: Twigs deep carmine above, green below, stem bark light gray brown, with age finely fissured.

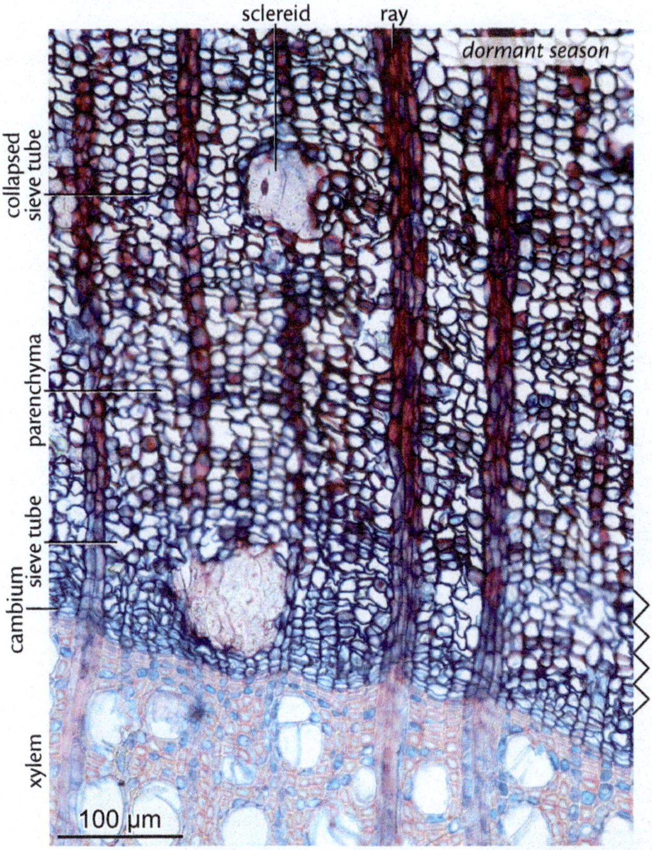

Bark anatomy: Soft structure of sieve tubes and parenchyma with a few round groups of sclerenchyma. Homogeneous tissue formed by angular sieve tubes and round to oval parenchyma cells. Sclereids frequent, in isolated, large groups. Fibers absent. Rays straight, not dilated. Prismatic crystals frequent in all cell types.

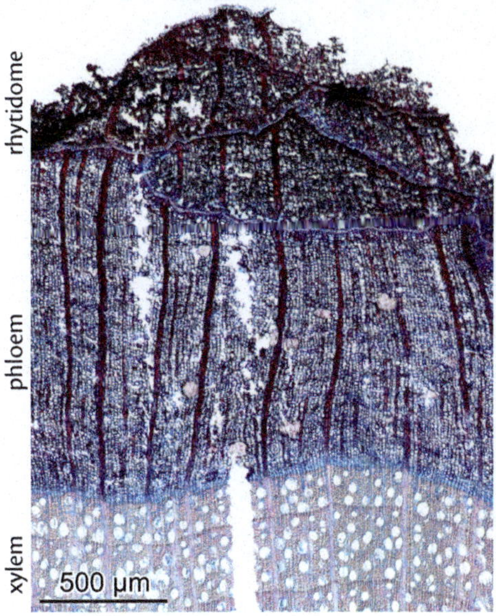

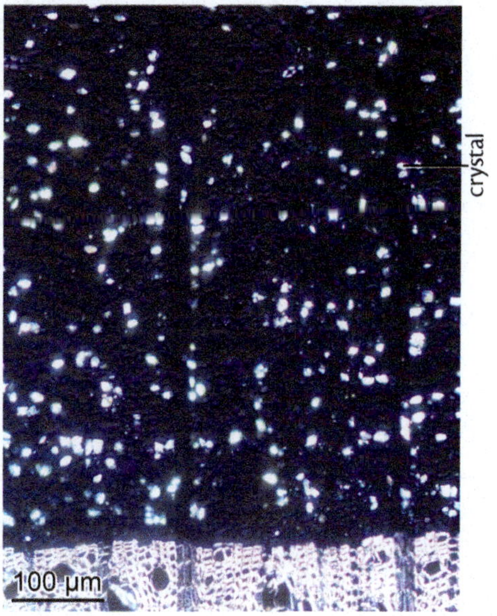

Corylus avellana L.

Betulaceae

hazel, Haselstrauch, noisetier, nocciòlo comune, avellano común

Form: Multi-stemmed, upright shrub, rarely a tree, to 15 m tall.

Features: Leaves deciduous, alternate, round ovate, felty hairy, margin double-toothed, tip pointed, to 10 cm long. Flowers monoecious, female catkins very small, deep red, concealed with only the styles visible, to 3 mm, male cone hanging, yellow, to 10 cm long. Fruit a globular nut with a smooth shell, kernel of the seed is edible.

Habitat: Europe, from the Mediterranean mountains northeast to Scotland, Scandinavia, the Ural Mountains, south to the Caucasus, northern Turkey and the Balkans and Apennines.

On dry to moist soils, to 1500 m.

Bark: Reddish gray and smooth, with horizontal white lenticels.

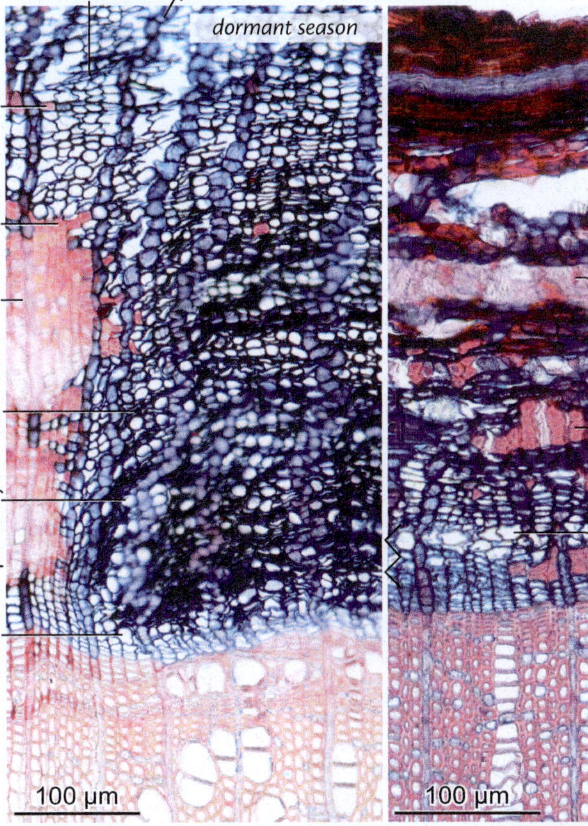

Bark anatomy: Dense structure with large groups of sclerenchyma.

Sieve tubes irregular shaped, collapsing very soon after formation. Parenchyma cells roundish. Fibers in large groups in continuation of aggregate rays. Sclereids and fibers in irregular and tangential groups. Uniseriate rays dilating. Prismatic crystals and crystal druses in all cell types.

Periderm consisting of many layers of rectangular, large, thin- to thick-walled cells.

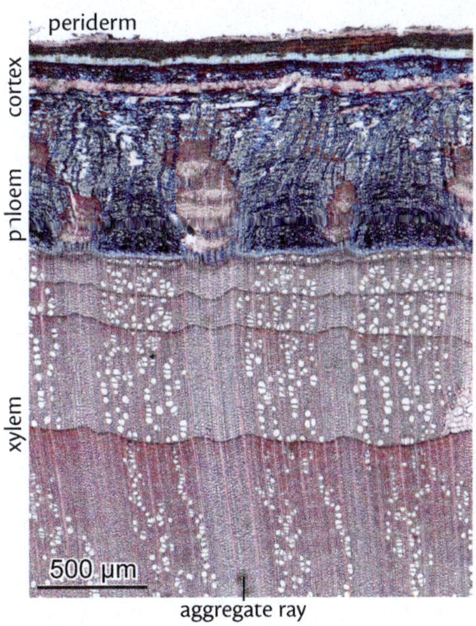

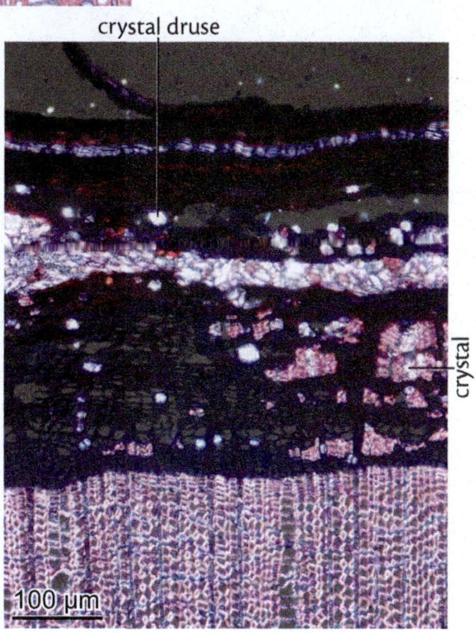

Cotoneaster integerrimus Medik. Rosaceae

wild cotoneaster, Gemeine Steinmispel, cotonéaster commun, cotognastro minore, griñolera

Form: Irregular shaped shrub, often multi-stemmed, to 3 m tall.

Features: Leaves deciduous, alternate, ovate, white hairy below, margin entire, to 4 cm long. Flowers small, white, bell-shaped, hanging, in corymbs of one to three. Fruits berry-like, red, to 8 mm wide.

Habitat: Europe and Southwest Asia, from Northern Spain northeast to Scandinavia, southeast to Northern Iran, the Caucasus, the Balkans and Southern Italy.

On dry, stony soils in full sun, to 2800 m.

Bark: Twigs first red brown, later gray, stem bark gray and smooth.

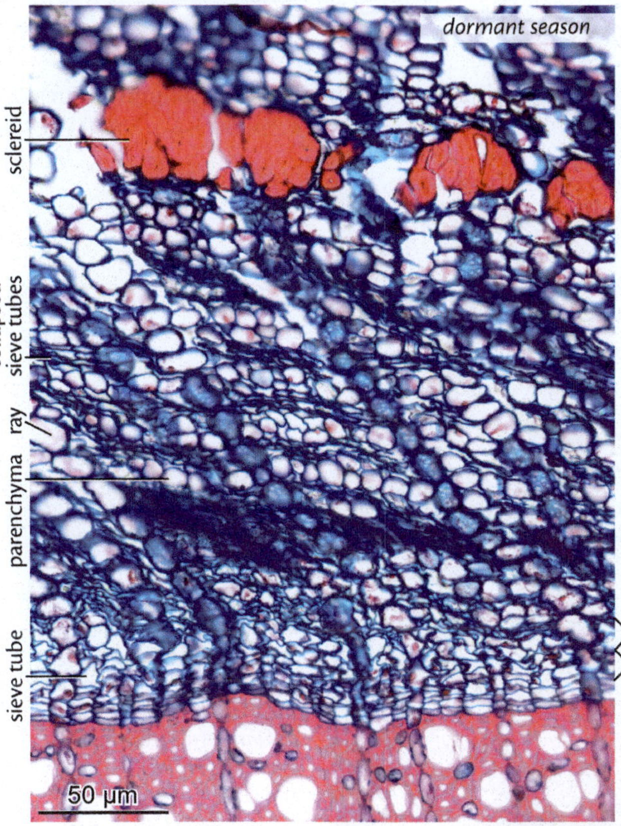

Bark anatomy: Structure with some groups of sclereids. Sieve tubes and parenchyma cells tangentially arranged. Sieve tubes collapsing soon after formation, collapsed sieve tubes appearing as unlignified, compact, tangential bands. Fibers absent. Long sclereids in many tangential bands, beginning in the cambial zone. Rays wavy. Dilatations absent. Prismatic crystals frequent in axial parenchyma cells and rays.

Phellem cells rectangular, thick-walled, filled with dark-staining substances.

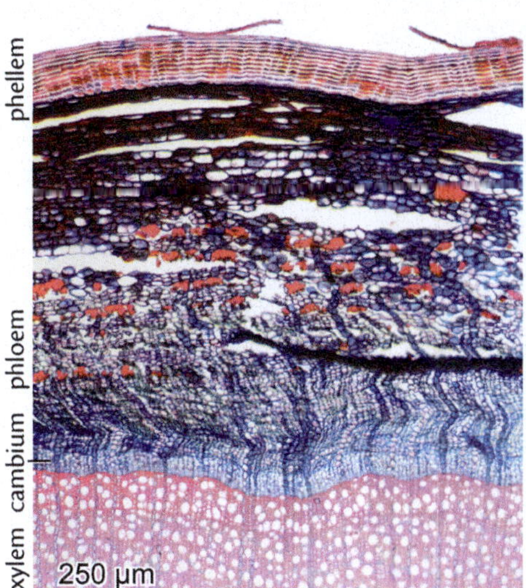

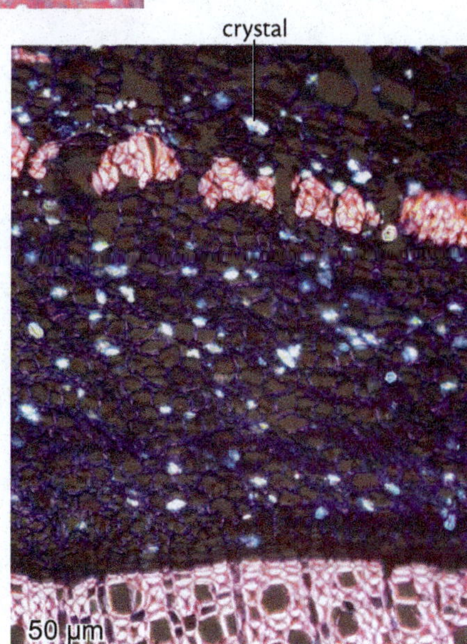

Crataegus monogyna Jacq. Rosaceae

hawthorn, Eingriffliger Weißdorn, aubépine à un style, biancospino comune, majuelo

Form: Irregular shaped shrub, often multi-stemmed, thorny, to 6 m tall.

Features: Leaves deciduous, alternate, sharply lobed, margin toothed, to 4 cm long. Flowers white, five-parted, with one style. Fruit a berry-like pome, dark red, to 8 mm wide.

Habitat: Europe and Western Asia, from the Iberian Peninsula northeast to Southern Scandinavia and the Volga River, southeast to Northern Iran, Turkey, Greece, Sicily and North Africa.

On dry to moist soils, to 1500 m.

Bark: Young smooth, light gray, with age cracking into small, rectangular plates.

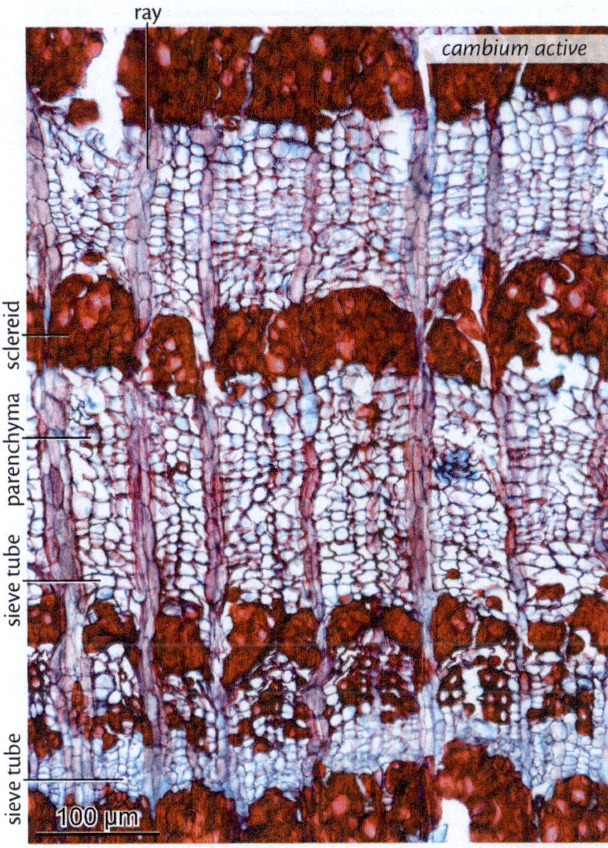

Bark anatomy: Tangentially layered structure of sieve tubes, parenchyma cells and groups of sclereids. Fibers absent. Sclereids solitary and in groups in tangential layers. Rays straight, dilatations rare. Prismatic crystals frequent, mainly in axial parenchyma cells.

Phellem cells rectangular.

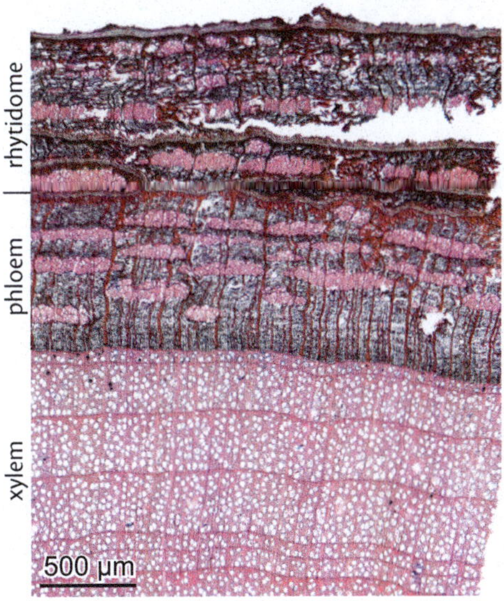

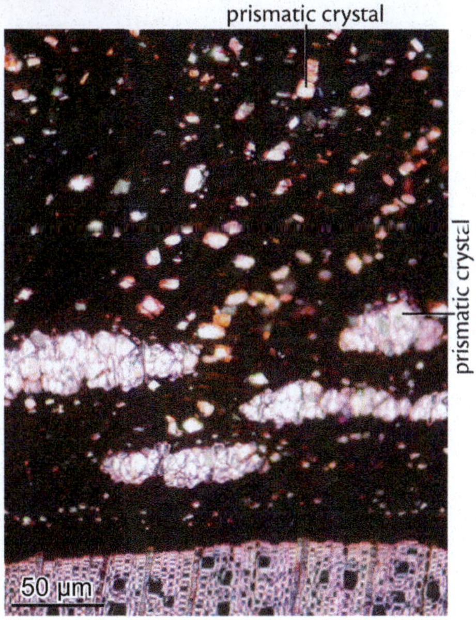

Euonymus europaeus L. Celastraceae

spindle tree, Gemeines Pfaffenhütchen, fusain d'Europe, berretto da prete, bonetero

Form: Upright shrub, rarely a small tree, twigs green, square in cross section, to 5 m tall.

Features: Leaves deciduous, opposite, long ovate, margin finely and regularly toothed, to 8 cm long. Flowers whitish yellow green, four-parted. Fruits unmistakable, with a ribbed, purple pink, fleshy capsule enclosing an orange seed to 1.5 cm wide.

Habitat: Europe, from Southern France northeast to Southern Scandinavia and the Dnieper River, isolated in the Caucasus, south to Greece and Sicily.

On moist to wet, often loamy soils, to 1400 m.

Bark: Twigs green, square in cross section, stem bark light gray brown, with age fissured.

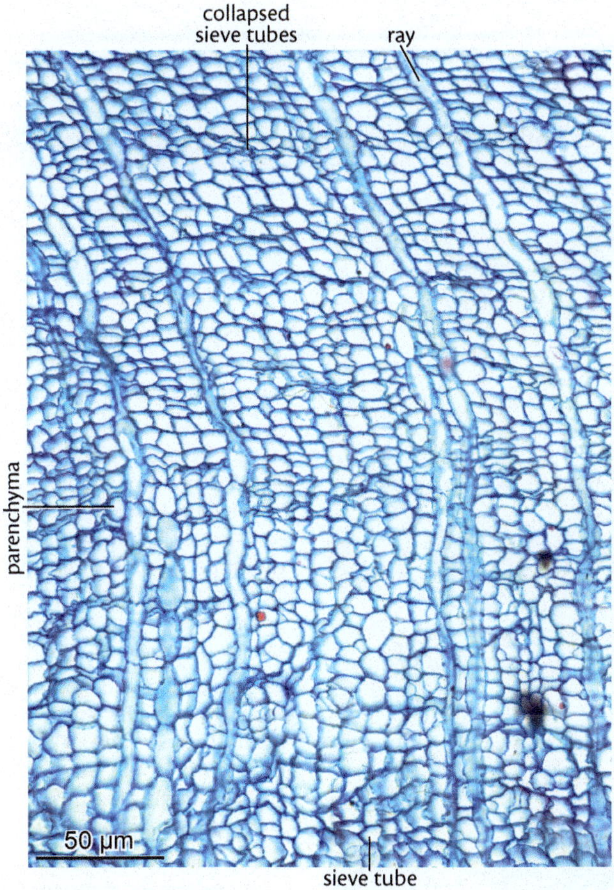

Bark anatomy: Slightly tangentially structured tissue of sieve tubes and parenchyma, without any lignified elements. Sieve tubes angular, accompanied by round to oval parenchyma cells. Fibers and sclereids absent. With extremely large dilatations. Many large crystal druses in cells of dilatations.

Rhytidome consisting of some layers of irregular zones of large, thin-walled cells.

Holdheide 1951 identified the angular cells in the latewood as unlignified fibers (Hohlfasern). His conclusion can not be confirmed on permanent slides embedded in Canada Balsam.

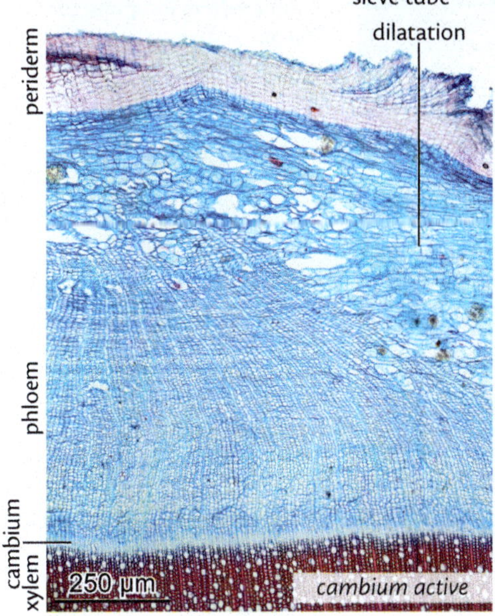

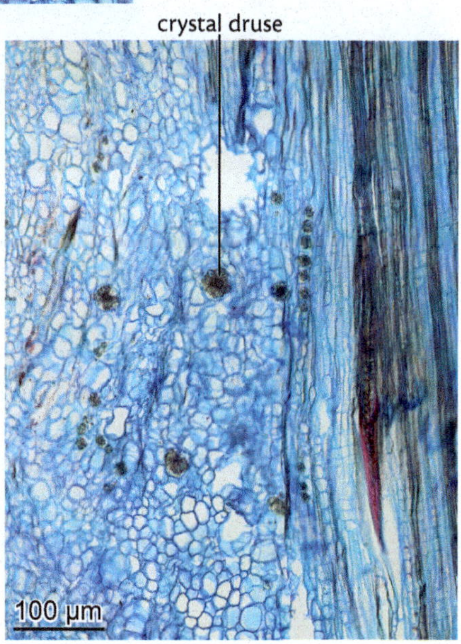

Euonymus verrucosus Scop. Celastraceae

warty-barked spindle, Warziges Pfaffenhütchen, fusain verrue, berretto porroso, bonetero verrugoso

Form: Spreading shrub, twigs densely covered in cork warts, to 2.5 m tall.

Features: Leaves deciduous, opposite, long ovate, margin finely and regularly toothed, to 5 cm long. Flowers brown red to greenish, four-parted. Fruits unmistakable, with a ribbed, carmine pink, fleshy capsule enclosing an orange seed, to 2 cm wide.

Habitat: Central and Eastern Europe, from the Eastern Alps northeast to Poland and the Ural Mountains, southeast to the Caucasus, Crimea and Albania.

On dry, calcareous soils, in summer-warm locations, to 700 m.

Bark: Green to light gray, densely covered with dark brown to black corky warts.

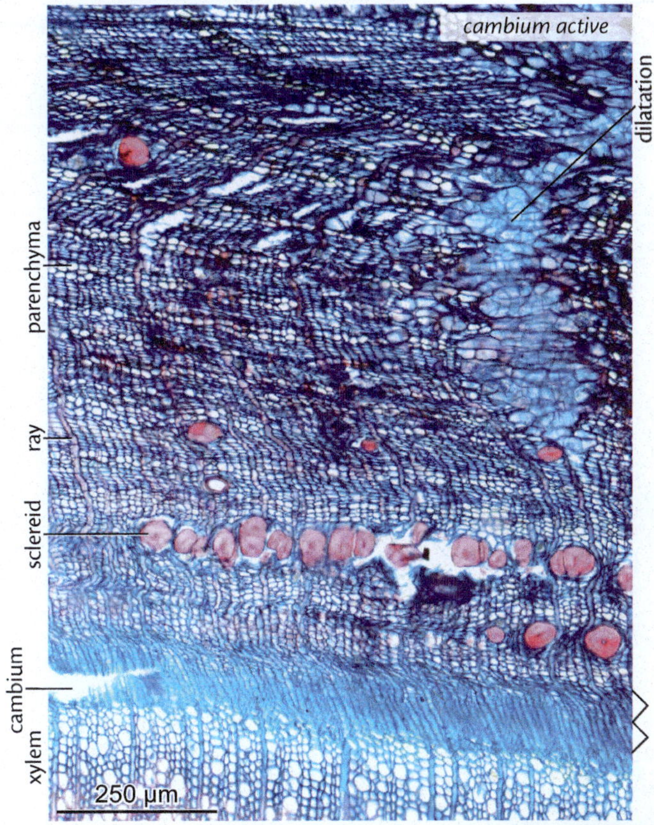

Bark anatomy: Slightly tangentially structured tissue of sieve tubes and parenchyma, with a few isolated sclereids. Sieve tubes angular, accompanied by round to oval parenchyma cells. Very large, thick-walled fibers in tangential rows. With intensive dilatations. Many large crystal druses in all cell types.

Rhytidome consisting of some layers of irregular zones of large, thin-walled cells.

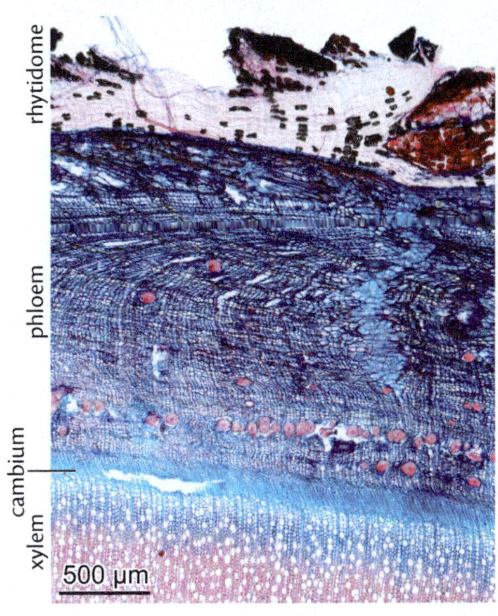

Fagus sylvatica L. Fagaceae

European beech, Rotbuche, hêtre, faggio comune, haya común

Form: Upright tree, crown regular shaped, to 40 m tall.

Features: Leaves deciduous, alternate, ovate, margin entire, to 10 cm long. Flowers monoecious, female and male flower similar, a small, pale yellow green, hanging tuft, to 1 cm long. Seed an ovate, sharply triangular nut, embedded in a four-valved cup, to 2.5 cm wide.

Habitat: Europe, from Northern Spain northeast to South Sweden, western Poland and the Carpathians and Crimea, south to the Balkans, Sicily and Corsica.

On moist, deep soils, often forming pure stands, to 1700 m.

Bark: Light gray and smooth, also on old trees.

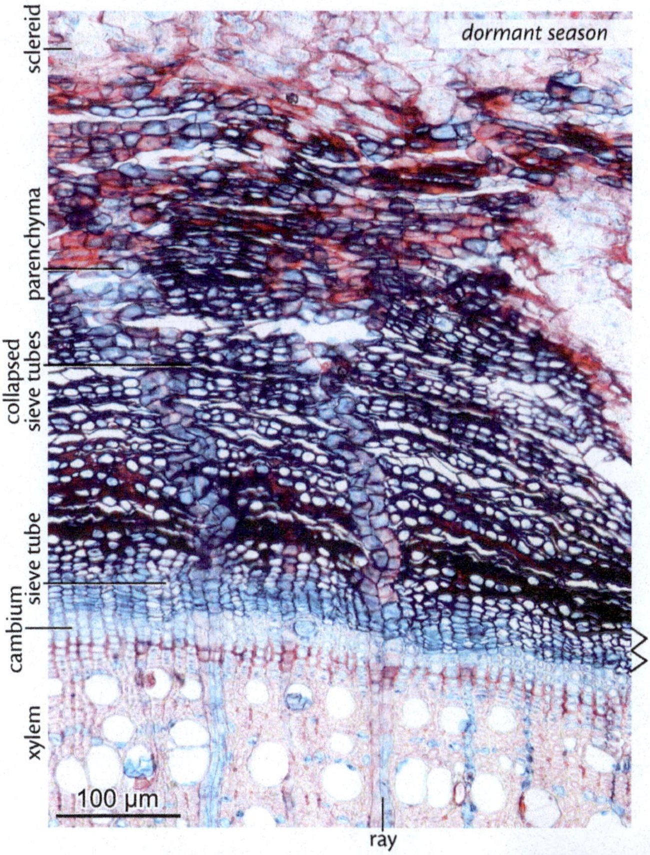

Bark anatomy: Very dense structure with large groups of sclerenchyma groups.

Sieve tubes irregular shaped, collapsing very soon after formation. Parenchyma cells roundish. Fibers rare. Sclereids in irregular and tangential groups and in dilatation zones of large rays. Multiseriate rays with extreme dilatations. Prismatic crystals and a few crystal druses only in sclereids.

Periderm consisting of many layers of large, rectangular, thick-walled cells.

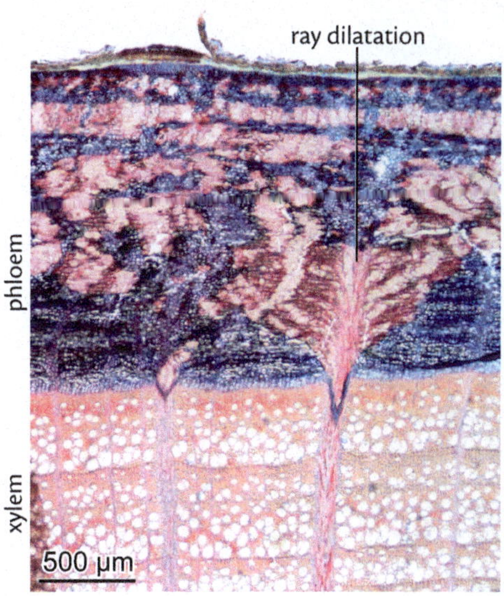

Fraxinus excelsior L. Oleaceae

ash, Gemeine Esche, frêne commun, frassino comune, fresno común

Form: Upright tree, crown regular shaped, to 40 m tall.

Features: Leaves deciduous, opposite, pinnate with 9–15 leaflets, margin toothed, to 40 cm long. Flowers in violet green, upright panicles, to 3 cm long. Fruit a flattened, winged nut, to 4 cm long, in dense, hanging tufts.

Habitat: Europe, from Northern Spain northeast to Southern Scandinavia, Lake Ladoga and the Volga River, Crimea and the Caucasus, south to northern Turkey and Greece, Sicily and Corsica.

On moist, deep soils, often riverine, to 1700 m.

Bark: Young smooth, light yellowish gray, with age markedly fissured.

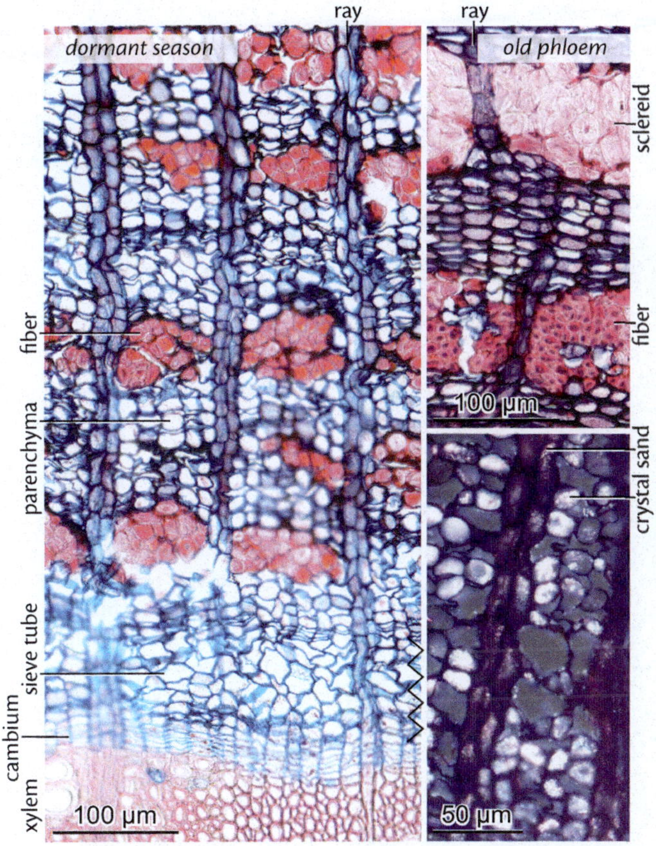

Bark anatomy: Regular tangential layers of bands of fibers and zones of sieve tubes and parenchyma.

Layers of angular sieve tubes and roundish parenchyma cells alternating with 2–4-seriate bands of thick-walled sclerenchyma. Fibers associated with sclereids. Rays mostly straight, present for many years, few rays dilating. A lot of crystal sand in axial parenchyma cells and rays.

Annual rings indistinct.

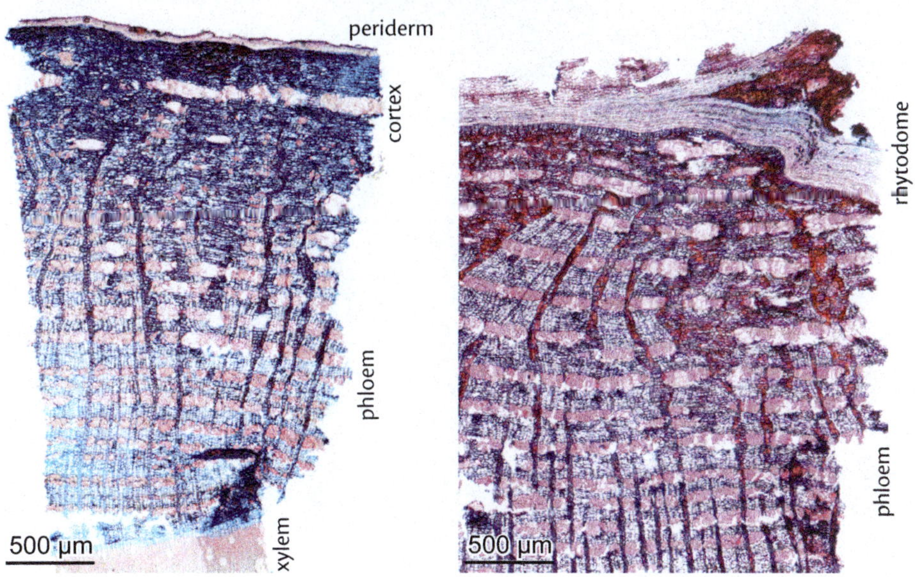

Hedera helix L.

Araliaceae

ivy, Efeu, lierre, edera, hiedra

Form: Upright, evergreen climber, often on trees, to 20 m tall.

Features: Leaves evergreen, leathery, alternate, changing from square-lobed to triangular and round ovate, margin entire, to 10 cm long. Flowers yellowish green in autumn, five-parted, to 7 mm long, in many-flowered, upright umbels, to 5 cm long. Fruit a globular, blue black drupe with a flattened upper side, in dense umbels during winter and spring.

Habitat: Europe, from Algeria northeast to Southern Scandinavia, Estonia, the Caucasus, south to northern Turkey, Crete and Malta.

On moist to dry soils, to 1400 m.

Bark: Young dull, gray, often corky, with age fissured or with transversal cracks, often bulging.

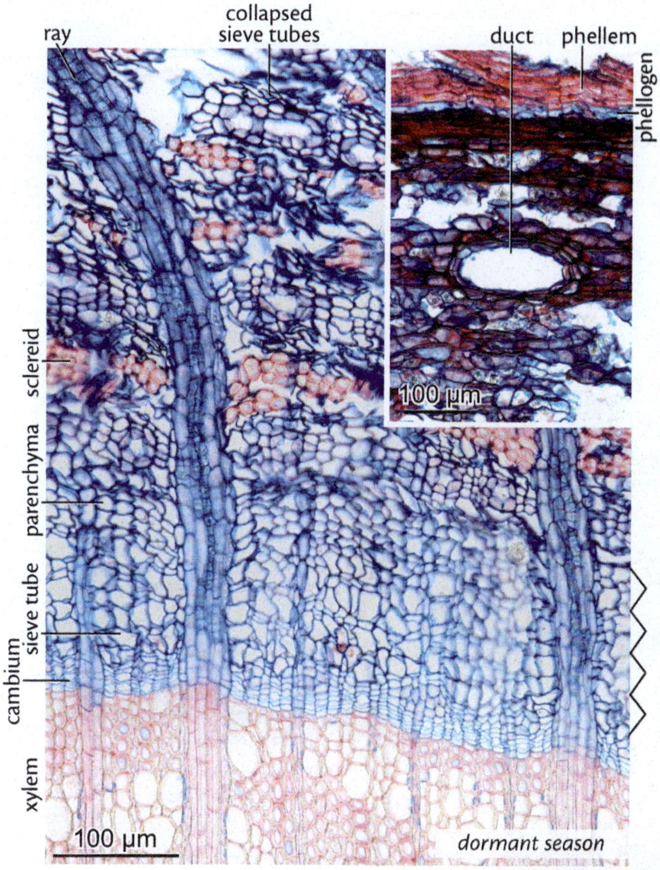

Bark anatomy: Tangential layers consisting of fiber bands and zones of sieve tubes and parenchyma and mostly irregular shaped groups of sclereids.

Sieve tubes angular, parenchyma roundish. Fibers absent. Sclereids in groups of various shapes. Rays slightly wavy. Few rays or parenchyma cells dilating. Crystal druses numerous in axial parenchyma cells and rays. Ducts occasionally present.

Annual ring boundaries recognizable in the first years. Phellem consisting of some layers of irregular zones of large, thin-walled cells.

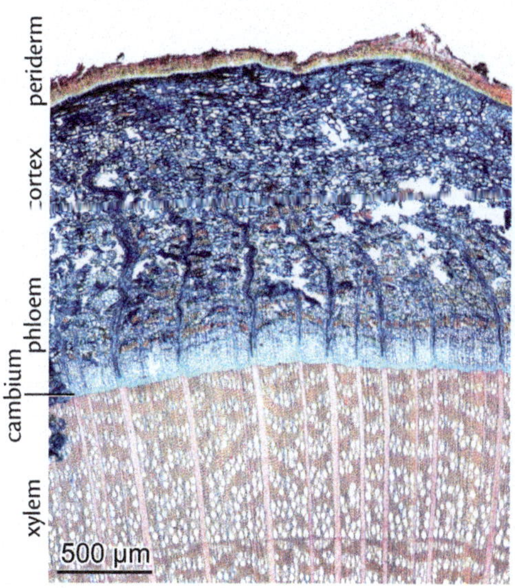

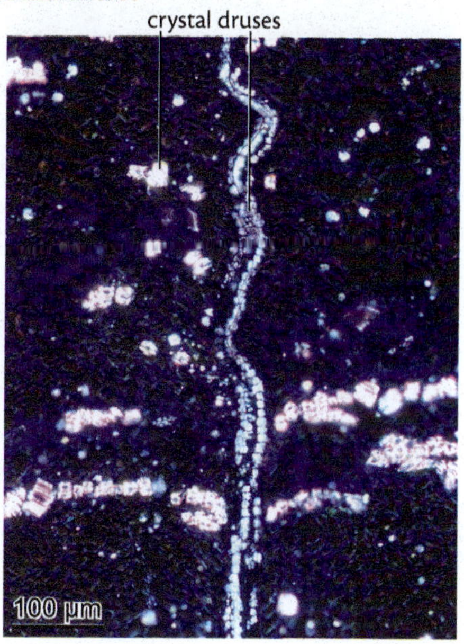

Hippocrepis emerus (L.) Lassen

Fabaceae

scorpion senna, Strauchwicke, hippocrépide émérus, cornetta dondolina, coletua

Form: Upright shrub, twigs angular in cross section, green, often hanging, to 2 m tall.

Features: Leaves deciduous, alternate, pinnate with 5–9 ovate leaflets, to 8 cm long. Flowers yellow, papilionaceous, in groups of two to three, to 1 cm long. Fruit a long, thin, gray brown pulse, to 10 cm long.

Habitat: Southern Europe and Western Asia, from Northern Spain northeast to Belgium, Hungary, Serbia, south to Lebanon, Sicily and Tunisia.

On dry, calcareous, often rocky soils, in warm locations, to 1800 m.

Bark: Light gray brown or olive green, often spotted.

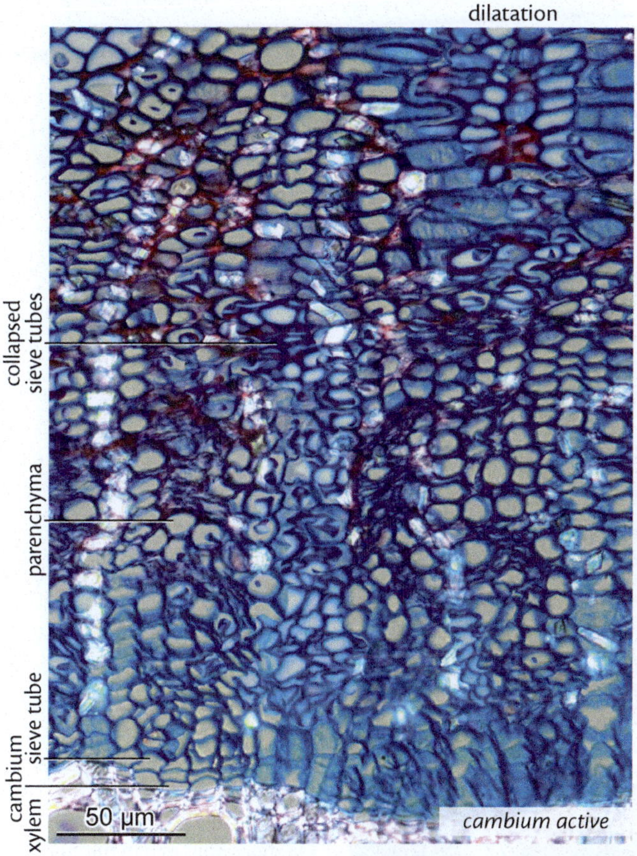

Bark anatomy: Structure of sieve tubes and parenchyma, with many rays. Young tissue formed by angular sieve tubes, accompanied by companion cells, and round and oval parenchyma. Fibers absent. A few small groups of sclereids in older parts. Most rays dilating. Many prismatic crystals (styloids) in parenchyma cells, rarely in rays.

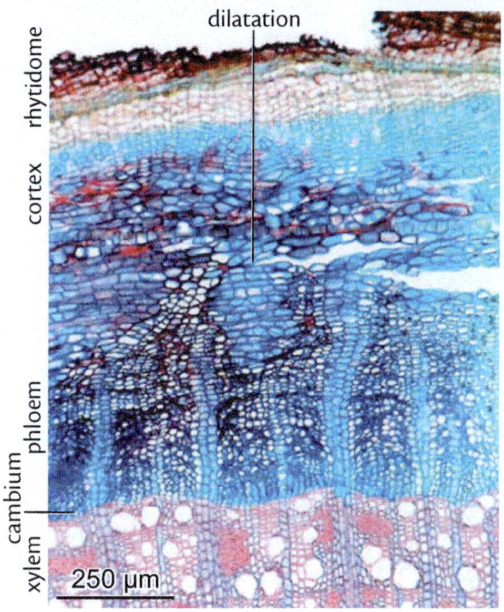

Ilex aquifolium L.

Aquifoliaceae

holly, Stechpalme, houx, agrifoglio, acebo

Form: Upright, evergreen tree or shrub, to 10 m tall.

Features: Leaves evergreen, leathery, alternate, long ovate, glossy above, margin pungently toothed or entire, to 8 cm long. Flowers dioecious, white, four-parted, to 8 mm long. Fruit a globular, deep red drupe on female plants, to 8 mm wide.

Habitat: Southwestern Europe, from Morocco northeast to England, Denmark, northeastern Germany, the Danube Delta and the Caucasus, south to western Turkey, Sicily and Algeria.

On moist soils rich in humus, in winter-mild sites, to 1300 m.

Bark: Dull gray, without fissures, sometimes with warty lenticels.

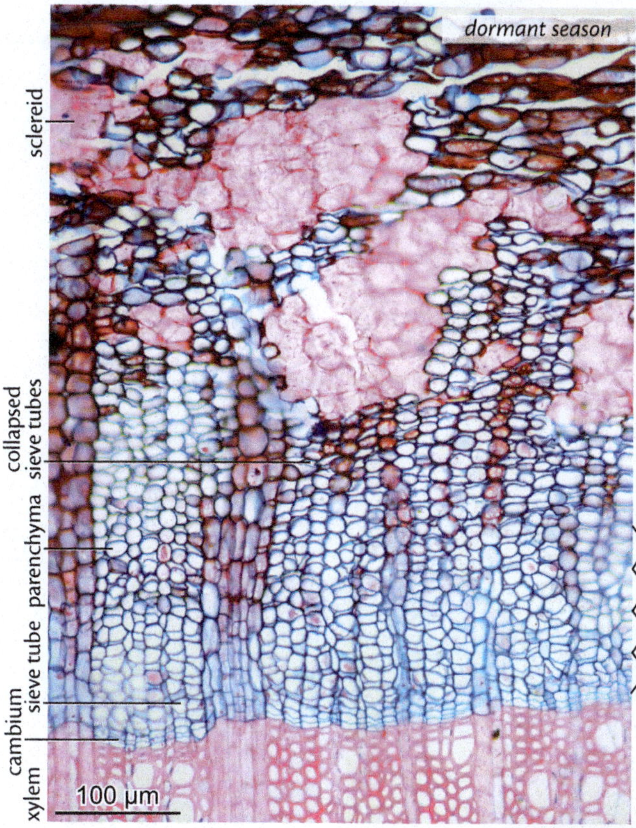

Bark anatomy: Unlayered zones of sieve tubes and parenchyma near the cambium, with many irregular groups of sclereids toward the periphery.

Sieve tubes angular, parenchyma cells roundish. Fibers absent. Sclereids in groups of various shapes. Ray straight. Few rays or parenchyma cells dilate. Many crystal druses and prismatic crystals in axial parenchyma cells, rays and sclereids.

Annual ring boundaries absent. Phellem consisting of some layers of large, thick-walled cells.

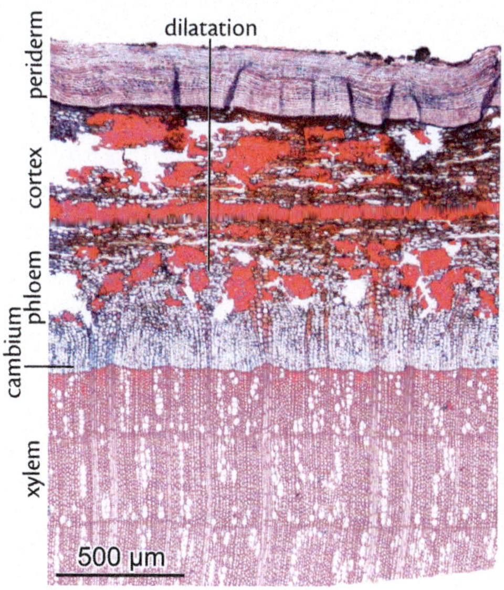

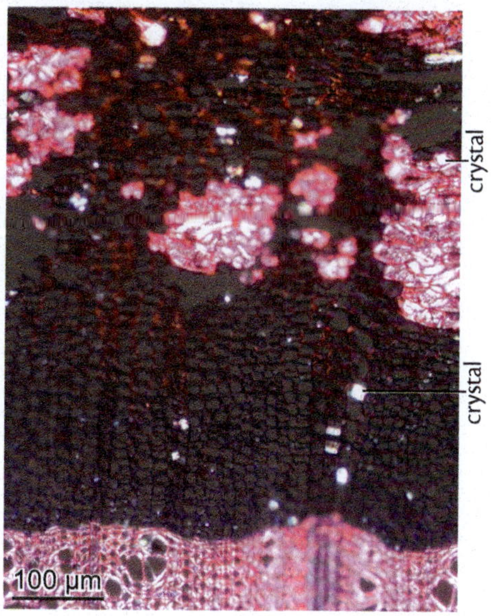

Ligustrum vulgare L. Oleaceae

wild privet, Liguster, troène commun, ligustro, aligustre

Form: Upright, densely branched shrub, to 5 m tall.

Features: Leaves deciduous, opposite, long ovate, margin entire, to 5 cm long. Flowers white, heavily scented, four-parted, to 5 mm long, in dense panicles. Fruit a globular, glossy black berry, to 1 cm wide.

Habitat: Europe, from Northern Spain northeast to Southern England, Central Germany, southwestern Ukraine and the Caucasus, south to northern Turkey, Albania and Southern Italy.

On moist, often loamy and calcareous soils, to 1400 m.

Bark: Light gray to gray brown, smooth, dotted with numerous lenticels.

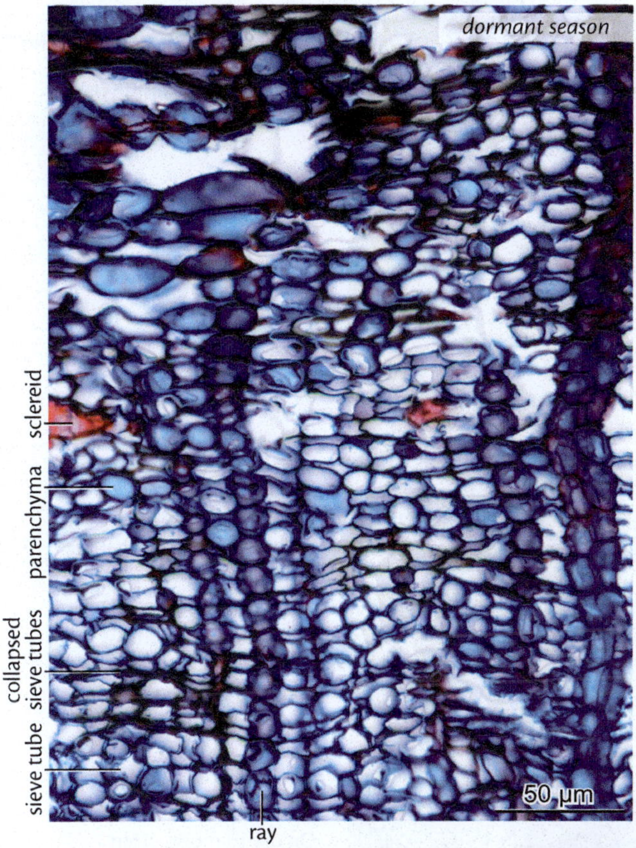

Bark anatomy: Indistinctly layered zones of sieve tubes and parenchyma near the cambium. With few irregular shaped groups of sclereids towards the periphery.

Sieve tubes angular, parenchyma cells roundish. Fibers absent. Groups of sclereids in loose tangential lines. Dilatations absent. Crystals absent.

Phellem consisting of some layers of large, thick-walled cells.

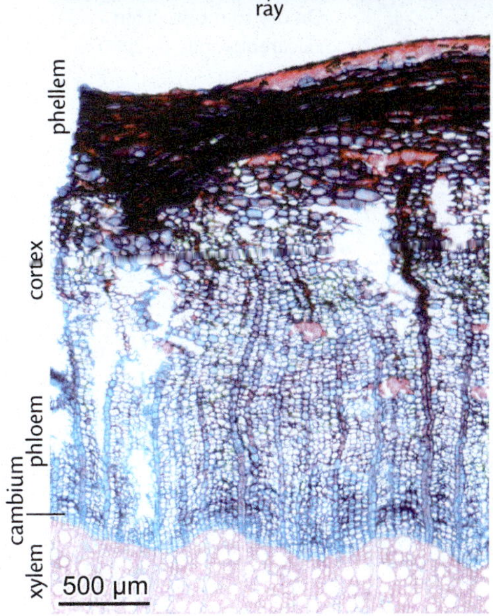

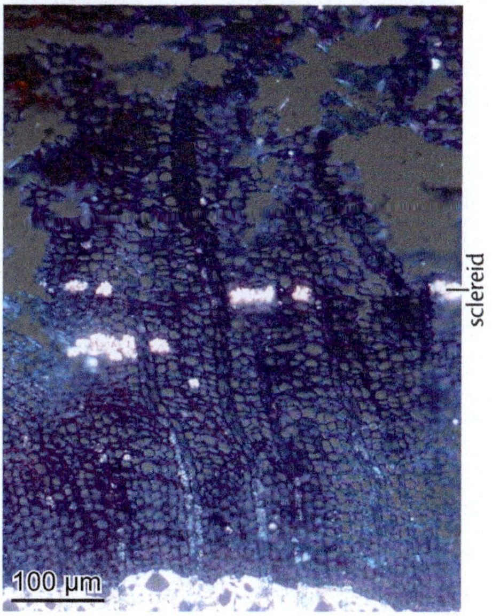

Lonicera xylosteum L. — Caprifoliaceae

fly honeysuckle, Rote Heckenkirsche, chèvrefeuille des haies, caprifoglio peloso, cerecillo

Form: Upright, densely branched shrub, to 2 m tall.

Features: Leaves deciduous, opposite, round ovate, margin entire, to 6 cm long. Flowers creamy white, in pairs, asymmetrically two-lipped, to 15 mm long. Fruit a globular, glossy red berry, in pairs, partly fused together, to 1 cm wide.

Habitat: Eurosiberian, from Middle Spain northeast to Southern Finland, Northern Russia to the Ob river, the Southern Ural and the Caucasus, south to Northern Turkey and Greece, Sicily and Southern France.

On dry to moist, often calcareous soils, to 2000 m.

Bark: Light gray to gray brown, young smooth, with age scaly fissured.

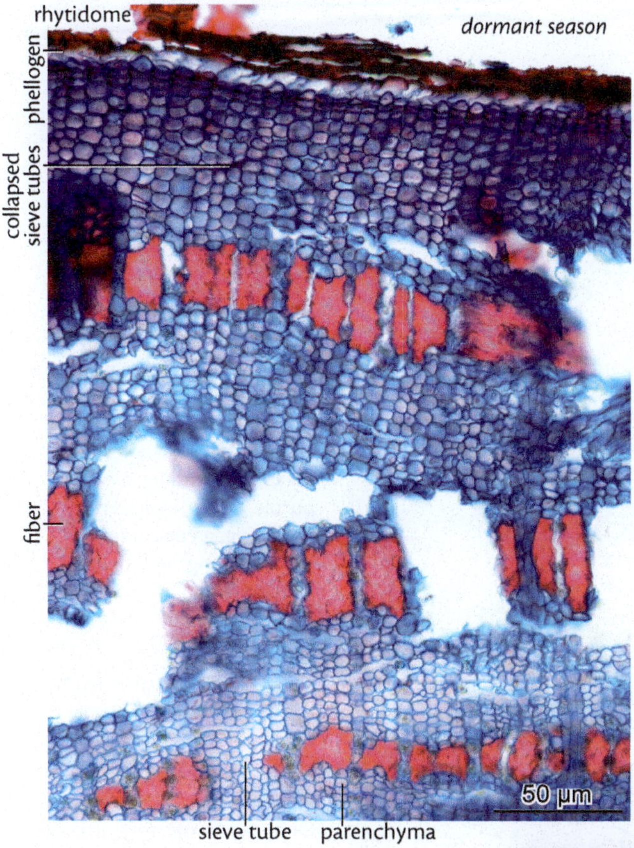

Bark anatomy: Tangentially layered structure, consisting of sieve tubes, parenchyma cells and fibers. Layers scaling off easily.

Sieve tubes angular, parenchyma cells round, both cell types irregularly mixed. Fibers in tangentially arranged rectangular blocks. Rays hardly recognizable in transmitting light, but clearly indicated by many crystal druses in polarized light.

Shed rhytidome consisting of a layer of living, unlignified square cells.

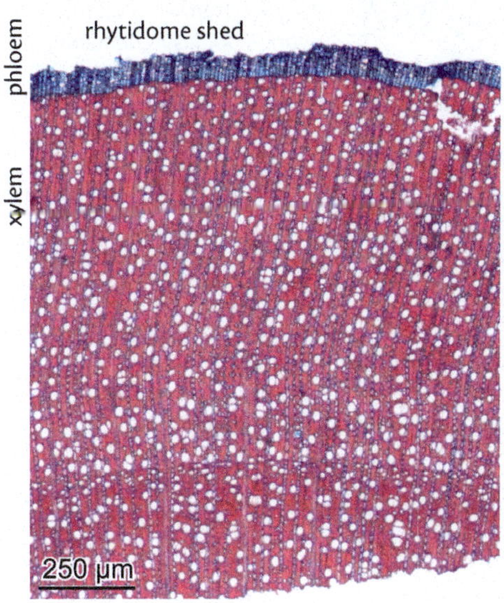

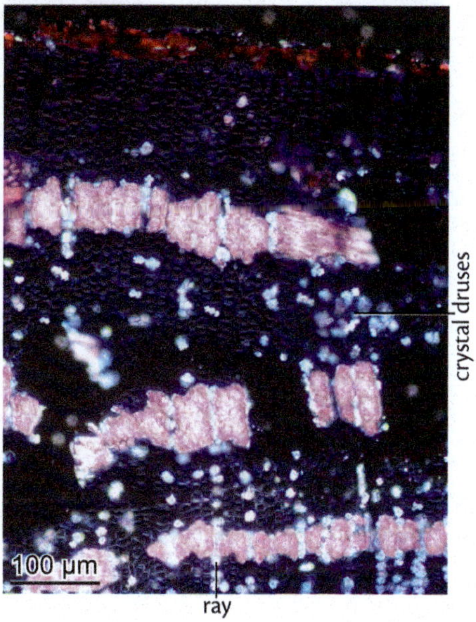

Populus nigra L.

Salicaceae

black poplar, Schwarzpappel, peuplier noir, pioppo nero, chopo negro

Form: Upright tree, crown irregular shaped, to 35 m tall.

Features: Leaves deciduous, alternate, triangular, rhombic or cordate, margin regularly finely toothed, to 7 cm long. Flowers dioecious in hanging catkins, female catkin thin, green, male catkin thick, purple, to 8 cm long. Fruit a hanging, green catkin, to 10 cm long, with many leathery capsules, the ripe, tiny seeds with abundant, white, woolly hairs.

Habitat: Eurasia, from North Africa northeast to Estonia, the southern Yenisei River and Central Asia, south to Turkey.

On moist to wet soils, strictly riverine in lowlands, to 800 m. Often replaced by the hybrid *Populus* x *canadensis*.

Bark: Young light gray, with age gray brown, roughly borky fissured, also with horizontal cracks.

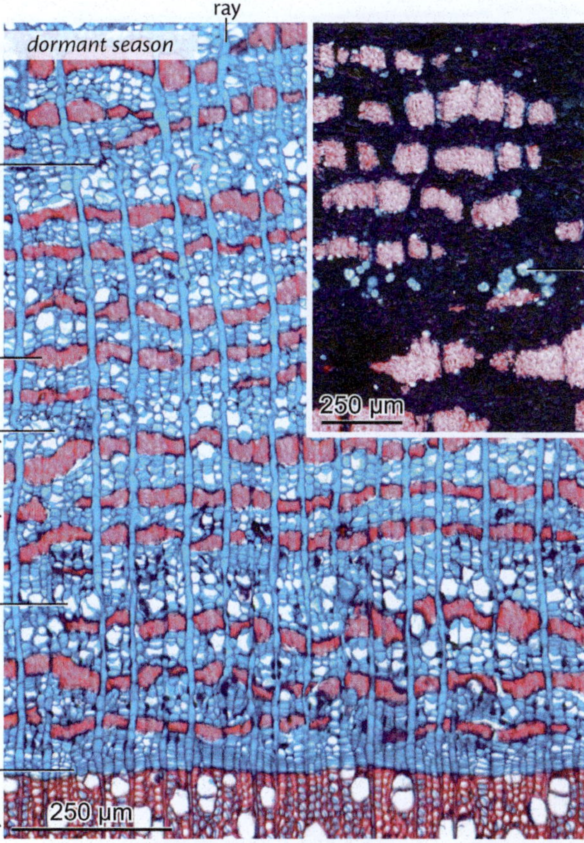

Bark anatomy: Regular tangential layers of parenchyma, sieve tubes and fibers.

Sieve tubes irregular shaped, companion cells small. Parenchyma cells roundish. Fibers in tangential layers. Sclereids in older parts of the phloem in groups surrounded by fibers. Prismatic crystals and crystal druses along fiber bands. Tangential layers not corresponding with annual rings. Annual rings visible at small magnifications.

Periderm tangentially irregularly layered, consisting of large, irregular shaped, thin-walled cells and small, rectangular, unlignified cells.

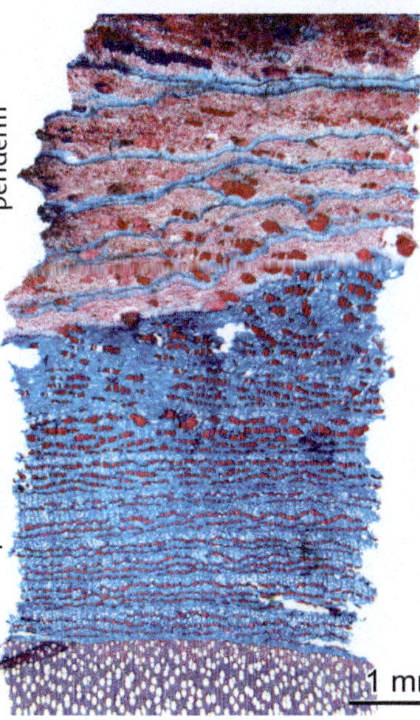

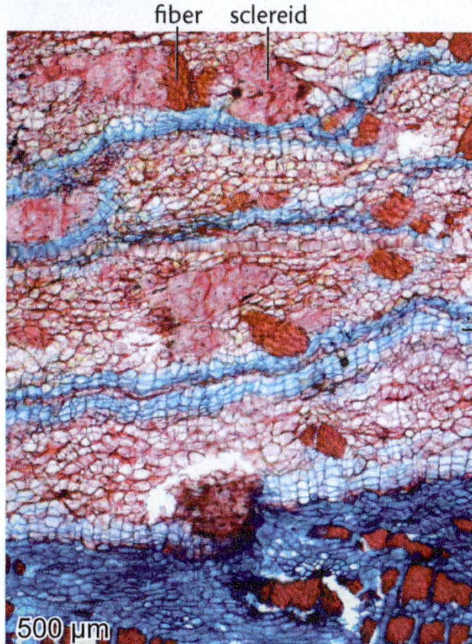

Prunus avium (L.) L. Rosaceae

wild cherry, Vogelkirsche, cerisier sauvage, ciliegio, cerezo silvestre

Form: Upright tree, crown regular shaped, trunk straight, to 25 m tall.

Features: Leaves deciduous, alternate, long ovate, margin regularly toothed, to 15 cm long. Flowers white, five-parted, to 2 cm long. Fruit a classical, glossy red to black cherry, to 2 cm wide.

Habitat: Europe and Western Asia, the whole of Europe except the northeast, east to Central Asia and Northern Iran south to northern Turkey and the Mediterranean Mountains.

On moist, deep and loamy, often calcareous soils, to 1700 m.

Bark: Glossy and smooth, gray to red brown, with conspicuous horizontal stripes.

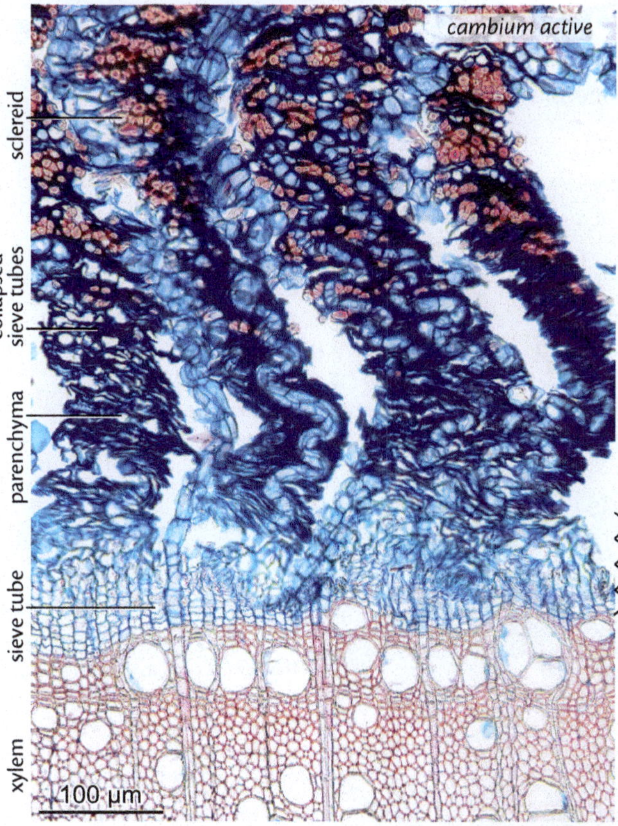

Bark anatomy: Radial structure with large intercellulars. Sieve tubes, parenchyma cells and many sclereids in intensively wavy, radial strips between large, bent rays. Fibers absent. Dilatations not observed. Structure chaotic in external parts. Many crystal druses and prismatic crystals, mainly in axial parenchyma cells.

Phellem consisting of a very large band of rectangular, thick- or thin-walled, elastic cells.

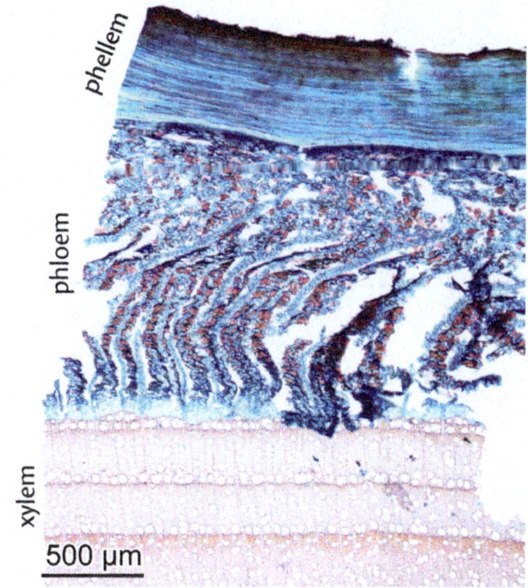

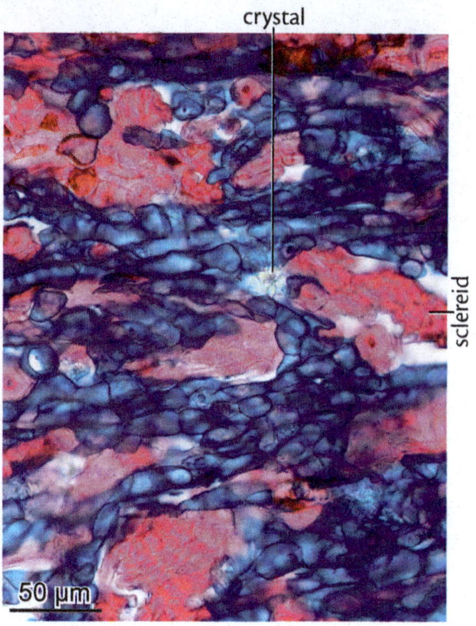

Prunus padus L. Rosaceae

bird cherry, Traubenkirsche, merisier à grappes, pado, cerezo aliso

Form: Tree or shrub, strictly upright, crown regular shaped, to 10 m tall.

Features: Leaves deciduous, alternate, long ovate, margin regularly toothed, to 15 cm long. Flowers white, five-parted, to 1 cm long, in upright racemes. Fruit a small, glossy black cherry, in hanging racemes, to 10 cm long.

Habitat: Eurasia, nearly the whole of Europe east to Siberia and Kamchatka, south to the Himalayas, Caucasus and Southern Italy.

On moist to wet, deeply loamy soils, to 2200 m.

Bark: Glossy and smooth, dark gray to black brown, with lenticels showing as short horizontal stripes.

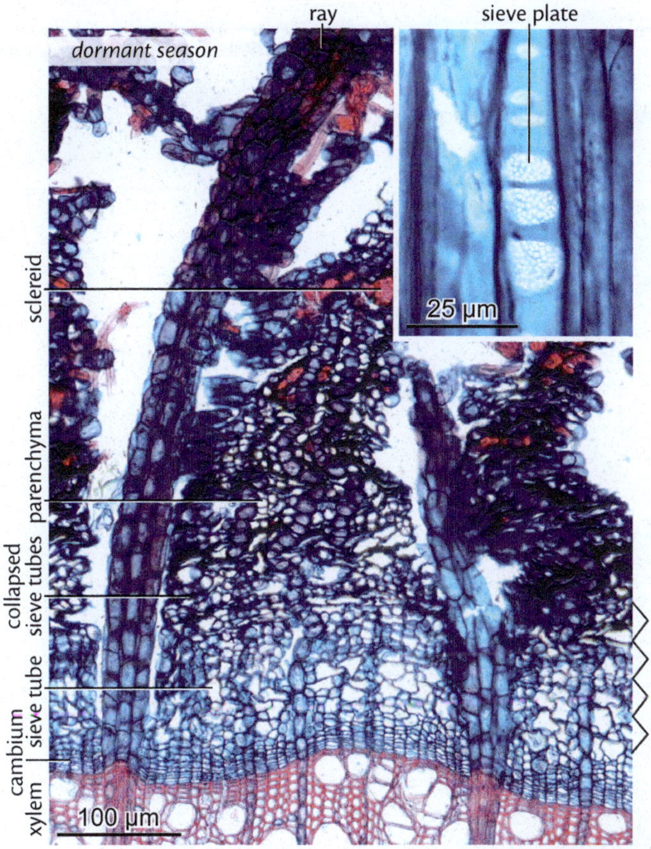

Bark anatomy: Radial structure with large intercellulars. Sieve tubes, parenchyma cells and sclereids in intensively wavy, radial strips between large, bent rays. Fibers absent. Dilatations not observed. Structure chaotic in external parts. Many crystal druses and a few prismatic crystals, mainly in axial parenchyma cells.

Phellem consisting of a small band of rectangular, unlignified, thin-walled, compressed cells.

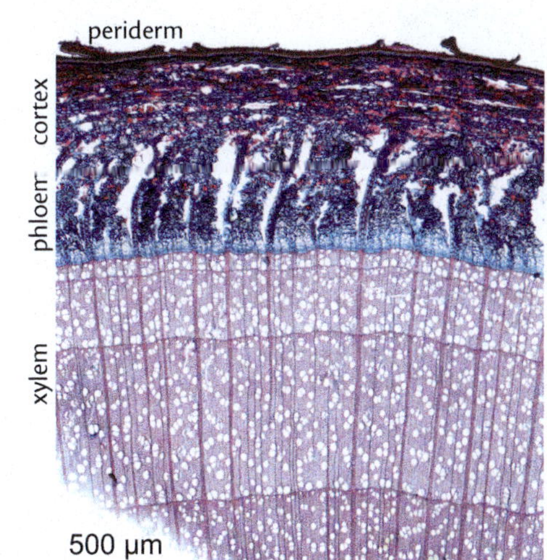

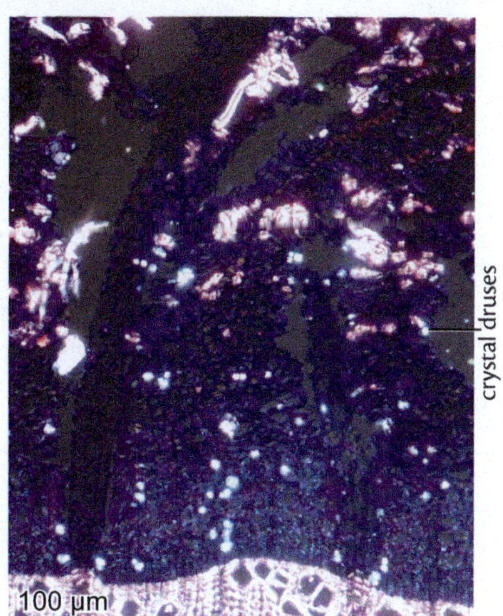

Prunus spinosa L. Rosaceae

blackthorn, Schwarzdorn, épine noire, prugnolo, endrino

Form: Dense, spiny shrub, to 4 m tall.

Features: Leaves deciduous, alternate, ovate, margin regularly toothed, to 4 cm long. Flowers white, five-parted, to 1 cm long. Fruit a globular, dull black drupe, frosted blue, to 1.5 cm wide.

Habitat: Europe and Western Asia, nearly the whole of Europe east to the Volga River, south to Iran, Kurdistan, Sicily and North Africa.

On dry, mostly calcareous soils, in warm, sunny positions, to 1800 m.

Bark: Young glossy and smooth, black brown, with short horizontal stripes, with age slightly fissured.

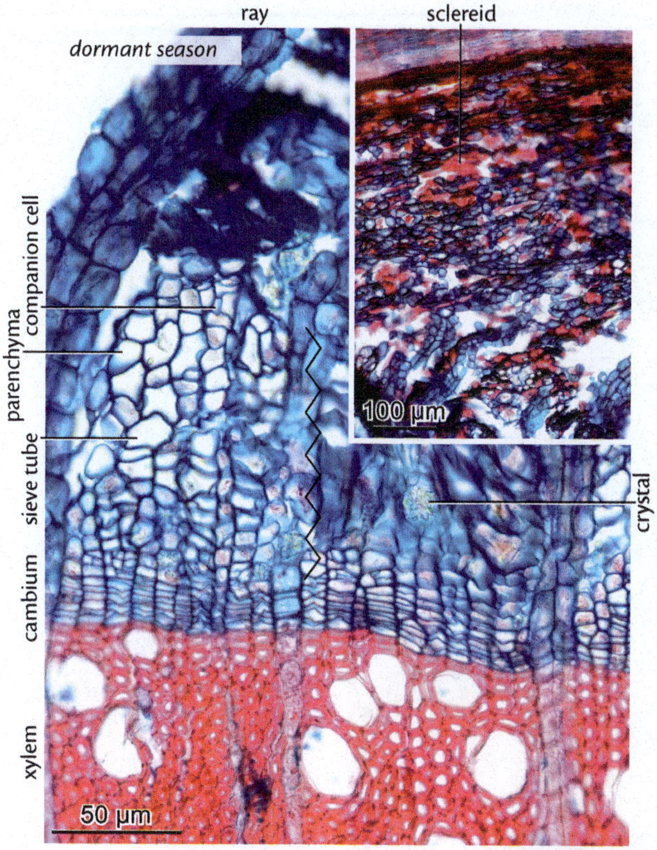

Bark anatomy: Radial structure with large intercellulars. Sieve tubes, parenchyma cells and many sclereids in intensively wavy, radial strips between large, bent rays. Fibers absent. Dilatations not observed. Structure chaotic in external parts. Many crystal druses and prismatic crystals, mainly in axial parenchyma cells.

Phellem consisting of a very large band of rectangular, thick- or thin-walled, elastic cells.

Structure almost identical to *Prunus avium*.

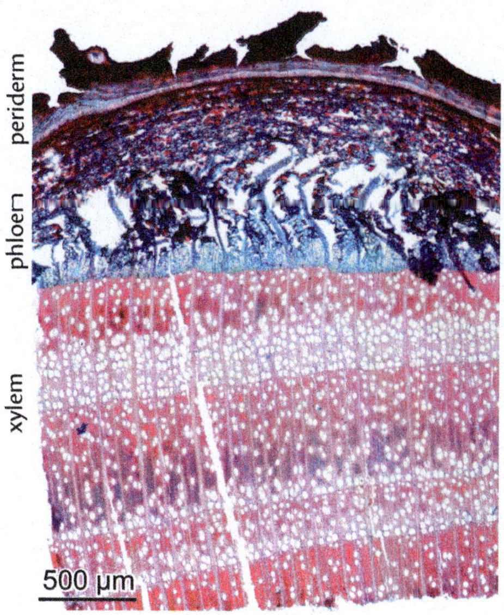

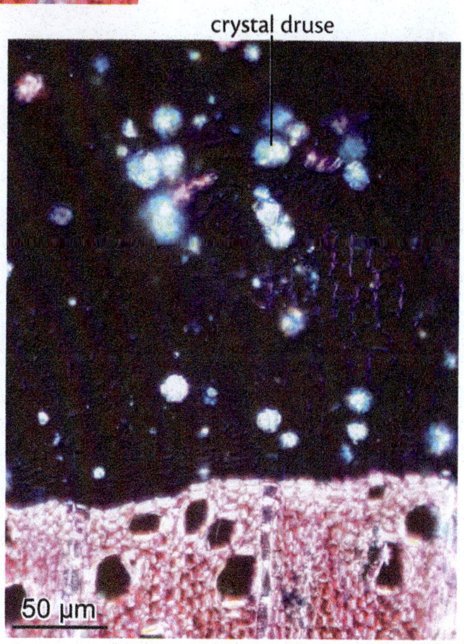

Pyrus pyraster (L.) Burgsd. Rosaceae

wild pear, Wildbirne, poirier sauvage, pero selvatico, peral sylvestre

Form: Upright tree or shrub, to 20 m tall.

Features: Leaves deciduous, alternate, ovate, margin regularly toothed or entire, to 6 cm long. Flowers white, five-parted, to 4 cm long. Fruits globular or pear-shaped, yellow, to 7 cm long.

Habitat: Europe and Western Asia, Europe without Scandinavia, southeast to Iran, the Caucasus, Turkey and Sicily.

On dry, stony, mostly calcareous soils, in warm, sunny positions, to 1300 m.

Bark: Young glossy and smooth, black brown, with short horizontal stripes, with age slightly fissured.

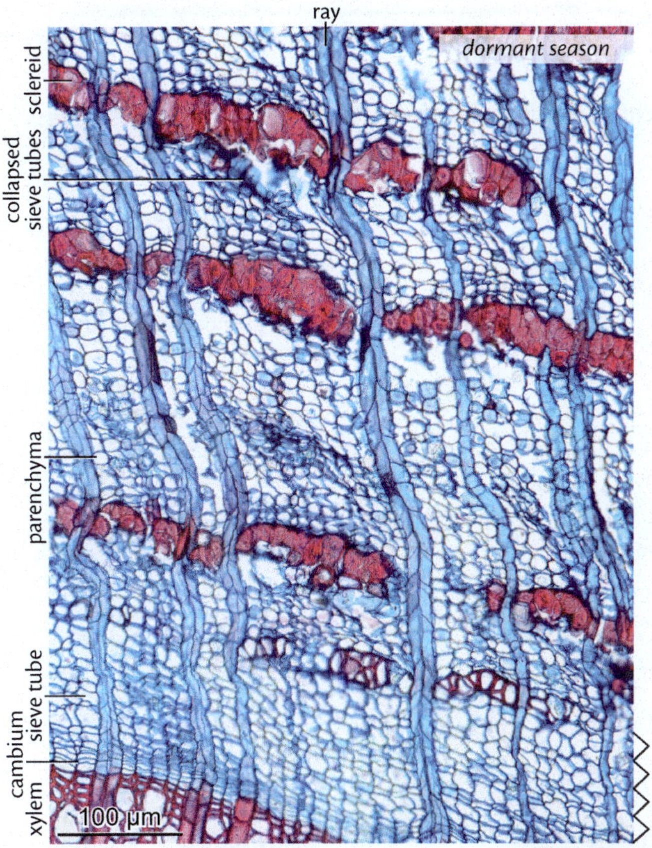

Bark anatomy: Structure tangentially layered. Sieve tubes and parenchyma cells indistinctly tangentially arranged. Sieve tubes collapsing soon after formation. Fibers absent. Groups of sclereids in tangential layers. Rays slightly wavy. Dilatations not observed. Many small crystal druses in the axial parenchyma cells. Crystals absent in rays.

Phellem cells rectangular, thick-walled.

Structure almost identical to *Malus × domestica*.

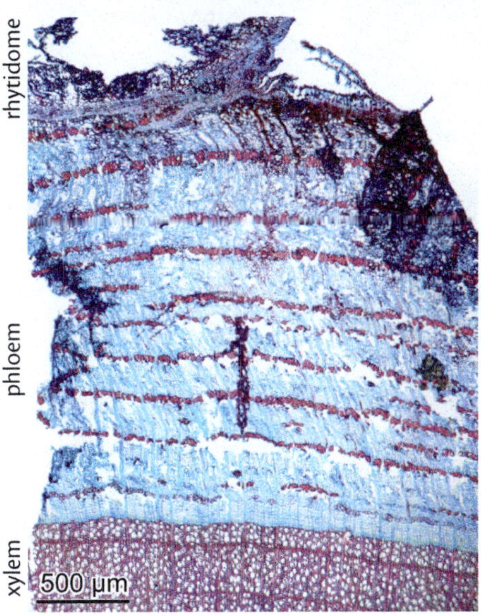

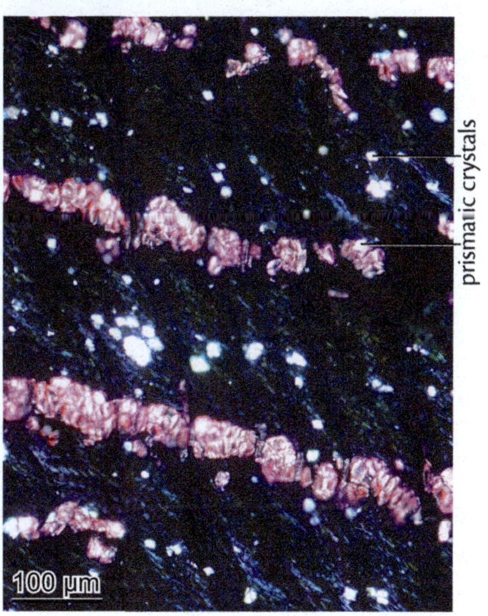

Quercus petraea (Matt.) Liebl. Fagaceae

sessile oak, Traubeneiche, chêne sessile, quercia rovere, roble albar

Form: Upright tree, crown large, irregular shaped, to 40 m tall.

Features: Leaves deciduous, alternate, stalked, margin roundly lobed, to 16 cm long. Flowers monoecious, female flower very small, reddish, grouped in clusters, male flowers in hanging, green yellow catkins, to 8 cm long. Fruit an ovate acorn, to 2.5 cm long, sitting in a stalkless cup, in clusters.

Habitat: Europe, from Central Spain northeast to Southern Sweden, the Carpathians and the Caucasus, south to northern Turkey, the Balkans and Sicily.

On dry, stony soils, to 1600 m.

Bark: Young smooth, dark to brown gray, soon deeply fissured.

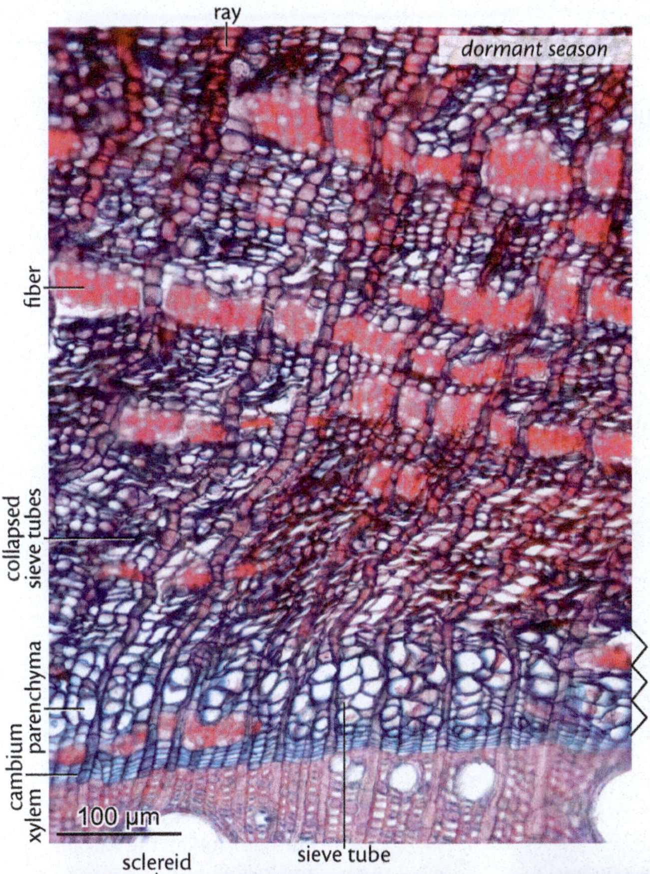

Bark anatomy: Structure tangentially layered. Sieve tubes angular, parenchyma cells round, radially arranged. Sieve tubes collapsing soon after formation. Fiber bands alternating with zones of sieve tubes and parenchyma cells. Sclereids in round groups, dominating large rays. Rays uniseriate, straight or wavy. Some large rays dilating. Prismatic crystals surrounding fiber bands. Crystal druses in parenchymatic zones, many small crystal druses in axial parenchyma cells.

Phellem formed by layers of rectangular, thin-walled cork cells.

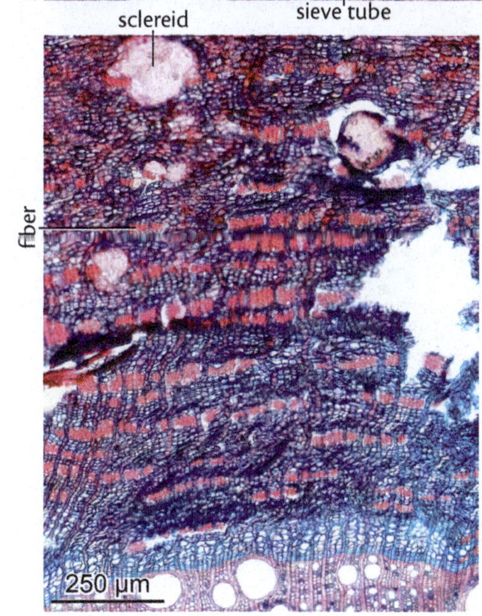

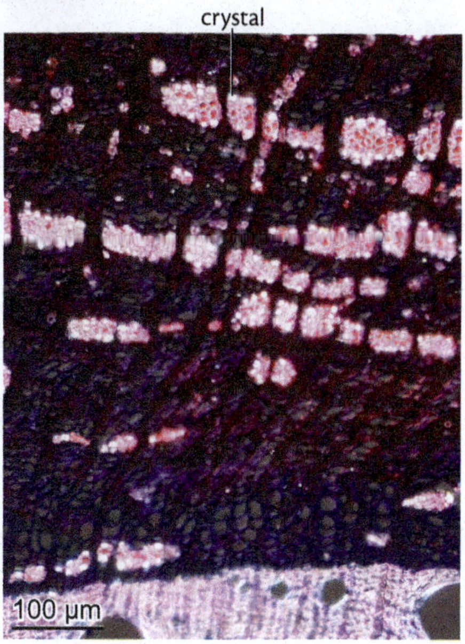

Quercus robur L. Fagaceae

common oak, Stieleiche, chêne pédonculé, quercia comune, roble carballo

Form: Upright tree, crown large, irregular shaped, to 50 m tall.

Features: Leaves deciduous, alternate, sessile, margin roundly lobed, to 16 cm long. Flowers monoecious, female flower, very small, reddish, grouped in clusters, hanging, male flowers in hanging, green yellow catkins, to 8 cm long. Fruit an ovate acorn, to 2.5 cm long, sitting in a long-stalked, pendulous cup, in clusters.

Habitat: Europe, from Portugal northeast to Southern Sweden, the Ladoga Lake and the Volga River, the Caucasus, south to northern Turkey, the Balkans and Italy.

On moist to wet, deep and loamy soils, to 1200 m.

Bark: Young smooth, dark to brown gray, soon deeply fissured.

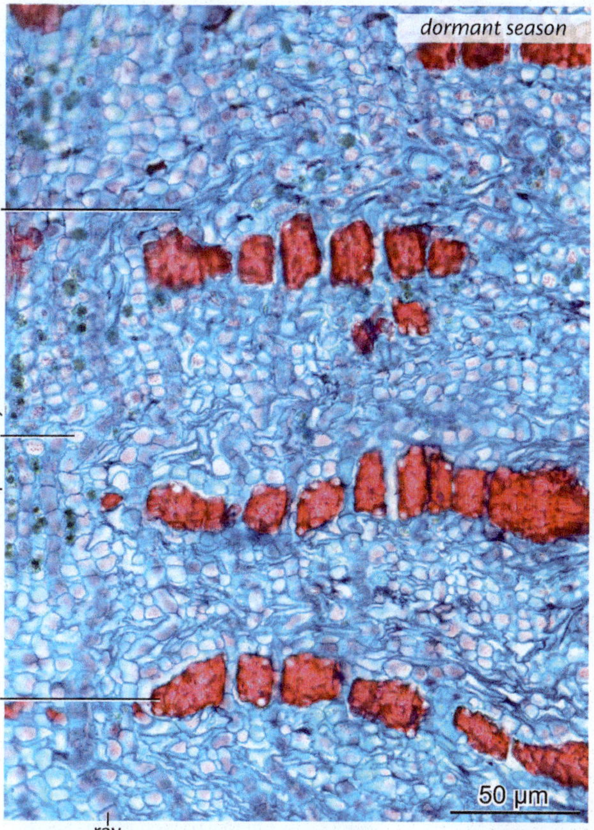

Bark anatomy: Structure tangentially layered. Sieve tubes angular, parenchyma cells round, radially arranged. Sieve tubes collapsing soon after formation. Fiber bands alternating with zones of sieve tubes and parenchyma cells. Sclereids in round groups, dominating large rays. Rays uniseriate, straight or wavy. Some large rays dilating. Prismatic crystals surrounding fiber bands. Crystal druses in parenchymatic zones, many small crystal druses in axial parenchyma cells.

Phellem formed by layers of rectangular, thin-walled cork cells.

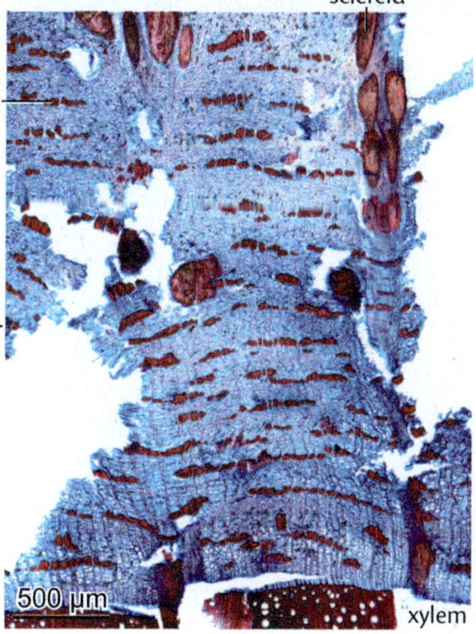

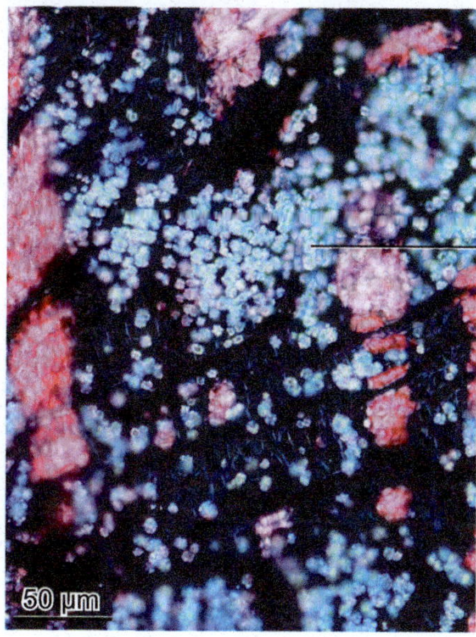

Rhamnus cathartica L. Rhamnaceae

buckthorn, Kreuzdorn, nerprun purgatif, spino cervino, espino cerval

Form: Upright shrub, densely branched, to 3 m tall.

Features: Leaves deciduous, alternate or opposite, round ovate, margin regularly toothed, to 8 cm long. Flowers yellow green, four-parted, to 6 mm long. Fruit a globular, berry-like, glossy black drupe, to 8 mm wide.

Habitat: Europe and Western Asia, from Central Spain northeast to Denmark, Estonia, east to Kirghistan, Northern Iran, south to Crimea, Albania and Sicily, isolated in Algeria and the Atlas Mountains.

On dry to wet, often calcareous soils, to 1600 m.

Bark: Gray brown, dotted with lighter-colored lenticels.

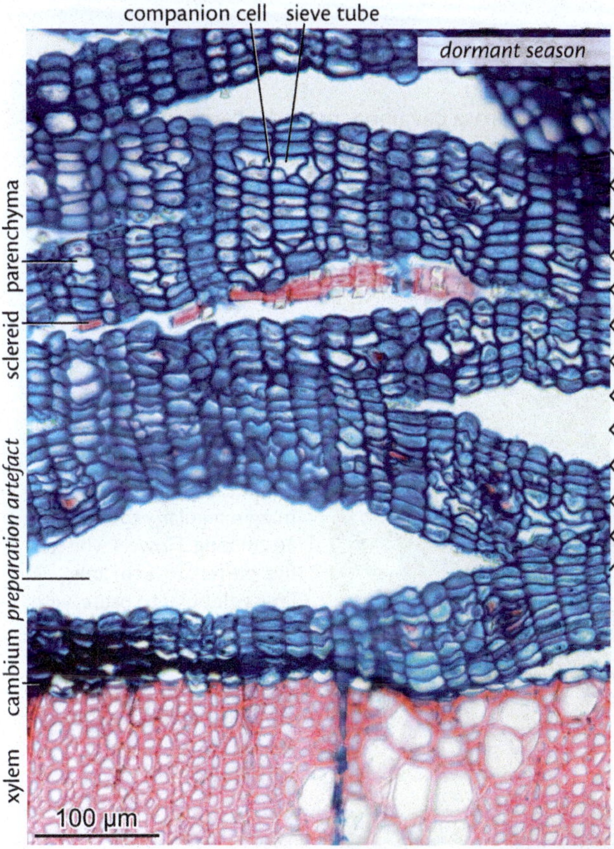

Bark anatomy: Irregular, unlignified tissue with many oval groups of fibers. Sieve tubes angular, parenchyma cells roundish. Fibers in tangentially arranged, oval groups, loosely connected with tissue of sieve tubes and parenchyma. (Holes are preparation artifacts.) Sclereids rare, in groups, or absent. Rays straight, dilating. Prismatic crystals around fiber groups. Crystal druses in most axial parenchyma and ray cells.

Rhytidome layered, cells thick-walled.

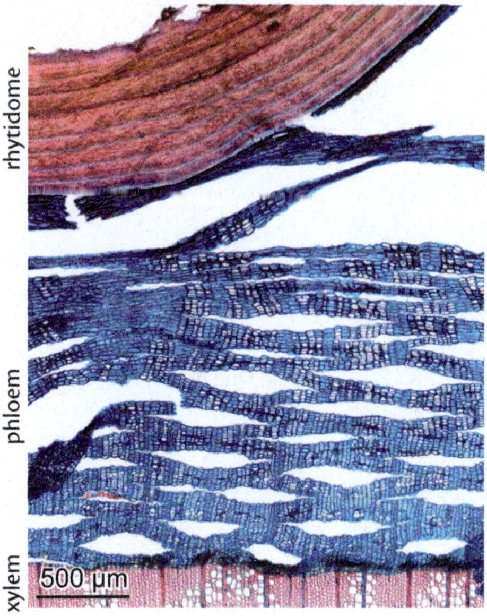

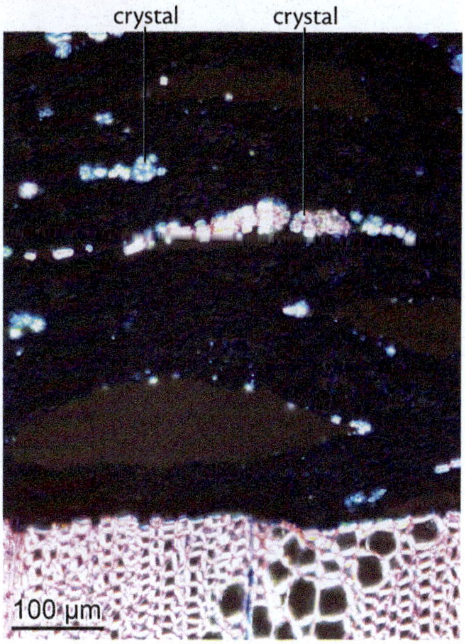

Rosa arvensis Huds. — Rosaceae

field rose, Feldrose, rosier des champs, rosa cavallina, rosal del campo

Form: Low shrub, densely branched and scrambling, twigs green, spiny, to 1 m tall.

Features: Leaves deciduous, pinnate, leaflets ovate, margin regularly toothed, to 10 cm long. Flowers white, five-parted, to 4 cm long. Fruits globular to ovate, with a permanent style on top, red, to 2 cm long.

Habitat: Europe, from France northeast to Ireland, Belgium, Slovakia, the Black Sea, Turkey and Italy.

On loamy, calcareous soils, to 1700 m.

Bark: Dull green, with bent spines.

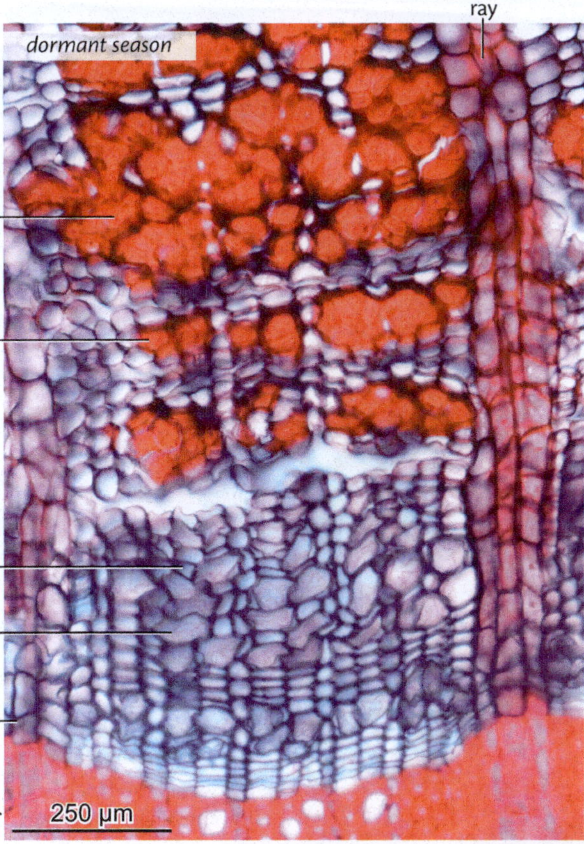

Bark anatomy: Structure tangentially layered. Sieve tubes collapsing during sclereid formation. Fibers absent. Sclereids in small tangential layers. Rays straight, large rays often dilating. Many prismatic crystals, mainly in axial parenchyma cells. Large rays parallelled by very large prismatic crystals.

Phellem layered, cells rectangular, thick-walled, filled with dark-staining substances.

No anatomical differences in bark of *Rosa* species.

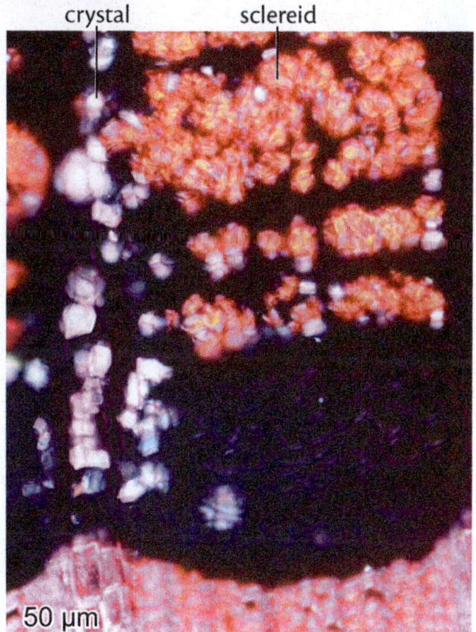

Rosa elliptica Tausch

Rosaceae

elliptic-leaved rose, Keilblättrige Rose, rosier à feuilles elliptiques, rosa ellittica, eglantina roja

Form: Densely branched, spiny shrub, twigs reddish, to 2.5 m tall.

Features: Leaves deciduous, pinnate, leaflets ovate, margin regularly toothed, to 10 cm long. Flowers white or pink, five-parted, to 3 cm long. Fruits globular to ovate, red, to 2 cm long.

Habitat: Central Europe, from the Pyrenees east to Central Germany, southern Poland east to the Don River.

On dry, stony and loamy, calcareous soils, to 1400 m.

Bark: Young reddish, later brown gray, with hooked spines.

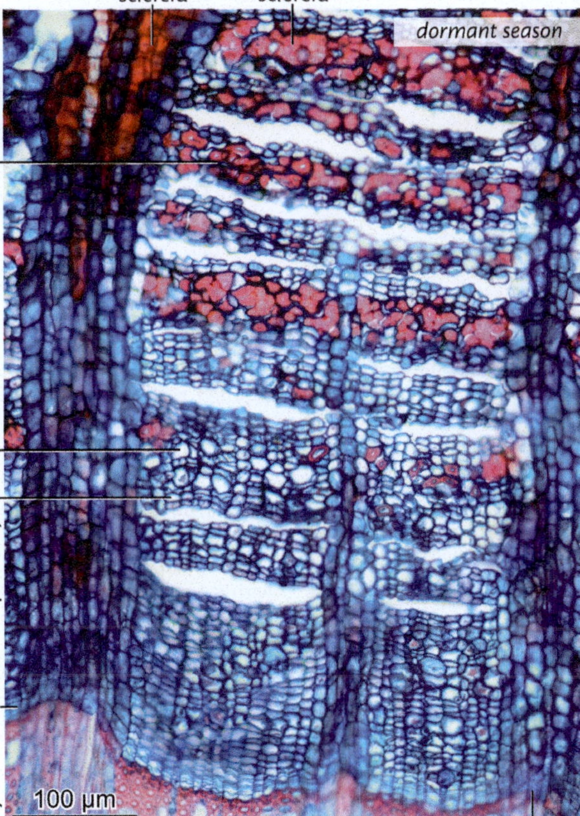

Bark anatomy: Structure tangentially layered. Sieve tubes collapsing during sclereid formation. Fibers absent. Sclereids in small tangential layers. Rays straight, large rays often dilating. Many prismatic crystals, mainly in axial parenchyma cells. Large rays paralleled by very large prismatic crystals.

Phellem layered, cells rectangular, thick-walled, filled with dark-staining substances.

No anatomical differences in bark of *Rosa* species.

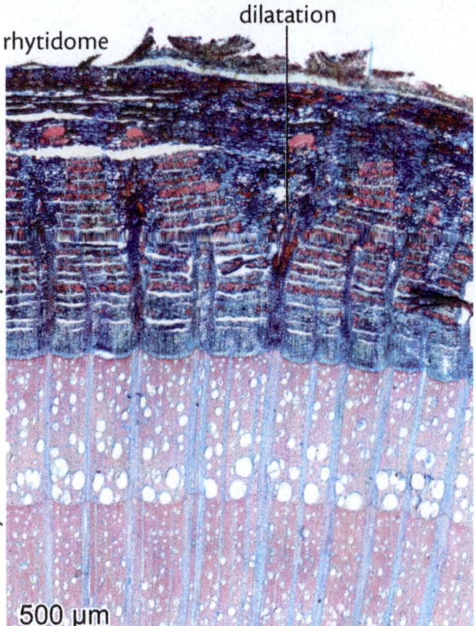

Salix alba L. Salicaceae

white willow, Silberweide, saule blanc, salice bianco, sauce blanco

Form: Upright tree, crown large, irregular shaped, to 35 m tall.

Features: Leaves deciduous, alternate, long lanceolate, lower side silvery, densely silky hairy, margin regularly toothed, to 12 cm long. Flowers monoecious, female flower an upright, thin, green catkin, male flower an upright, dense, yellow catkin, to 8 cm long. Fruit a hanging, green catkin, to 10 cm long, with many leathery capsules, the ripe, tiny seeds with abundant, white woolly hairs.

Habitat: Eurasia, from North Africa northeast to England, Northern Germany and Estonia, east to the Ob River and Northwest China, south to Iran and Turkey, the Balkans and Italy.

On moist to wet, strictly riverine soils, forming the zone of trees closest to the water, to 1000 m (China to 3000 m).

Bark: Twigs yellow to dark brown, stem bark young smooth, light gray, with age brown gray, soon deeply fissured.

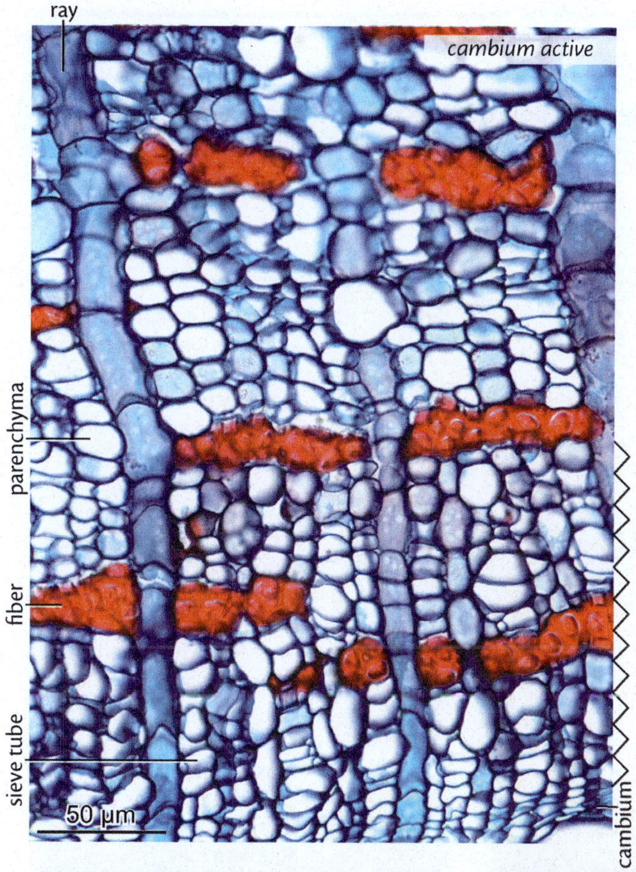

Bark anatomy: Structure of all willows principally the same. Occurrence of fibers reduced under poor growing conditions. Annual rings normally unrecognizable.

Regular tangential layers of parenchyma, sieve tubes and fibers in the juvenile phloem .

Sieve tubes irregular shaped, collapsing within a few years. Companion cells small. Parenchyma cells roundish. Fibers in tangential layers, not corresponding with annual rings. Groups of sclereids in older parts of the phloem. Prismatic crystals and a few crystal druses along fiber bands.

Periderm of old stems irregularly tangentially layered, consisting of large, rectangular, thin-walled cells and small, thick-walled cells.

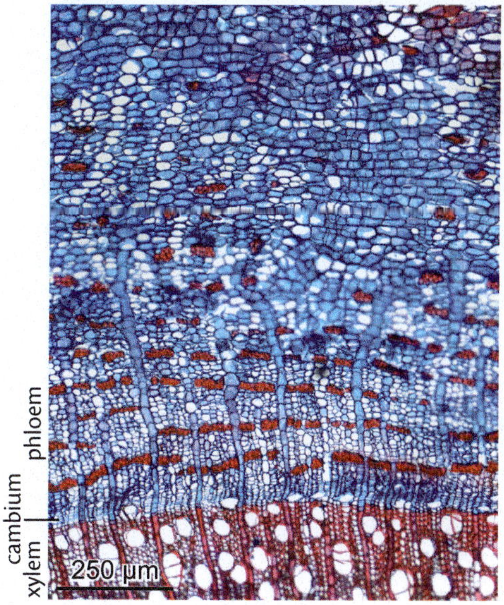

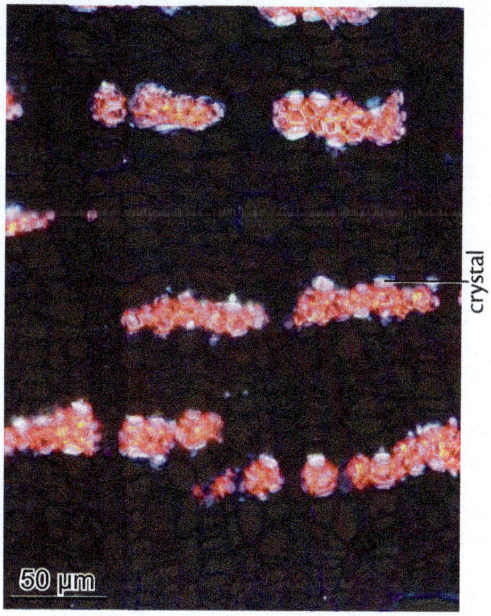

Salix purpurea L. — Salicaceae

purple willow, Purpurweide, saule pourpre, salice rosso, sauce colorado

Form: Upright shrub, twigs deep purple, to 6 m tall.

Features: Leaves deciduous, alternate, long lanceolate, lower side bluish, margin regularly toothed, to 12 cm long. Flowers monoecious, female flower an upright, thin, green catkin, male flower an upright, purple catkin, to 6 cm long. Fruit a hanging, green catkin, to 5 cm long, with many leathery capsules, the ripe, tiny seeds with abundant, white woolly hairs.

Habitat: Eurasia, from North Africa northeast to Scotland, Denmark and Estonia, east to the Ob River, south to Crimea, Albania and Southern Italy.

On moist to wet, sandy to gravelly riverine soils, to 2200 m.

Bark: Smooth, light gray, with age still almost without fissures.

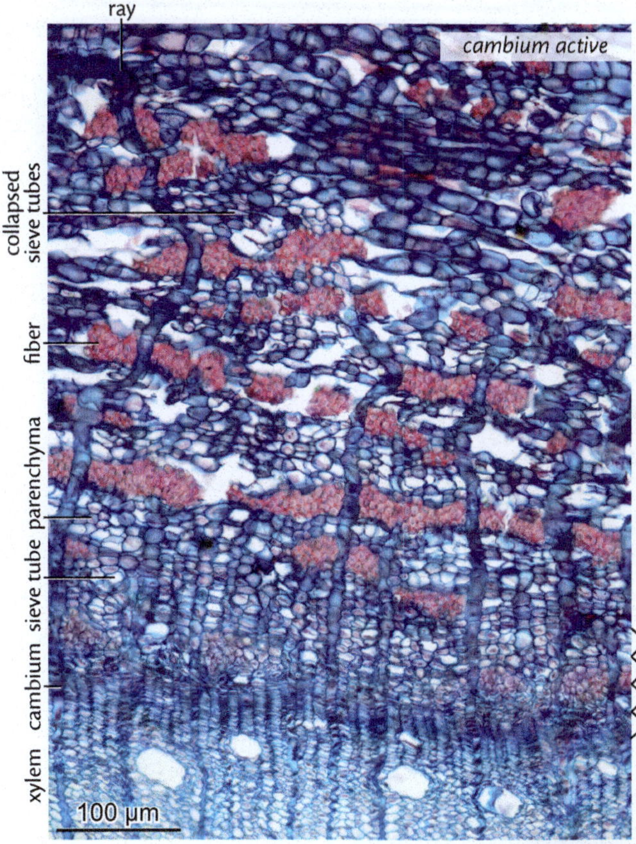

Bark anatomy: Structure of all willows principally the same. Occurrence of fibers reduced under poor growing conditions. Annual rings normally unrecognizable.

Regular tangential layers of parenchyma, sieve tubes and fibers. Sieve tubes irregular shaped, collapsing within a few years. Companion cells small. Parenchyma cells roundish. Fibers in tangential layers, not corresponding with annual rings. Groups of sclereids in older parts of the phloem. Prismatic crystals and a few crystal druses along fiber bands.

Periderm of old stems irregularly tangentially layered, consisting of large, rectangular, thin-walled cells and small, thick-walled cells.

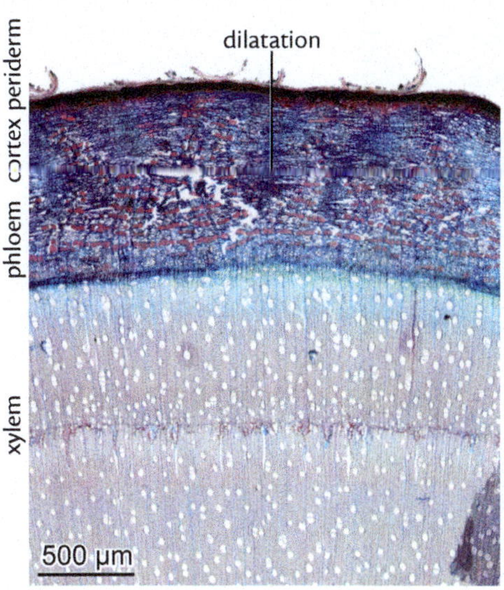

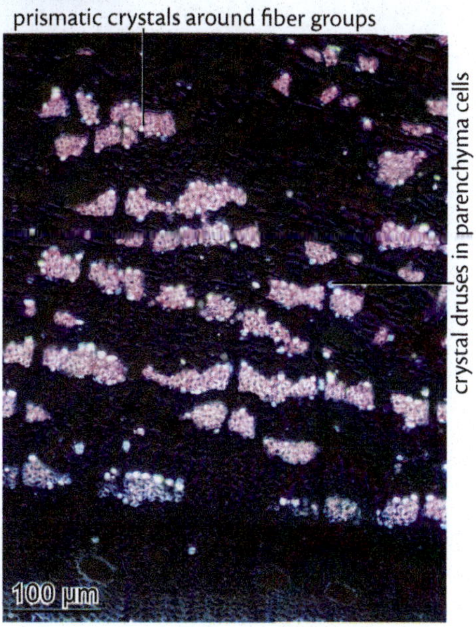

Sambucus nigra L. Adoxaceae

elder, Schwarzer Holunder, sureau noir, sambuco nero, saúco negro

Form: Upright shrub or small tree, to 7 m tall.

Features: Leaves deciduous, opposite, pinnate, usually with five leaflets, margin toothed, to 30 cm long. Flowers creamy white, scented, in pseudo-umbels, to 25 cm across. Fruit a globular, glossy black, berry-like drupe, to 7 mm long, in pseudo-umbels, to 25 cm across.

Habitat: Europe, from Southern Spain northeast to Southern Sweden, Estonia, Ukraine and the Caucasus, south to northern Turkey, Greece and Sicily.

On moist, deeply loamy soils, to 1600 m.

Bark: Twigs light to brown gray, bark gray brown, with age deeply fissured.

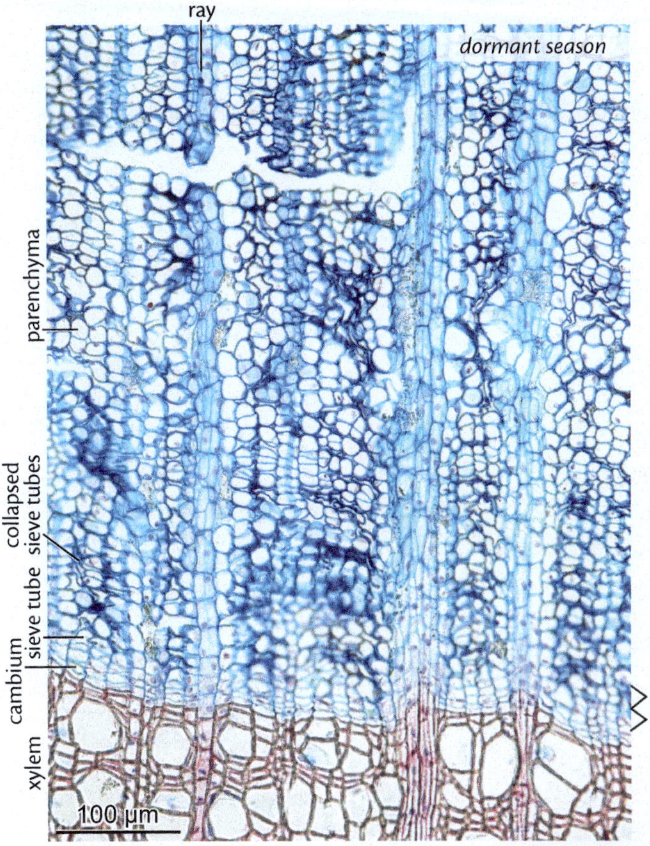

Bark anatomy: Soft tissue without lignification. Sieve tubes angular, parenchyma cells round. Sieve tubes collapsing within a short time, indicated by tangential and radial lines. Fibers absent. Sclereids often absent or in small tangential lines. Rays straight, dilatations absent. Crystal sand in radial rows.

Phellem layered, cells rectangular or square, thin-walled.

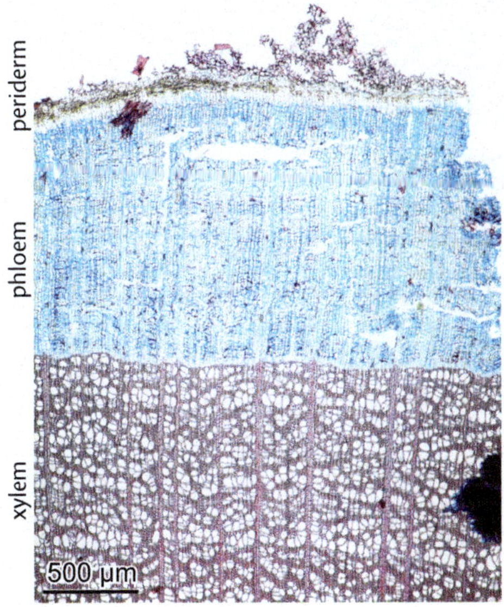

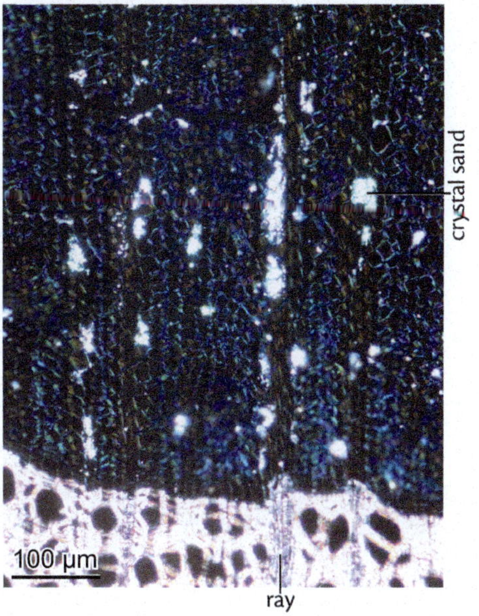

Sambucus racemosa L. Adoxaceae

red-berried elder, Roter Holunder, sureau à grappes, sambuco rosso, saúco rojo

Form: Upright shrub or small tree, to 4 m tall.

Features: Leaves deciduous, opposite, pinnate, usually with five leaflets, margin toothed, to 30 cm long. Flowers greenish yellow cream, in upright racemes, to 15 cm across. Fruit a globular, glossy red, berry-like drupe, to 5 mm long, in racemes, to 15 cm across.

Habitat: Europe, from the Pyrenees northeast to Northern Germany, Estonia, and the Carpathians, south to Bulgaria and the Central Apennines.

On moist, deeply loamy, often lime-poor soils, to 2400 m.

Bark: Gray brown, with conspicuous lenticels, with age deeply fissured.

Bark anatomy: Tissue soft, with some groups of sclereids. Sieve tubes angular, parenchyma cells round. Sieve tubes collapsing within a short time, indicated by tangential and radial lines. Fibers absent. Sclereids often absent or in small tangential lines. Rays straight, dilatations absent. Crystal sand in radial rows.

Phellem layered, cells rectangular or square, thin-walled.

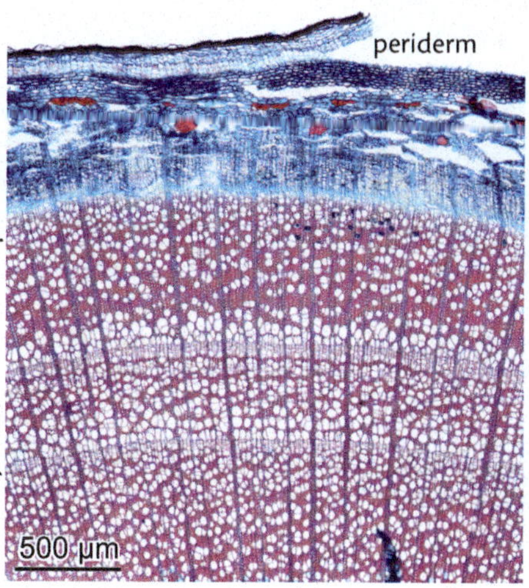

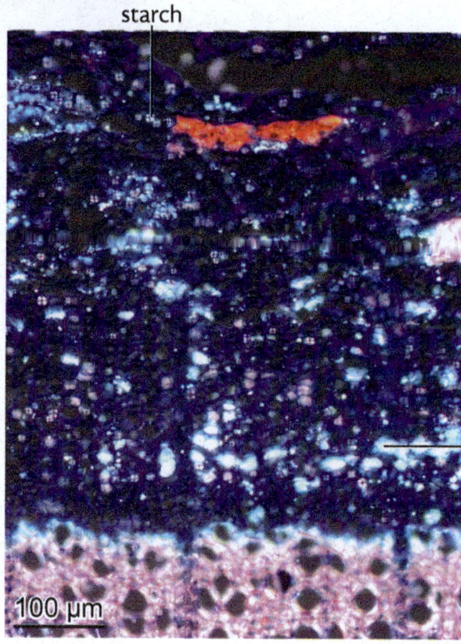

Sorbus torminalis (L.) Crantz

Rosaceae

wild service tree, Elsbeere, alisier torminal, sorbo torminale, mostajo

Form: Upright tree, crown regular shaped, to 25 m tall.

Features: Leaves deciduous, alternate, with irregular, sharply pointed lobes, margin regularly toothed, to 7 cm long. Flowers white, five-parted, to 5 mm long, in pseudo-umbels. Fruit a dull brown, ovate or globular pome, to 2 cm wide, in hanging pseudo-umbels.

Habitat: Europe and Western Asia, from North Africa northeast to Denmark and Poland, east to Northern Iran, south to Lebanon and Sicily.

On dry, often loamy and calcareous soils, to 1400 m.

Bark: Young smooth, gray to reddish gray, often with conspicuous horizontal diamonds, with age splitting into small, square plates.

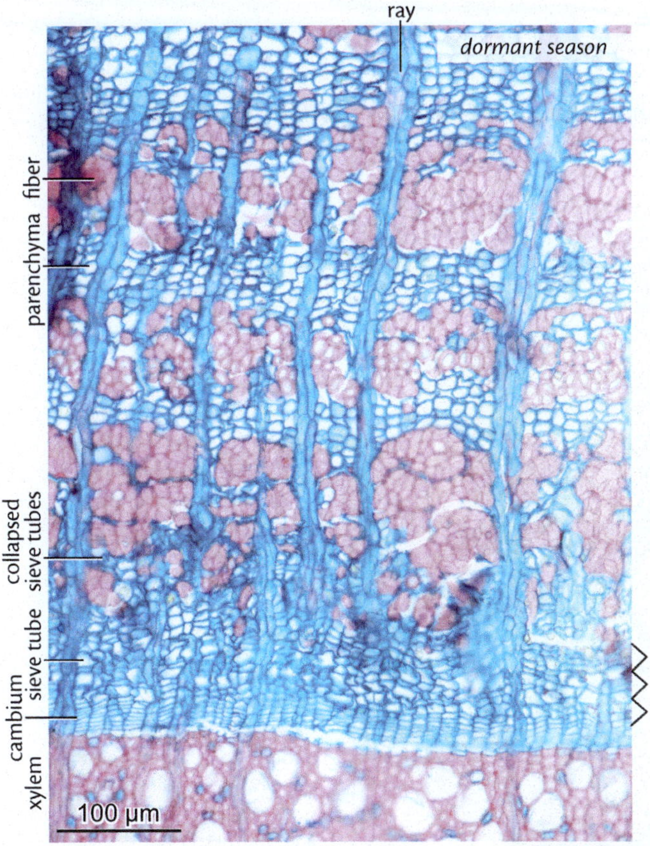

Bark anatomy: Structure regular, tangentially layered. Sieve tubes collapsing before sclereid formation. Fibers absent. Tangential bands of sclereids in tangential layers. Rays straight in the phloem. Dilatations not observed. Many small and large prismatic crystals in axial parenchyma cells.

Layers of rectangular, thick-walled phellem cells separating living from dead bark (rhytidome).

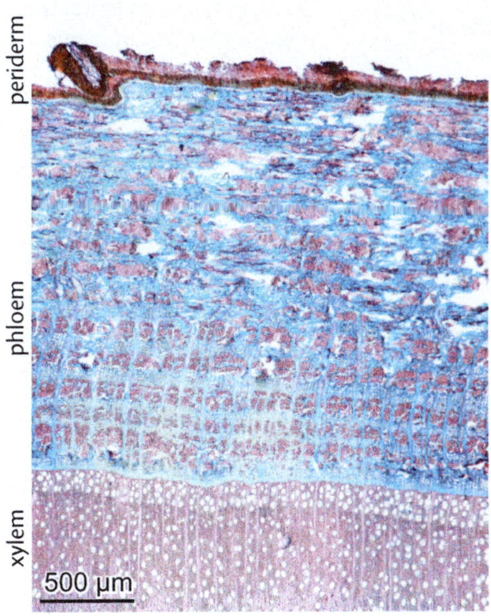

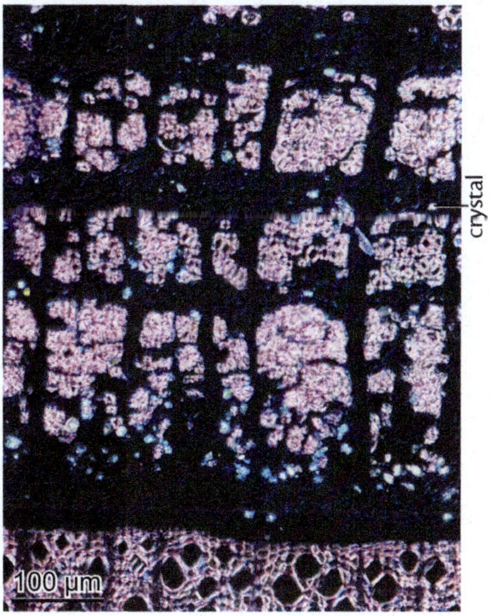

Tilia cordata Mill. Malvaceae

small-leaved lime, Winterlinde, tilleul à petites feuilles, tiglio selvatico, tilo de hoja pequeña

Form: Upright tree, crown regular, heart-shaped, to 30 m tall.

Features: Leaves deciduous, alternate, distinctly heart-shaped, with a pointy tip, margin regularly toothed, to 8 cm long. Flowers yellow white, scented, to 8 mm long, grouped in three to ten beneath a wing-like bract. Fruit a dull brown, capsular nut, to 6 mm wide, hanging beneath the bract.

Habitat: Europe, from Northern Spain northeast to Southern Scandinavia and the Lake Ladoga, east to the Ural Mountains and the Caucasus, south to Albania and Southern Italy.

On dry, often stony limestone soils, on warm escarpments, to 1600 m.

Bark: Twigs red or brown, in shadowy positions also green, stem bark smooth, gray, with age markedly fissured.

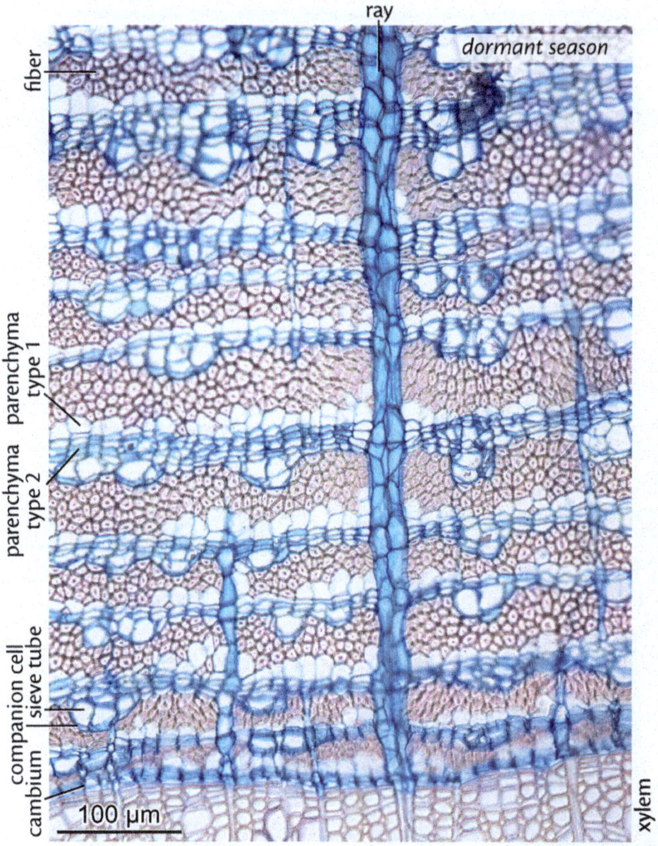

Bark anatomy: Structure regular, tangentially layered. Sieve tubes angular, arranged in groups, accompanied by companion cells. Parenchyma cells in two types: large, unlignified cells in uniseriate rows adjacent to fiber bands and small cells in layers of 2–4 rows. Tangential fiber bands of variable width in tangential layers. Sclereids absent. Rays straight, often distinctly dilating. Very few large crystals in large parenchyma cells and cortex. Slime occasionally in older parts of dilatations.

Phellem layered, cells rectangular, thick-walled.

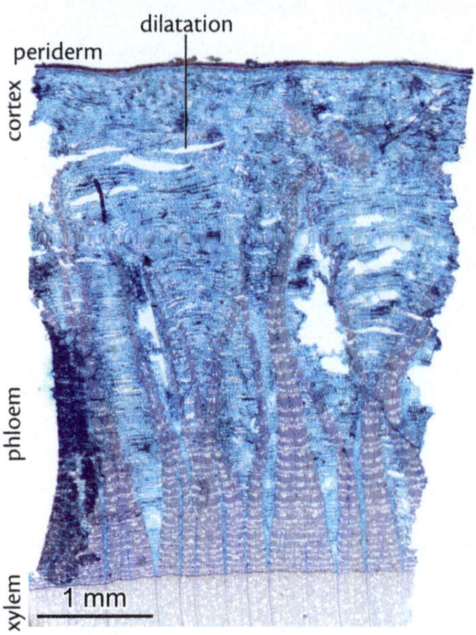

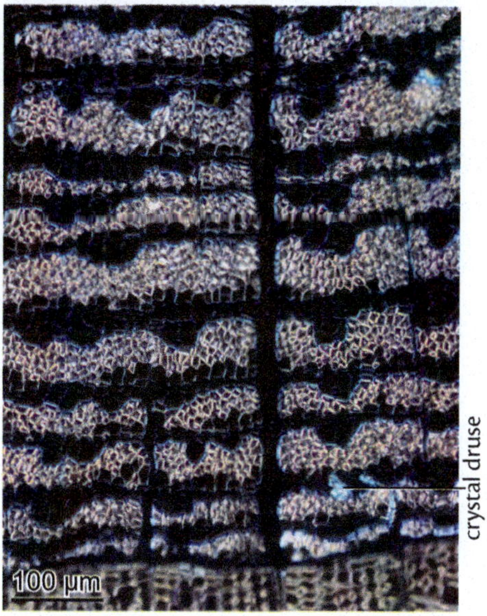

Ulmus glabra Huds. Ulmaceae

wych elm, Bergulme, orme montagnard, olmo di montagna, olmo de montaña

Form: Upright tree, crown regular shaped, broad, to 40 m tall.

Features: Leaves deciduous, alternate, long ovate, with one or several tips, leaf base always asymmetrical, margin double-toothed, to 20 cm long. Flowers reddish, emerging early, to 3 mm long. Fruit a one-seeded nut, in the center of a round wing, first light green, maturing to light brown.

Habitat: Europe, from Northern Spain northeast to coastal Norway and the Lake Ladoga, east to the Ural Mountains and the Caucasus, south to Greece and Southern Italy, isolated in the Taurus Mountains.

On moist, calcareous, often stony or loamy soils, to 1800 m.

Bark: Young smooth, gray, with age gray brown and markedly fissured.

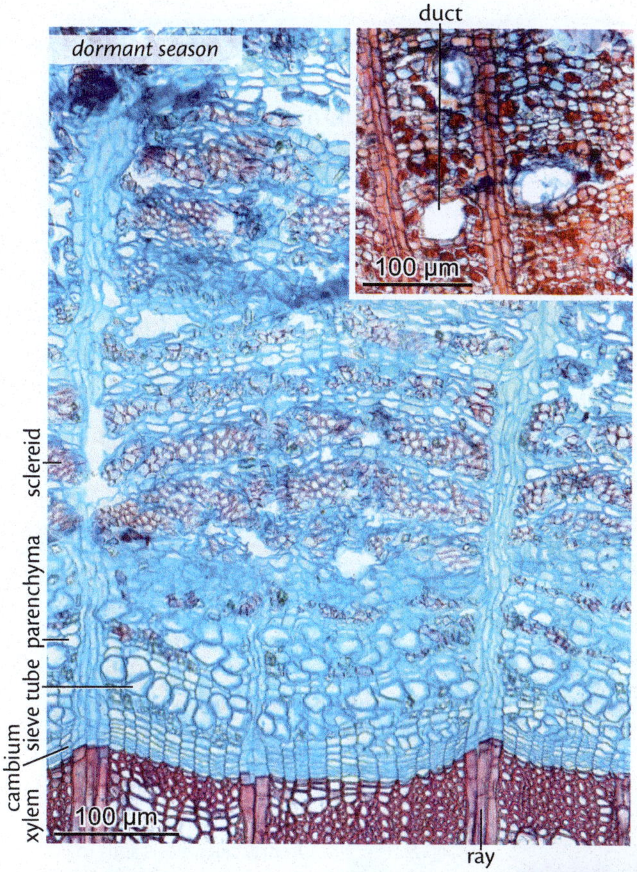

Bark anatomy: Structure regular, tangentially layered. Early phloem consisting of a zone of angular sieve tubes with many companion cells. Late phloem composed of rectangular parenchyma cells. Sieve tubes collapsing after two seasons. Tangential bands of fiber-like sclereids alternating with zones of parenchyma cells and sieve tubes. Rays straight in the beginning, later wavy, dilatations mostly absent. Many large prismatic crystals in axial parenchyma cells. Characteristic slime ducts with a surrounding layer of excretion cells and slime (stained blue).

Phellem layered, cells square or irregular shaped.

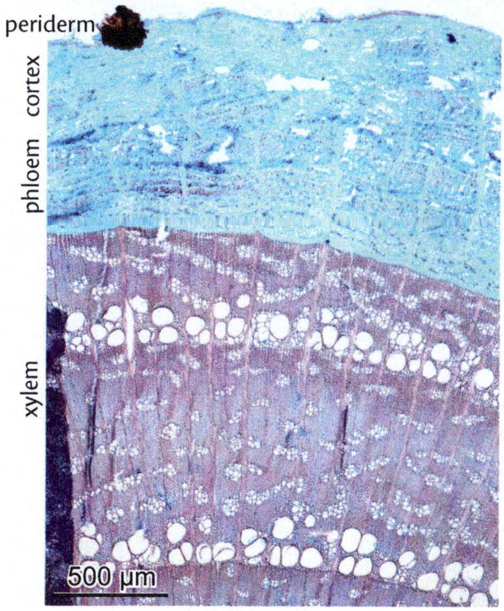

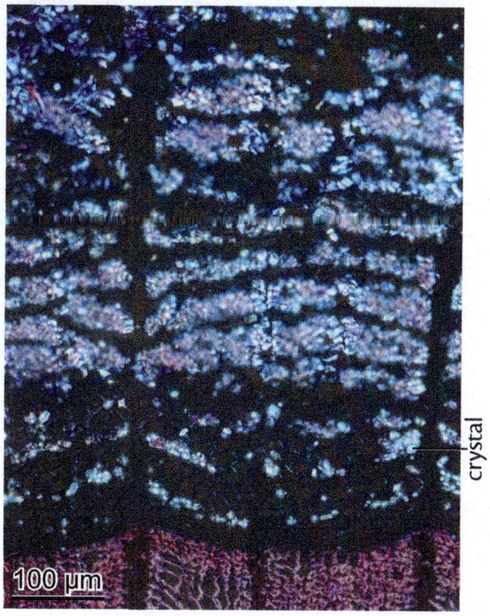

Viburnum lantana L. Adoxaceae

wayfaring tree, Wolliger Schneeball, viorne lantane, viburno lantana, morrionera

Form: Upright, densely branched shrub, to 5 m tall.

Features: Leaves deciduous, opposite, round ovate, felty hairy, margin regularly toothed, to 12 cm long. Flowers white, in flat pseudo-umbels, to 10 cm across. Fruit a globular, berry-like drupe, first glossy red, later maturing to blue black, to 1 cm long, in pseudo-umbels.

Habitat: Europe, from the Atlas Mountains northeast to Czechia, Ukraine, the Caucasus, south to northern Turkey, Northern Greece and Central Italy.

On dry, calcareous soils in sunny sites, to 1600 m.

Bark: Light brown to brown gray, dotted with many lenticels, with age fissured.

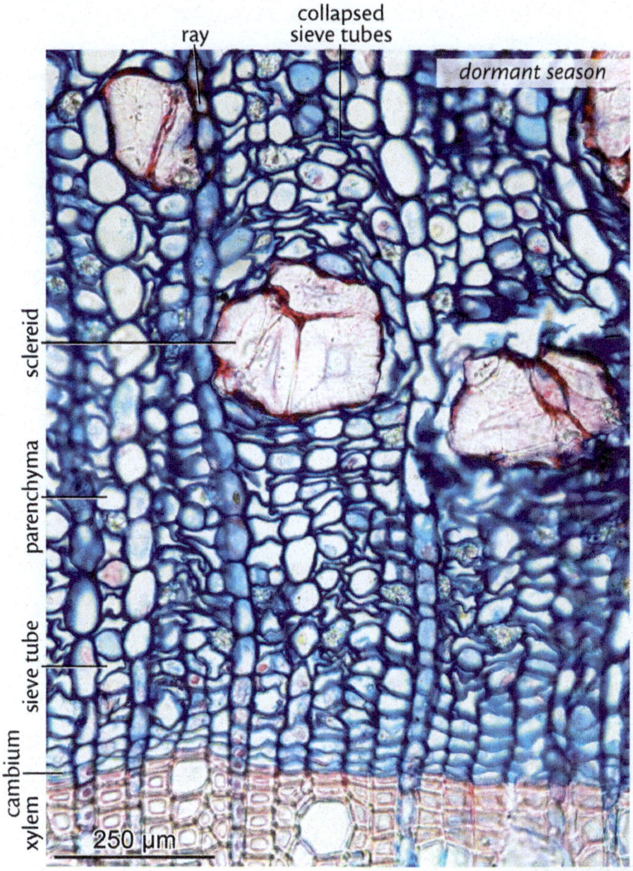

Bark anatomy: Structure irregular, tissue unlignified, with many round to oval groups of sclereids.

Sieve tubes angular, collapsing soon after formation. Parenchyma cells roundish. Fibers absent. Sclereids in round groups. Rays straight, dilatations absent. Many crystal druses in axial parenchyma cells. Annual ring boundaries not recognizable.

Phellem layered, consisting of 2–4 rows of thin-walled, square cells and one row of rectangular cells.

Structure almost identical to *Viburnum opulus*.

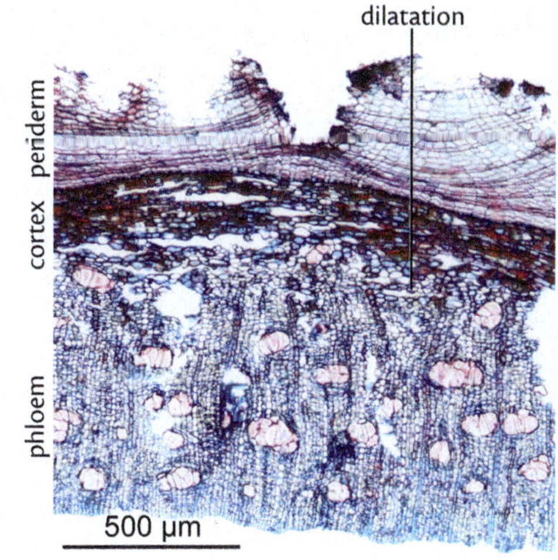

Viburnum opulus L. Adoxaceae

guelder rose, Gemeiner Schneeball, viorne obier, oppio, mundillo

Form: Upright, densely branched shrub, to 4 m tall.

Features: Leaves deciduous, opposite, three-lobed, margin irregularly acutely toothed, to 12 cm long. Flowers white, in flat pseudo-umbels with enlarged outer flowers, umbels to 10 cm across. Fruit a globular, glossy red, berry-like drupe, to 1 cm wide, in hanging pseudo-umbels, often holding through winter.

Habitat: Eurasia, from Northwest Africa northeast to Scotland, Southern Scandinavia, the Ob River, the Altai Mountains and the Tien Shan Mountains, south to northern Turkey, Greece and Central Italy.

On moist, deeply loamy soils, often riverine, to 1500 m.

Bark: Gray brown, with age finely fissured.

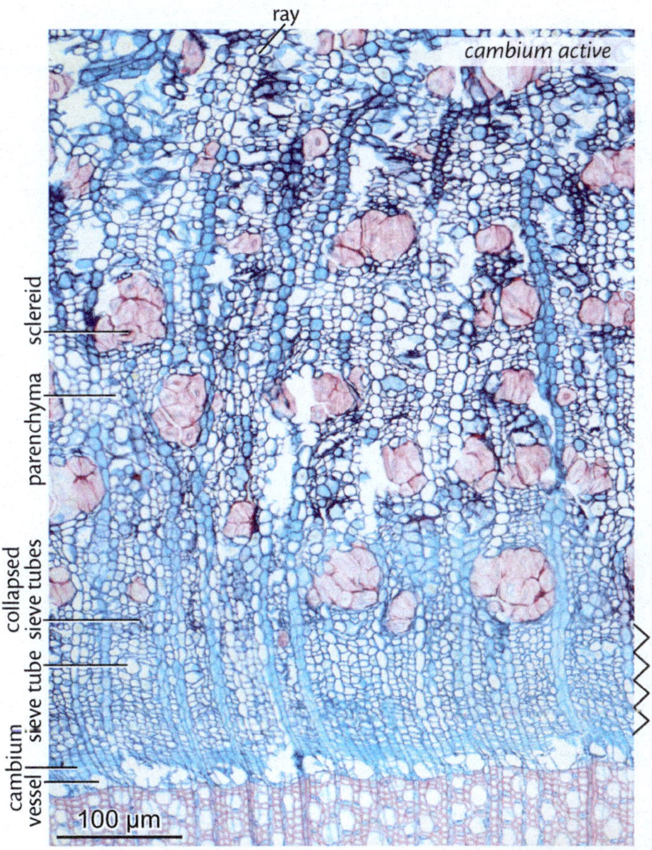

Bark anatomy: Structure irregular, tissue unlignified, with many round to oval groups of sclereids.

Sieve tubes angular, collapsing soon after formation. Parenchyma cells roundish. Fibers absent. Sclereids in round groups. Rays straight, dilatations absent. Many crystal druses in axial parenchyma cells.

Annual ring boundaries unrecognizable. Phellem layered, consisting of 2–4 rows of thin-walled, square cells and one row of rectangular cells.

Structure almost identical to *Viburnum lantana*.

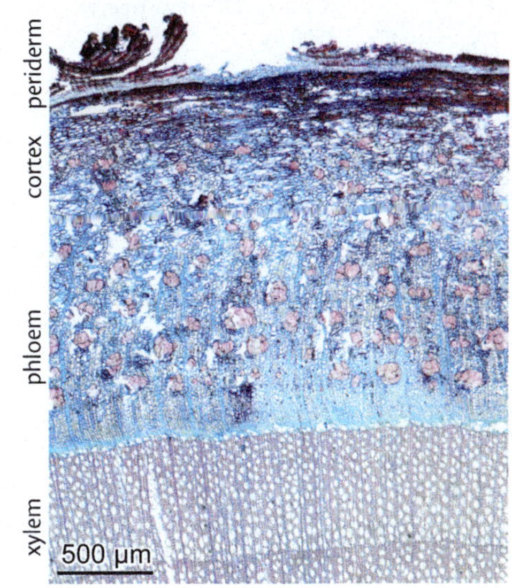

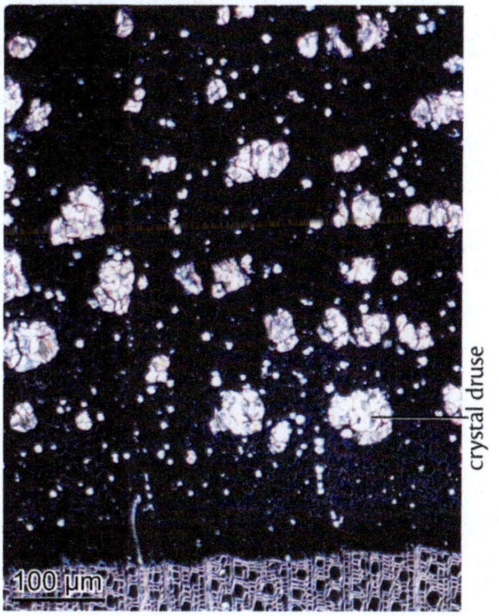

Viscum album L. Santalaceae

misteltoe, Mistel, gui, vischio comune, muérdago

Form: Globular shrub, densely branched, strictly epiphytic, twigs green, to 1 m in diameter.

Features: Leaves evergreen, leathery, opposite, long ovate, margin entire, to 5 cm long. Flowers dioecious, small, greenish, four-parted, to 4 mm long. Fruit a globular, fleshy, white, berry-like drupe, to 1 cm long.

Habitat: Eurasia, from Algeria northeast to Southern Scandinavia and Lithuania, east to Ukraine and the Caucasus, south to the Mediterranean Sea, isolated in the Himalayas, Northeast China, Japan and Southeast Asia.

Exclusively in the crowns of the specific host tree, to 1500 m.

Bark: Twigs remain green to olive yellow-green.

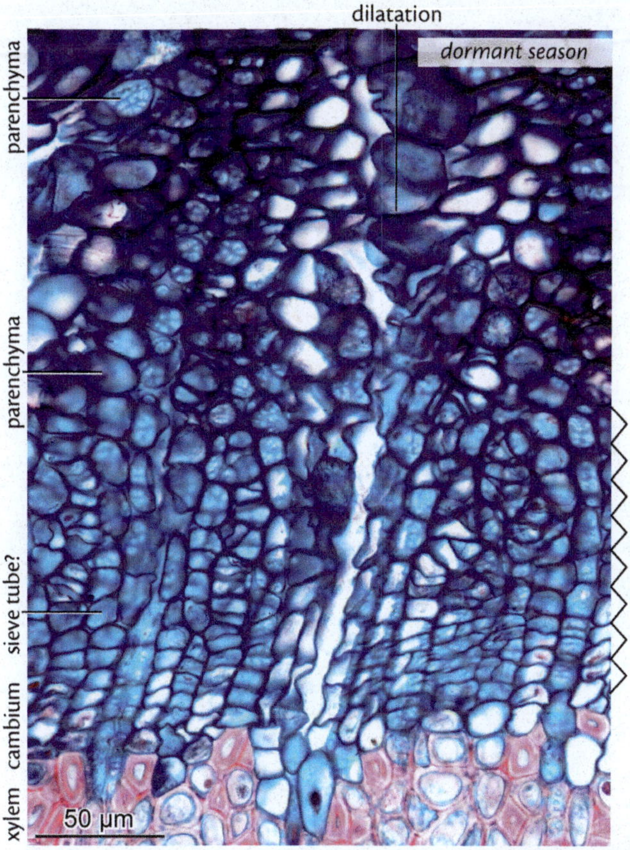

Bark anatomy: Structure with a small phloem, large cortex and epidermis with a thick cuticula.

Sieve tubes and roundish parenchyma cells difficult to distinguish. Sieve tubes not collapsing. Fibers absent. Groups of sclereids in the cortex. Rays dilating. Many crystal druses in axial parenchyma.

Phellem absent.

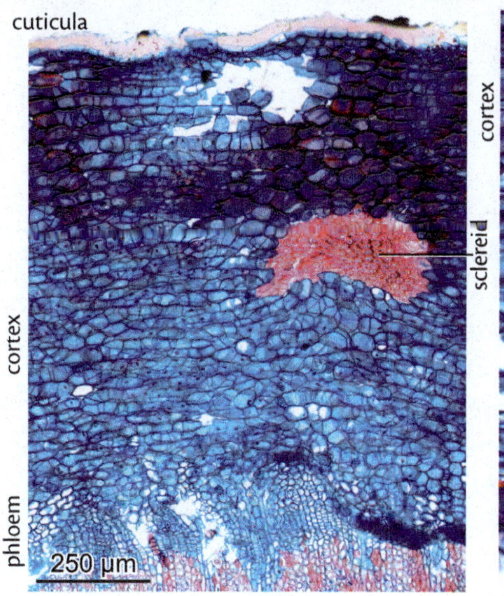

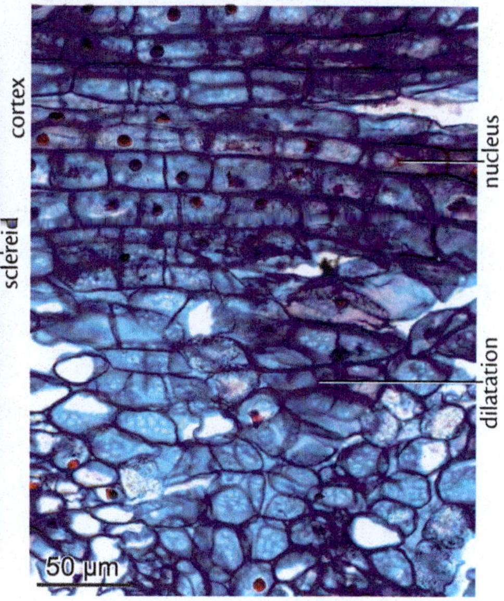

Geographically and ecologically very close and difficult to separate from the temperate forest is the biome of the **deciduous forest of submediterranean Europe**, centered in the higher, frost-influenced areas of the Mediterranean, the Balkans and the Southern Alps, with a few thermically favored islands in Central Europe, such as the Pannonian Basin. These forests are usually more species-rich than those in Central Europe, and are often dominated by deciduous oaks, *Castanea sativa*, *Ostrya carpinifolia* and *Fraxinus ornus*.

F. H. Schweingruber et al., *Bark Anatomy of Trees and Shrubs in the Temperate Northern Hemisphere*, https://doi.org/10.1007/978-3-030-14056-4_6

6 Deciduous forest of submediterranean Europe

Acer monspessulanum L. Sapindaceae

Montpellier maple, Französicher Ahorn, érable de Montpellier, acero minore, afre

Form: Upright tree, crown round, to 10 m tall.

Features: Leaves deciduous, opposite, three-lobed, margin entire, to 6 cm long. Flowers yellow green, in pseudo-umbels. Fruits splitting, with two wings at an acute angle.

Habitat: Mediterranean, northeast to the French Jura, Central Germany, the southern Balkans and Turkey.

On dry, stony and calcareous soils, to 1500 m.

Bark: Young smooth, light gray brown, later finely fissured.

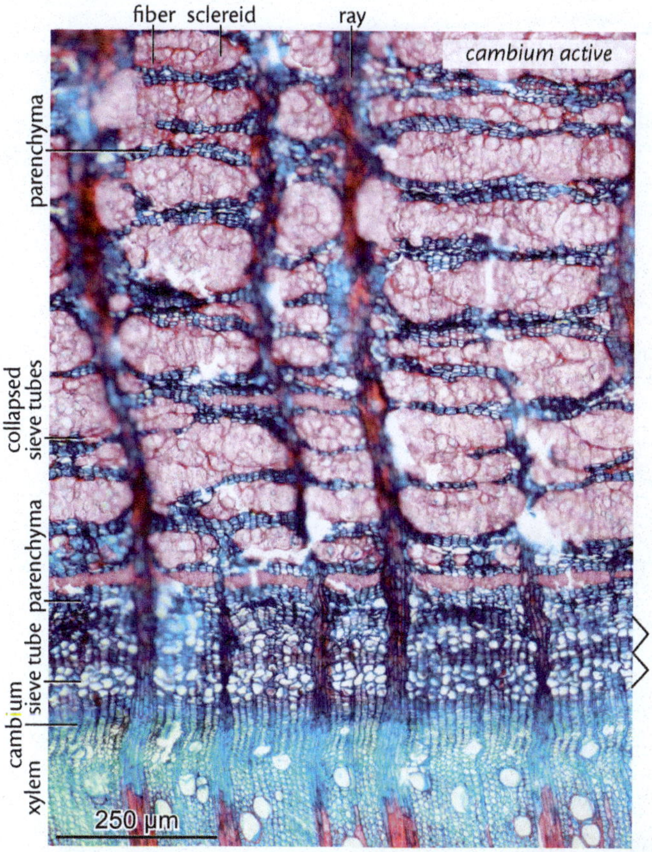

Bark anatomy: Regular tangential layers of sclereids.

Sieve tubes irregular shaped, companion cells small. Parenchyma cells roundish. Sieve tubes collapsing within short time after formation. Fibers very rare. Sclereids in large groups. Rays large, dilating after some years. Prismatic crystals mainly within fibers and around fiber zones. Crystals absent in rays.

Annual rings only visible in young phloem. Regular tangential structures becoming indistinct due to dilatation. Periderm tangentially layered, alternating between rectangular, thin-walled cells and thick-walled cells.

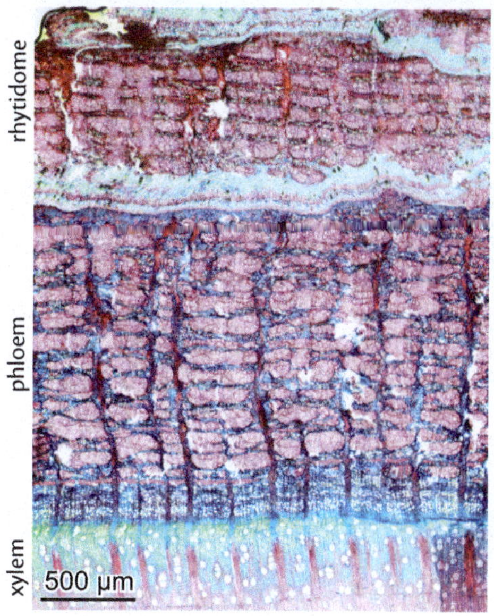

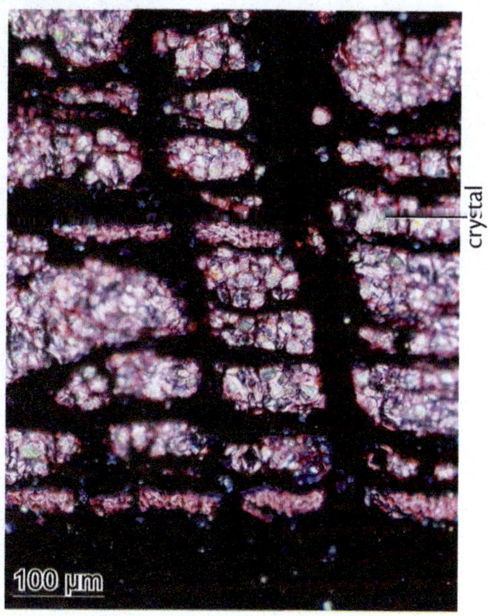

Acer opalus Mill. Sapindaceae

Italian maple, Schneeballblättriger Ahorn, érable à feuilles d'obier, acero alpino, acirón

Form: Upright tree, crown round or pyramidal, to 15 m tall.

Features: Leaves deciduous, opposite, shallowly three-lobed, margin weakly toothed, to 10 cm long. Flowers yellow green, in panicles. Fruits splitting, with two wings at an acute angle.

Habitat: Western Mediterranean, northeast to the Swiss Jura, the Valais, the North Apennines.

On dry, stony and calcareous soils, to 1400 m.

Bark: Young smooth, light gray brown, later dark gray, fissured and cracking into small plates.

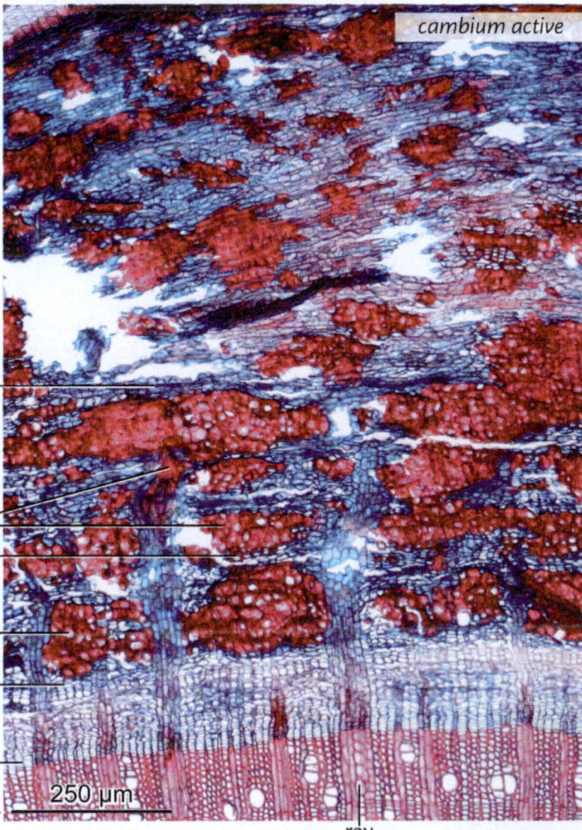

Bark anatomy: Regular tangential layers, consisting of sclereids in the young phloem.

Sieve tubes irregular shaped, companion cells small. Parenchyma cells roundish. Sieve tubes collapsing within short time after formation. Fibers very rare. Sclereids in large groups. Prismatic crystals mainly within fibers and around fiber zones. Rays disappearing after some years due to dilation. Crystals absent in rays.

Annual rings only visible in young phloem. Regular tangential structures becoming indistinct due to dilatation. Periderm tangentially layered, alternating between rectangular, thin-walled cells and thick-walled cells.

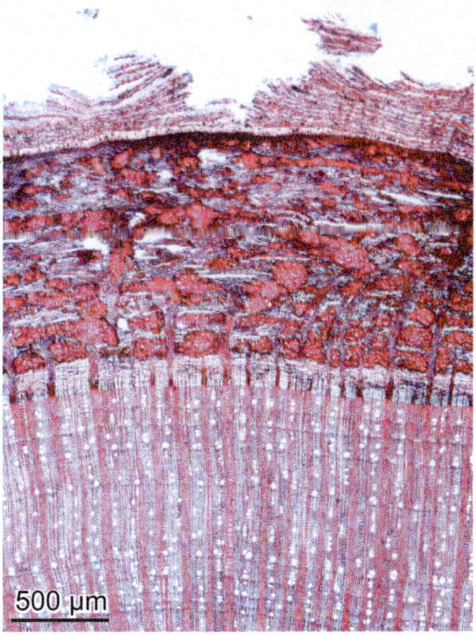

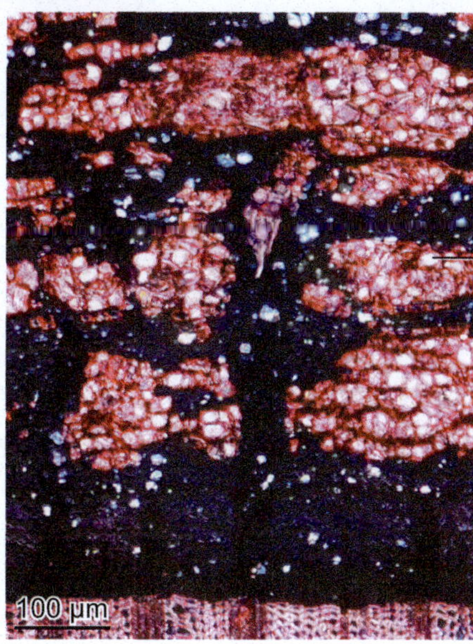

Aesculus hippocastanum L. Sapindaceae

horse chestnut, Rosskastanie, marronnier, ippocastano, castaño de Indias

Form: Upright tree, crown round or pyramidal, to 30 m tall.

Features: Leaves deciduous, opposite, digitate with five to nine leaflets, each to 20 cm long, margin double toothed. Flowers white, scented, irregularly five-parted, in upright panicles to 20 cm long. Fruit a globular, spiny, green capsule, containing one or two glossy brown, hard seeds, to 5 cm wide.

Habitat: Southeastern Europe and Western Asia, a few isolated localities in Albania, Macedonia and Northern Greece, from Turkey east to the Caucasus, Iran and the Himalayas; widely planted.

Naturally on moist, stony and mostly calcareous soils, to 1800 m.

Bark: Young dull, smooth, light gray, later gray brown, fissured and cracking into small plates.

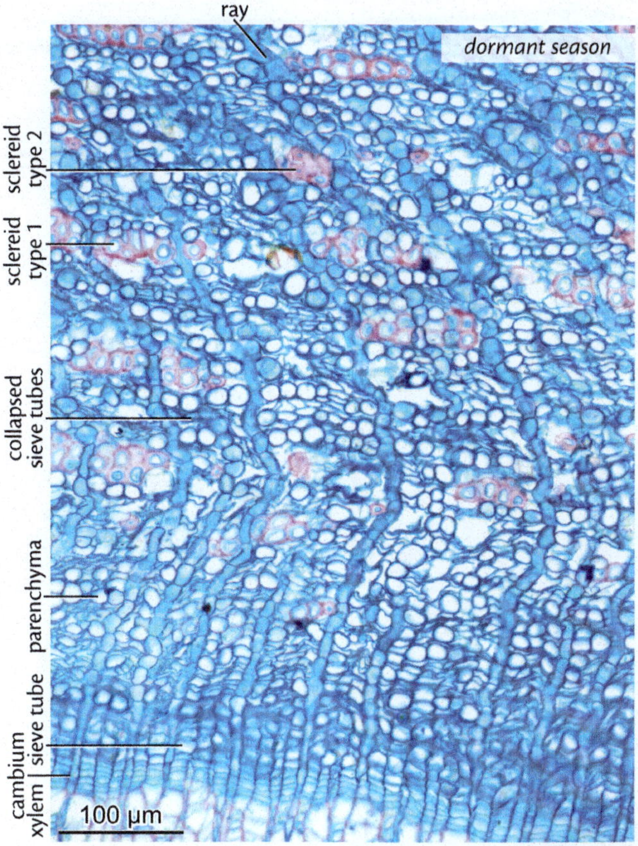

Bark anatomy: Poorly lignified structure with small spots of sclereids.

Sieve tubes collapsing within short time after formation. Parenchyma cells round, indistinctly tangentially layered. Fibers absent. Sclereids most abundant in form of thick-walled cells with a large lumen and unlignified tertiary walls (type 1), rarely as thick-walled cells with a very small lumen (type 2). Rays slightly wavy and bent. Ray dilatation only in thick bark. Prismatic crystals and crystal druses very large.

Phellem cells alternating between rectangular, thin-walled cells and often crumpled forms.

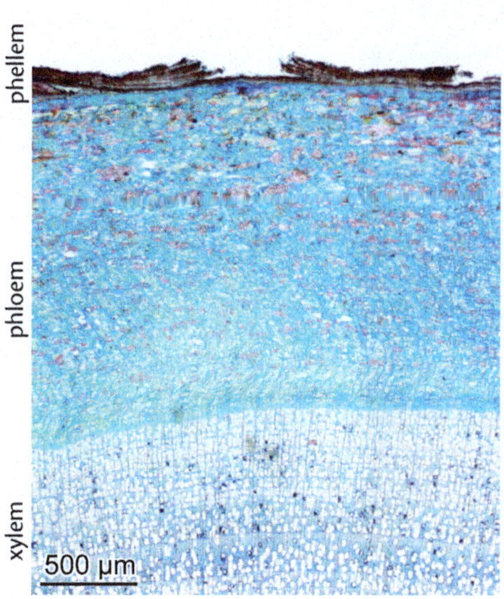

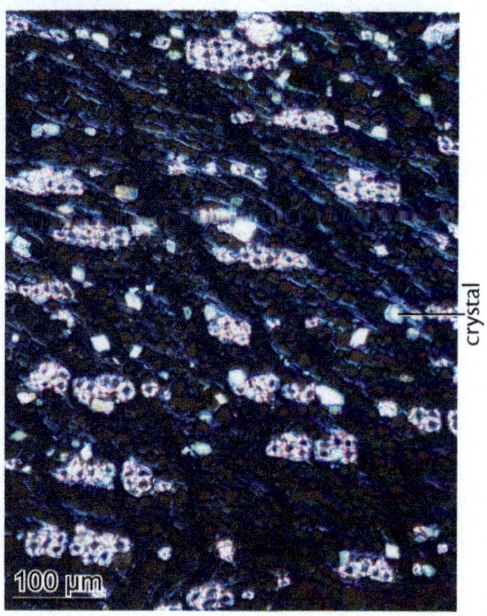

Buxus sempervirens L.

Buxaceae

box, Buchs, buis, bosso comune, boj

Form: Evergreen, small-leaved, upright shrub or small tree, to 5 m tall.

Features: Leaves evergreen, leathery, opposite, ovate, glossy green above, margin entire, to 2.5 cm long. Flowers creamy yellow, without petals, to 3 mm long. Fruit a three-valved, woody capsule, to 8 mm long.

Habitat: Southern Europe and Western Asia, northeast to Paris, Belgium, the Balkans south of the Danube River, the Caucasus and Northern Iran.

On dry, stony and calcareous soils, to 2100 m.

Bark: Dull, light gray brown, later finely fissured.

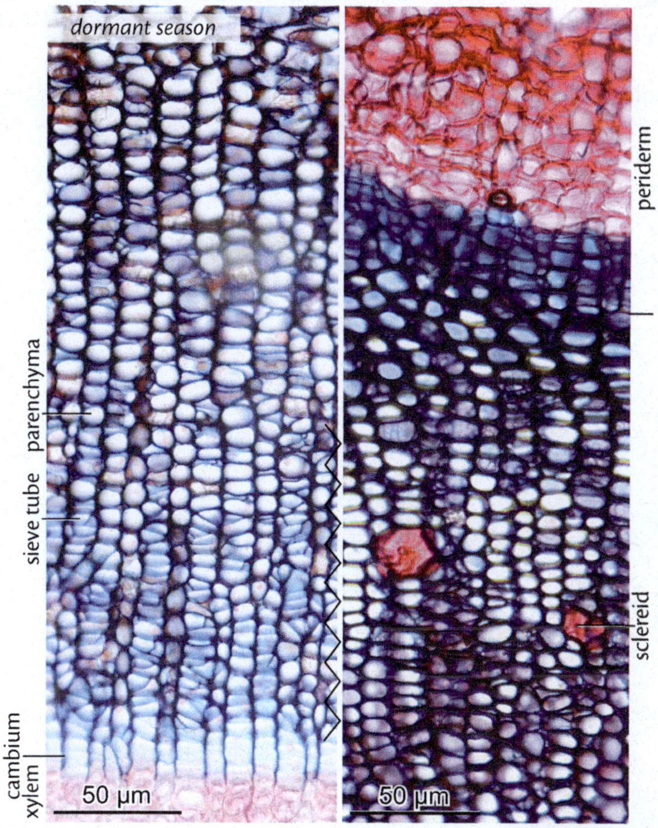

Bark anatomy: Uniform structure of sieve tubes and parenchyma. Sieve tubes rectangular, accompanied by small companion cells and round to oval parenchyma cells. Sclereids rare, isolated, large. Rays barely differentiated from axial tissue of sieve tubes and parenchyma. Many prismatic crystals in long axial cell chambers.

Periderm consisting of layers of large, rectangular, thin-walled cells and small, thick-walled cells.

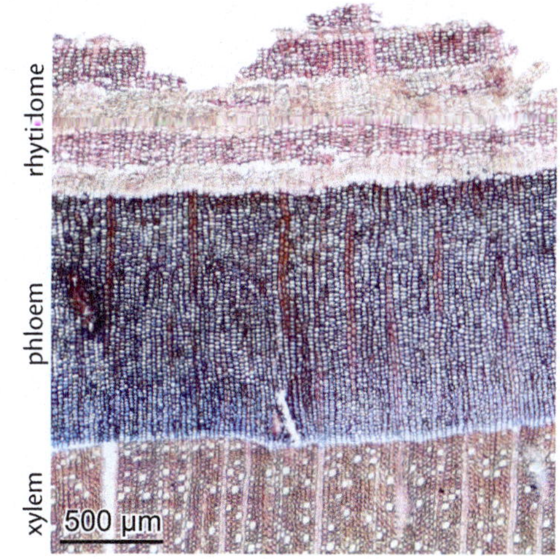

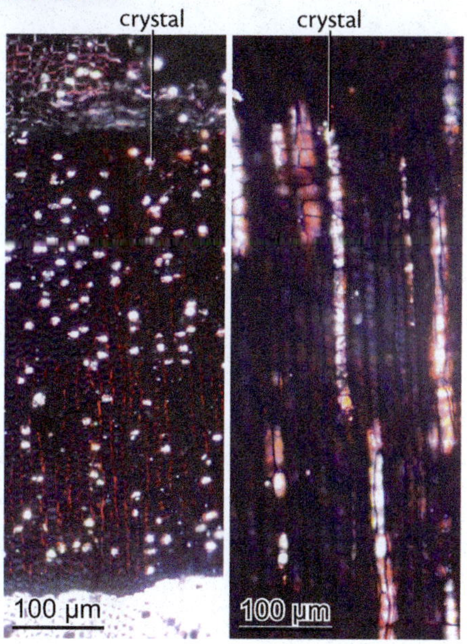

Castanea sativa Mill. Fagaceae

sweet chestnut, Edelkastanie, châtaignier cultivé, castagno comune, castaño

Form: Upright tree, crown large, irregular shaped, to 35 m tall.

Features: Leaves deciduous, alternate, long ovate, margin regularly sharply toothed, to 25 cm long. Flowers monoecious, female flower in the lower portion of the flower, male flowers in long, thin, white, scented catkins, to 20 cm long. Fruits hedgehog-like spiny cupules, to 6 cm wide, containing three red brown, edible nuts, called chestnuts.

Habitat: Originally Western Asia, from Turkey to the Caucasus and Iran, since Roman times widely planted in warmer parts of Europe.

On moist, permeable, acidic soils, to 1800 m.

Bark: Young smooth, silver gray, with age gray brown, deeply fissured.

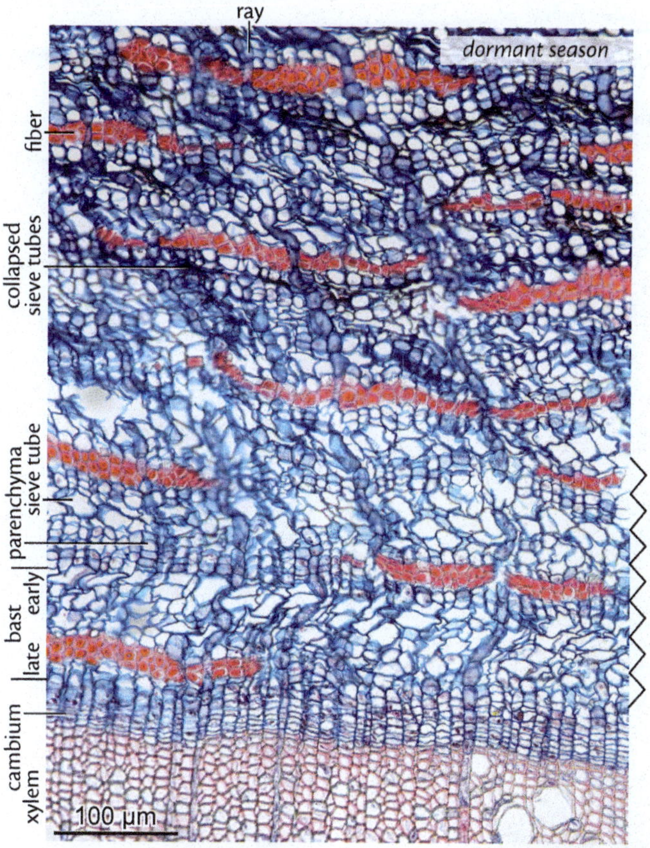

Bark anatomy: Regular tangential layers of fiber bands and zones of sieve tubes and parenchyma.

Early phloem formed by irregular shaped sieve tubes and roundish parenchyma cells. Sieve tubes collapsing within a few years after formation. Fibers in laterally interrupted tangential bands of 2–4 layers. Ray structure remains for many years. Prismatic crystals mainly within fibers and around fiber zones, crystals absent in rays.

Annual ring boundaries corresponding with fiber bands.

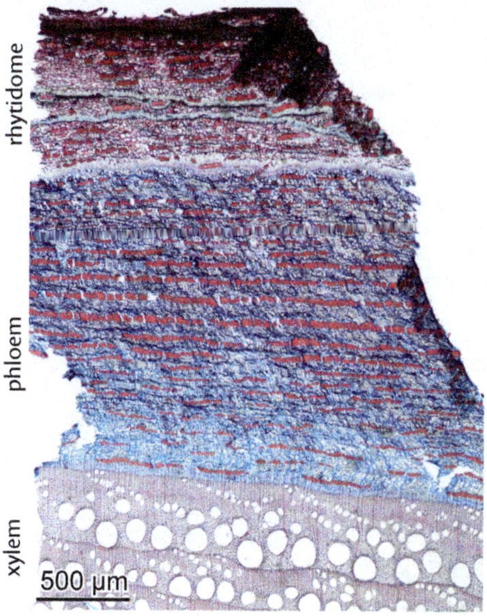

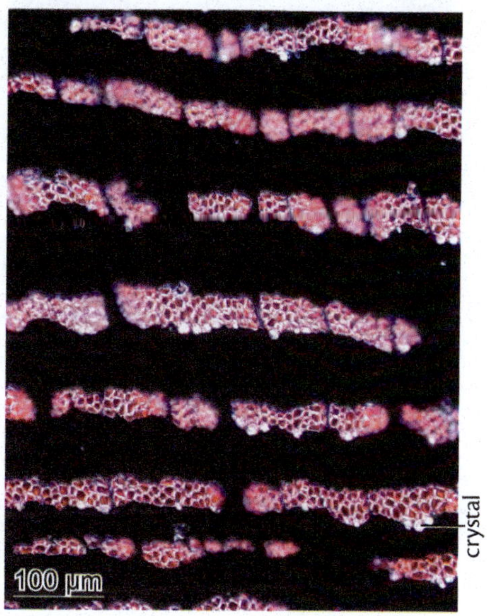

Celtis australis L. Cannabaceae

nettle tree, Zürgelbaum, micocoulier, bagolaro, almez

Form: Upright tree, crown regular shaped, broad, to 25 m tall.

Features: Leaves deciduous, alternate, long ovate and long-tipped, the leaf base always asymmetrical, margin regularly toothed, to 15 cm long. Flowers tiny, whitish green, to 3 mm long. Fruit a dull red brown, globular drupe, to 1 cm wide.

Habitat: Mediterranean, north to the Southern Alps and the southern Balkans, east to Syria and Israel.

On dry calcareous soils, mostly above limestone, to 1200 m.

Bark: Smooth, gray, occasionally with round, corky warts.

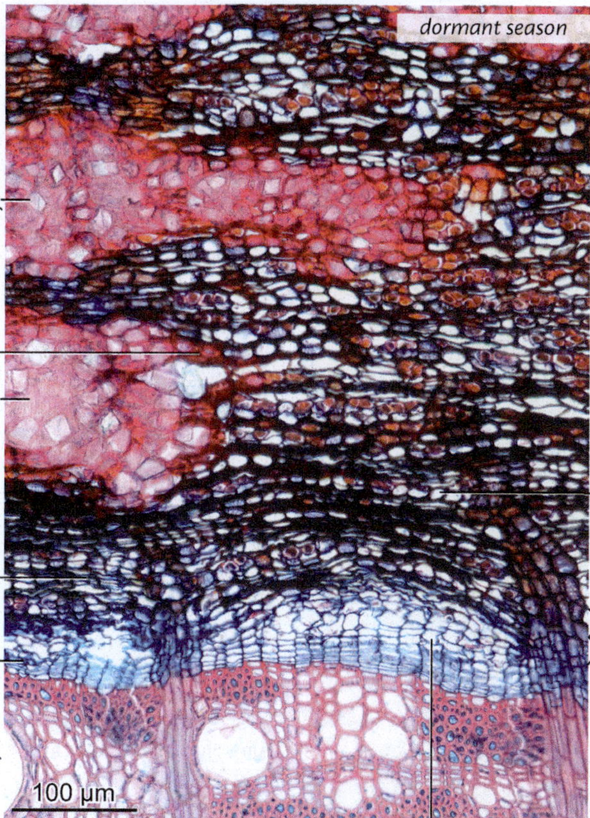

Bark anatomy: Dense structure with large zones of tangential sclerenchyma.

Sieve tubes irregular shaped, collapsing within a short time after formation. Parenchyma cells roundish. Sclereids dominating in the sclerenchymatic tissue. Fibers rare, around the sclereids. With indistinct dilatations. Prismatic crystals and crystal druses in sclerenchymatic zones.

Periderm consisting of many layers of large, rectangular, thin- to thick-walled cells.

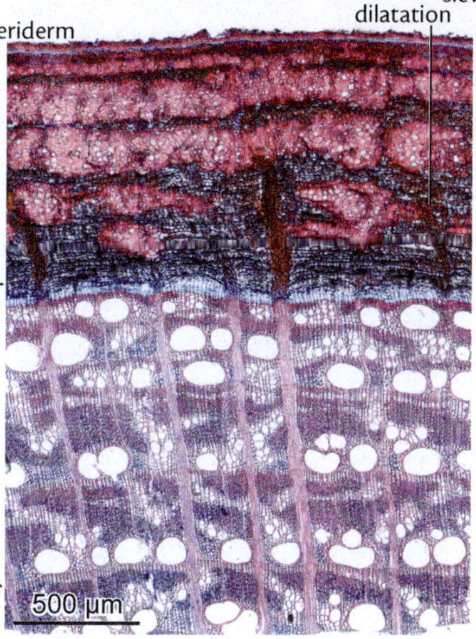

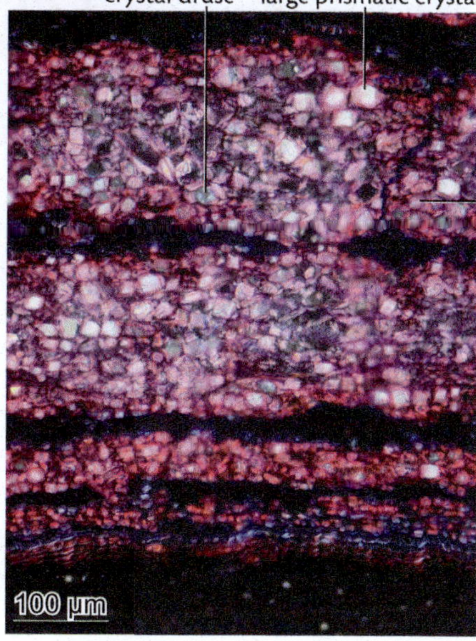

Cotinus coggygria Scop. Anacardiaceae

smoke tree, Perückenstrauch, arbre à perruque, sommacco selvatico, árbol de las pelucas

Form: Irregular shaped, densely branched, broad shrub, to 5 m tall.

Features: Leaves deciduous, alternate, round ovate, margin entire, to 8 cm long. Flower a tiny, yellowish, five-parted star, to 4 mm wide, in large inflorescences. Fruit a light brown, hard drupe, to 4 mm long, surrounded by wig-like, relaxed panicle of long-haired flower stalks, giving an unmistakable appearance.

Habitat: Mediterranean and Asia, north to the French Jura, the Valais, the Southern Alps, Hungary, Crimea and the Caucasus, isolated in the Western Himalayas and China.

On dry calcareous soils, on hot slopes with limestone, to 1000 m (in China to 2400 m).

Bark: Twigs red brown or brown, often dotted with white lenticels, stem bark light gray, fissured and dotted with lenticels.

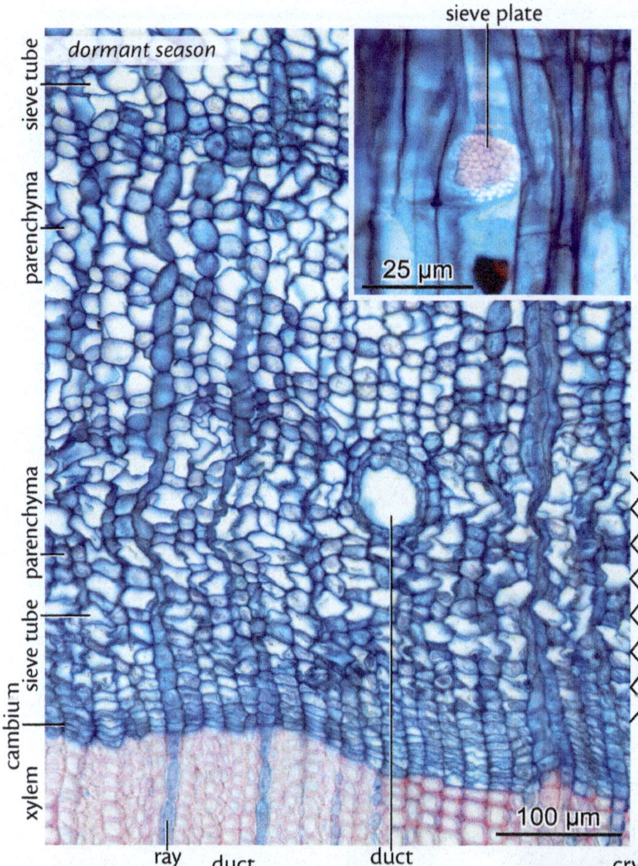

Bark anatomy: Slightly tangentially structured tissue of sieve tubes and parenchyma, with some gum ducts. Sieve tubes angular, accompanied by round to oval parenchyma cells. Sieve plates on radial walls. Sclereids isolated and in tangential bands. Fibers absent. Rays barely differentiated from axial tissue of sieve tubes and parenchyma. Many prismatic crystals and crystal druses of different sizes in axial cell chambers.

Rhytidome layered, consisting of irregular zones of large, thin-walled cells and small, thick-walled cells.

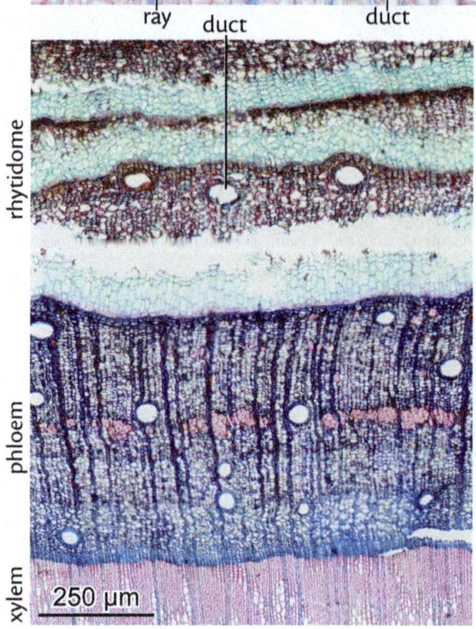

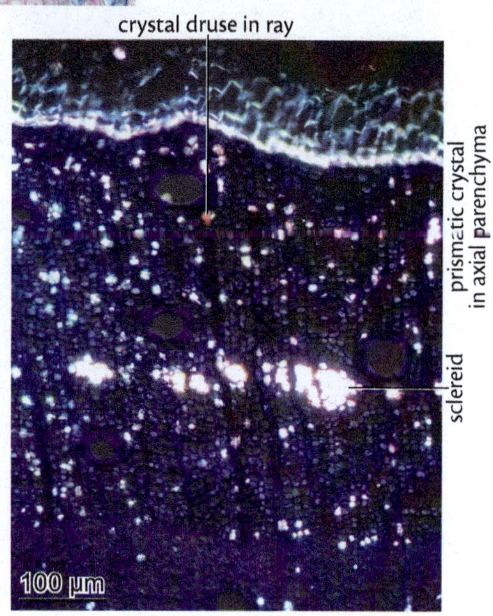

Cydonia oblonga Mill.

Rosaceae

quince, Quitte, cognassier, cotogno, membrillero

Form: Small tree or shrub, crown round, to 5 m tall.

Features: Leaves deciduous, alternate, ovate, margin entire, to 10 cm long. Flowers white pinkish, five-parted, to 5 cm across. Fruit a pear-shaped or globular, fleshy, yellow pome with fluffy hairs, to 12 cm long.

Habitat: Western and Central Asia, from Turkey, the Caucasus, Northern Iran to Turkmenistan and Kirghistan; widely planted.

On dry to moist, often calcareous soils, to 1700 m.

Bark: Young smooth, brown, with age gray to red brown, cracking into roundish plates, often developing parallel, fissured bulges.

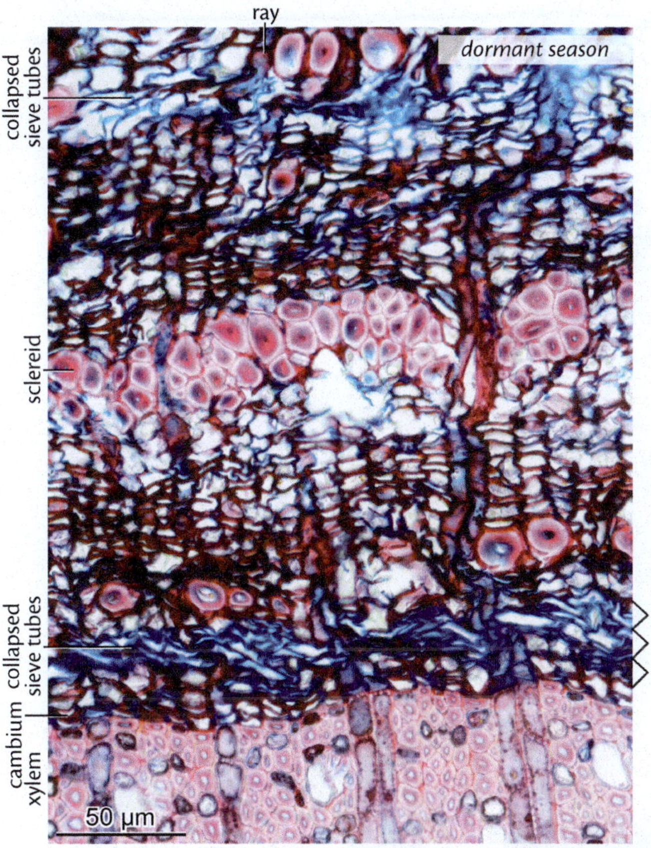

Bark anatomy: Tangentially layered structure of sieve tubes and parenchyma cells. Sieve tubes collapsing within a short time after formation. Fibers absent. Long sclereids in tangential groups, mainly in older tissues. Rays bent. Dilatations occurring after approximately four bands of sclereids. Many prismatic crystals in axial parenchyma cells and rays.

Phellem cells rectangular, thin- to thick-walled, filled with dark-staining substances.

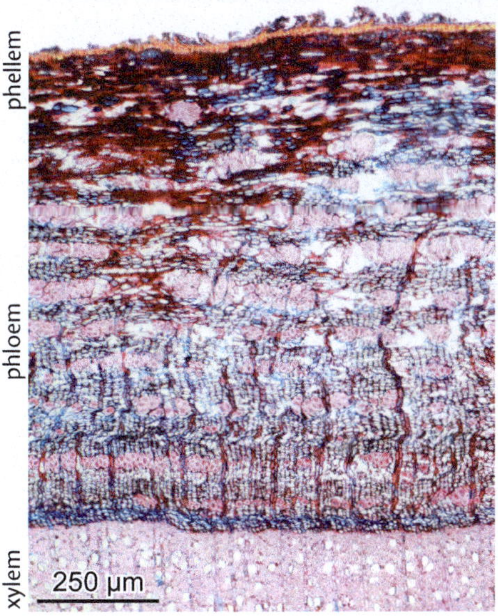

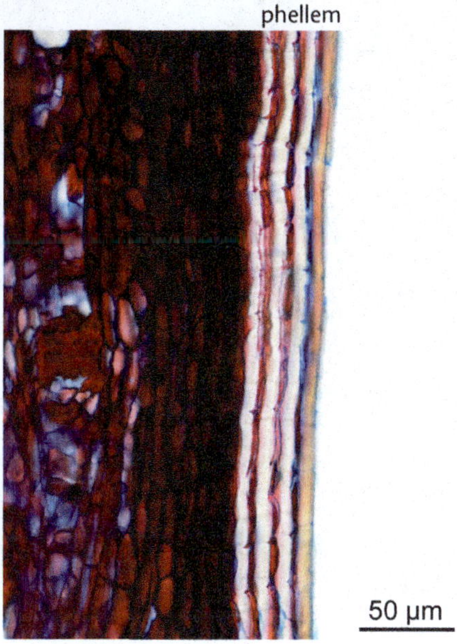

Daphne laureola L. Thymelaeaceae

spurge laurel, Lorbeer-Seidelbast, daphné lauréole, dafne laurella, adelfilla

Form: Small, evergreen, upright shrub, to 1 m tall.

Features: Leaves evergreen, leathery, alternate, long ovate, glossy green above, margin entire, to 14 cm long. Flowers yellow green, with four petals, to 1 cm long. Fruit an ovate, black, berry-like drupe, to 1 cm long.

Habitat: Southwestern Europe, from the Atlas Mountains northeast to Belgium, the Swiss Jura, the Southern Alps, isolated in the Vienna Basin and Hungary, south to Northern Greece, Sicily and Algeria.

On dry to moist, mostly calcareous soils, to 1500 m.

Bark: Dull light gray or slightly glossy reddish brown.

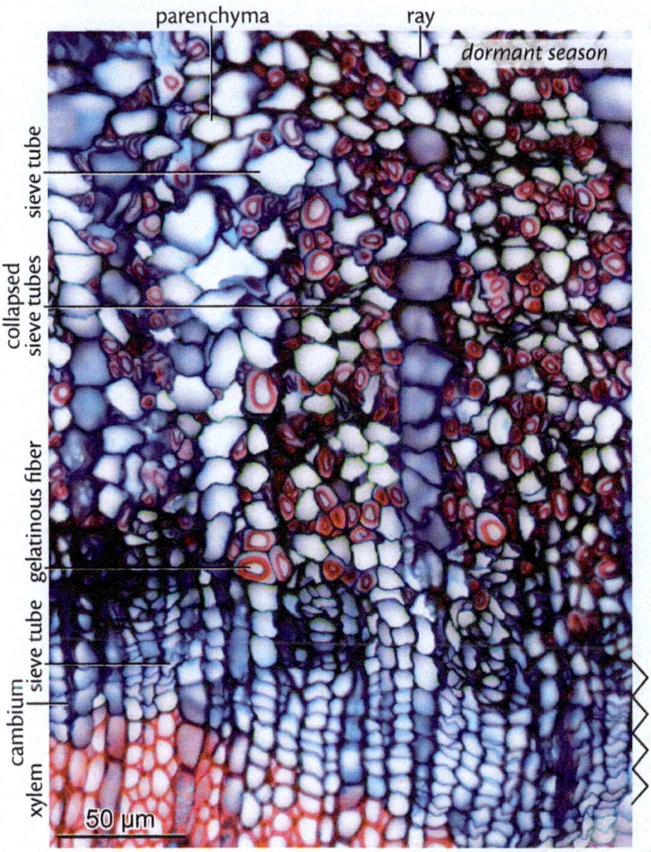

Bark anatomy: Slightly radially structured tissue of sieve tubes and parenchyma, without solitary fibers. Sieve tubes angular, accompanied by round to oval parenchyma cells. Fibers gelatinous.

Some rays and axial parenchyma cells dilating, separating the regular structure into fragments. Prismatic crystals rare.

Periderm cells rectangular, thin-walled.

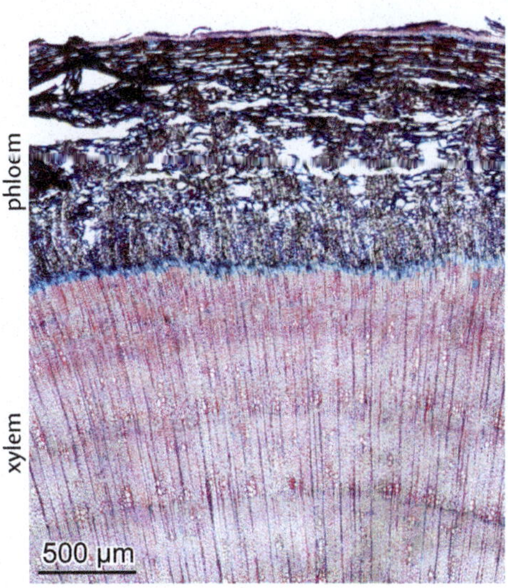

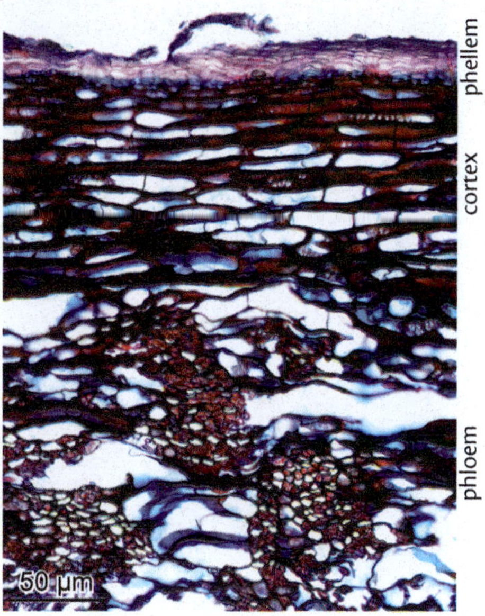

Fraxinus ornus L. Oleaceae

manna ash, Blumenesche, frêne à fleurs, orno, fresno de flor

Form: Upright tree, crown regular shaped, to 15 m tall.

Features: Leaves deciduous, opposite, pinnate with 5–11 leaflets, margin toothed, to 20 cm long. Flowers in white, scented, upright panicles, to 20 cm long. Fruit a flattened, winged nut, to 4 cm long, in dense, hanging tufts.

Habitat: Southeastern Europe, from the Southern Alps east to Hungary and Romania, south to the Balkans, southwestern Turkey and Greece, Sicily and Corsica, isolated near Valencia.

On dry, rocky limestone soil, to 1500 m.

Bark: Smooth, light or yellowish gray, remaining smooth with age.

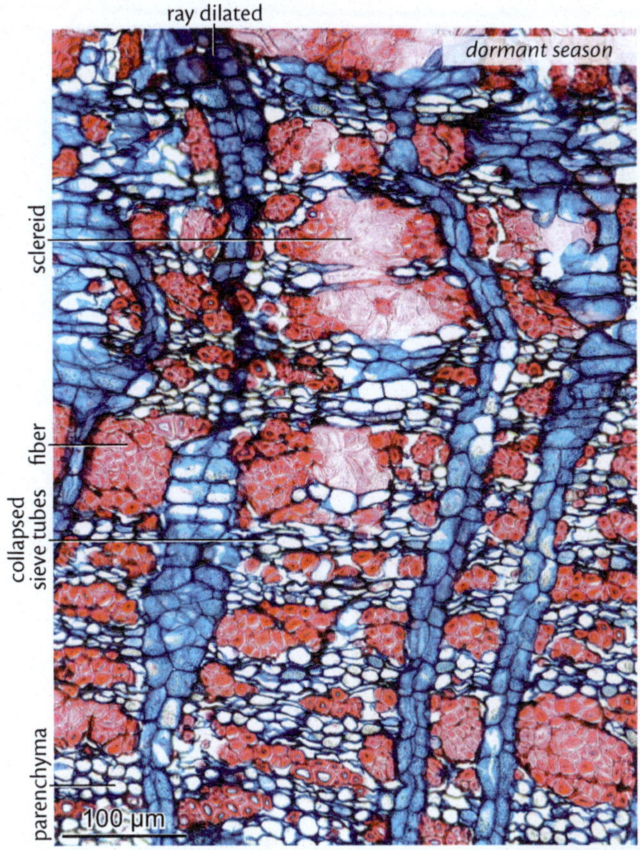

Bark anatomy: Regular tangential layers of fiber bands, sclereids and zones of sieve tubes and parenchyma.

Sieve tubes angular, parenchyma cells roundish, alternating in layers with 2–4-seriate bands of thick-walled sclerenchyma. Fibers associated with sclereids. Rays normally straight, persisting over many years. Few rays dilating. A lot of crystal sand in axial parenchyma cells and rays.

Annual ring boundaries recognizable only in most recently formed years.

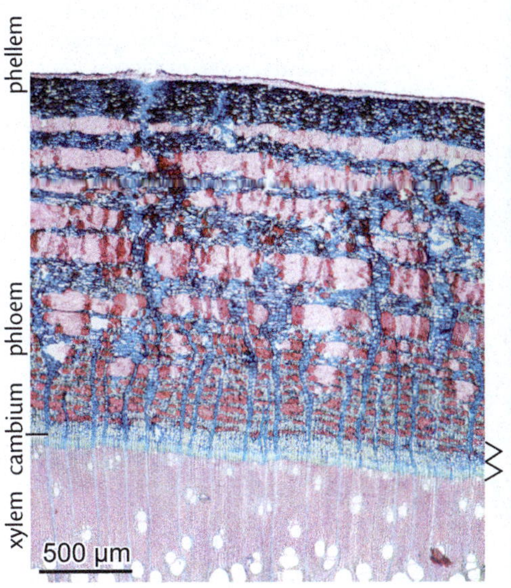

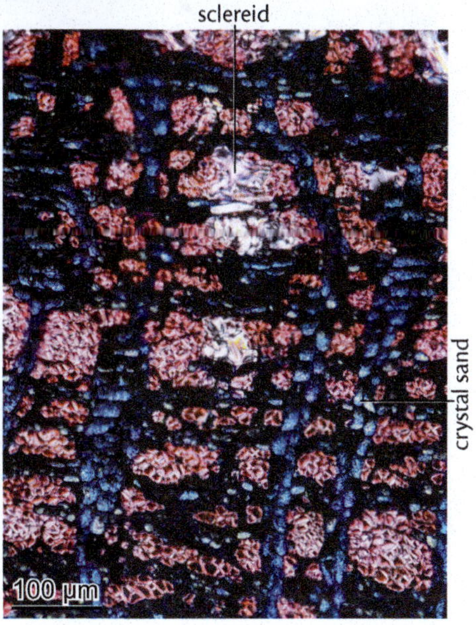

Juglans regia L.

Juglandaceae

walnut, Walnuss, noyer royal, noce comune, nogal común

Form: Upright tree, crown regular shaped, broad, to 25 m tall.

Features: Leaves deciduous, alternate, pinnate with four to ten leaflets, margin entire, leaflets to 15 cm long. Flowers monoecious, female flower very small, green, male flower a hanging, green catkin, to 6 cm long. Fruit an ovate, green drupe, to 5 cm across, containing the hard, beige brown nutshell with the edible seed, called walnut.

Habitat: Southeastern Europe and Asia, a few isolated localities in the Balkans, from Turkey east to Iran, Kirghistan and China; the natural distribution often hidden by intense planting.

Naturally on dry to moist, rocky soil, in moist gorges and ravines, to 1800 m.

Bark: Twigs olive green to gray brown, stem bark dull gray, young smooth, fissured with age.

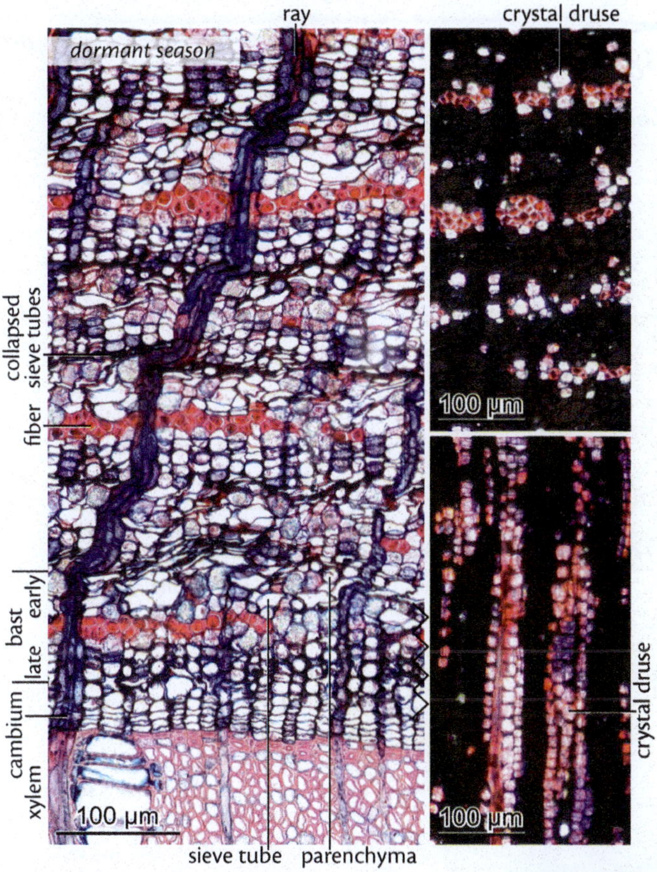

Bark anatomy: Regular tangential layers of fiber bands and zones of sieve tubes and parenchyma.

Sieve tubes angular, parenchyma cells roundish, alternating in layers with 2–4-seriate bands of thick-walled fibers. Sieve tubes collapsing within very short time after formation, collapsed cells indicated by a dark, dense band. Sclereids occasionally at the periphery of the phloem. Rays usually straight or slightly wavy. Few rays dilating. Many crystal druses in axial chambers.

Annual ring boundaries recognizable only in most recently formed years.

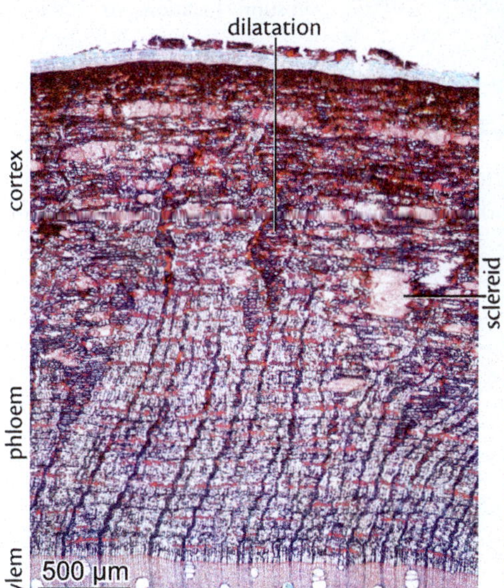

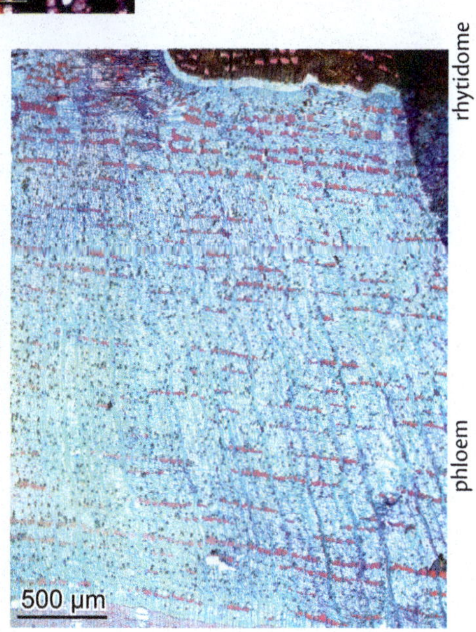

Laburnum anagyroides Medik. Fabaceae

golden chain, Gemeiner Goldregen, aubours faux anagyris, maggiacondolo comune, lluvia de oro

Form: Upright shrub or small tree, crown regular shaped, to 7 m tall.

Features: Leaves deciduous, alternate, three-parted with ovate leaflets, margin entire, to 8 cm long. Flowers lemon yellow, scented, pea-like asymmetrical, in many-flowered, hanging racemes, to 25 cm long. Fruit a dark gray brown, long ovate, hanging legume, to 5 cm long.

Habitat: Mountains of Southern Europe, the French Jura, the Southern Alps, the Apennines, the Balkans south to Bulgaria.

On dry, rocky limestone soil, in sunny locations, to 1200 m.

Bark: Smooth and glossy, olive green to brown, with many diamond-shaped lenticels.

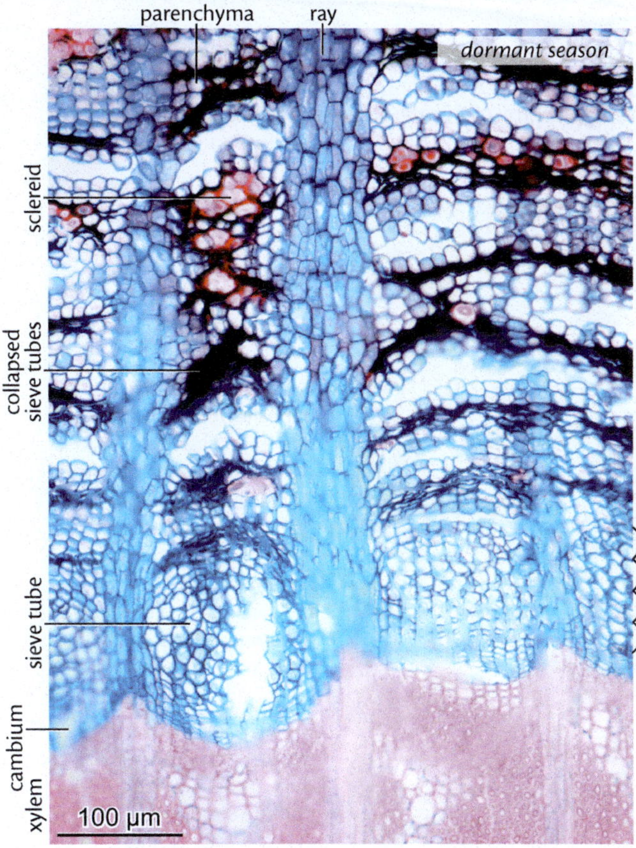

Bark anatomy: Regular tangential structure.

Sieve cells angular, companion cells small, in groups. Sieve tubes collapsing within a very short time after formation, collapsed cells indicated by dense, dark, curved bands. Parenchyma cells roundish. Fibers absent. Sclereids in young phloem in parenchymatic bands, as solitary cells or in groups. Sclereids in older phloem in irregular groups. Dilatations rare, in parenchyma between large rays. Crystals not observed.

Periderm large, tangentially layered, consisting of large, rectangular, thick-walled cells.

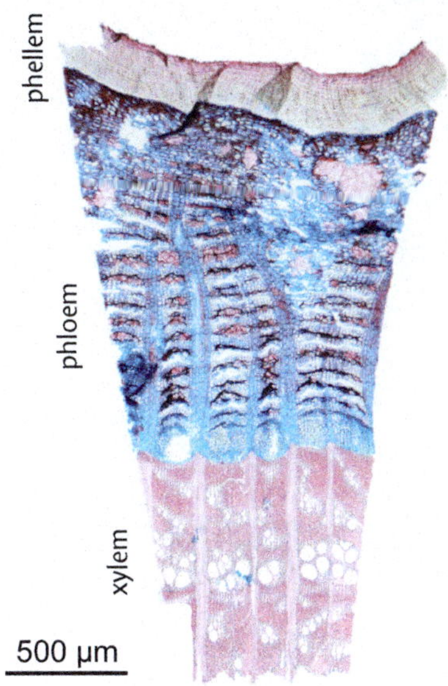

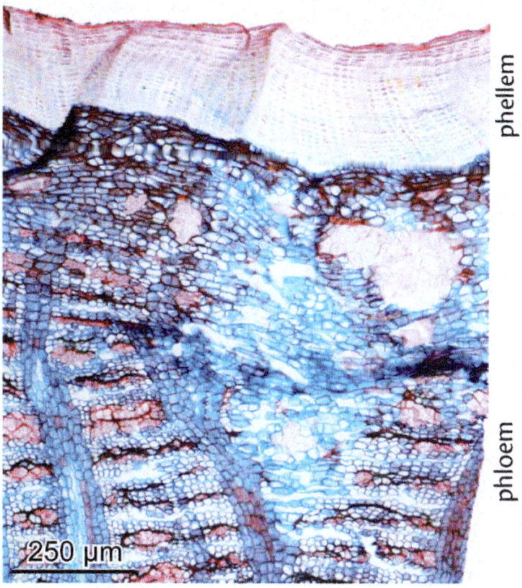

Lonicera caprifolium L. Caprifoliaceae

perfoliate honeysuckle, Gartengeißblatt, chèvrefeuille des jardins, caprifoglio comune, madreselva

Form: Winding vine, to 6 m tall.

Features: Leaves deciduous, opposite, round ovate, the uppermost leaves fused in pairs, margin entire, to 6 cm long. Flowers creamy white, night-scented, sitting at the end of twigs in groups of six, asymmetrically two-lipped, to 5 cm long. Fruit a globular, dull orange berry, to 1 cm wide, in groups at the end of twigs.

Habitat: Southern Europe, the Mediterranean and the Caucasus, north to the Danube Basin; often planted.

On dry to moist, often calcareous soils, to 1000 m.

Bark: Brown, fissured with age.

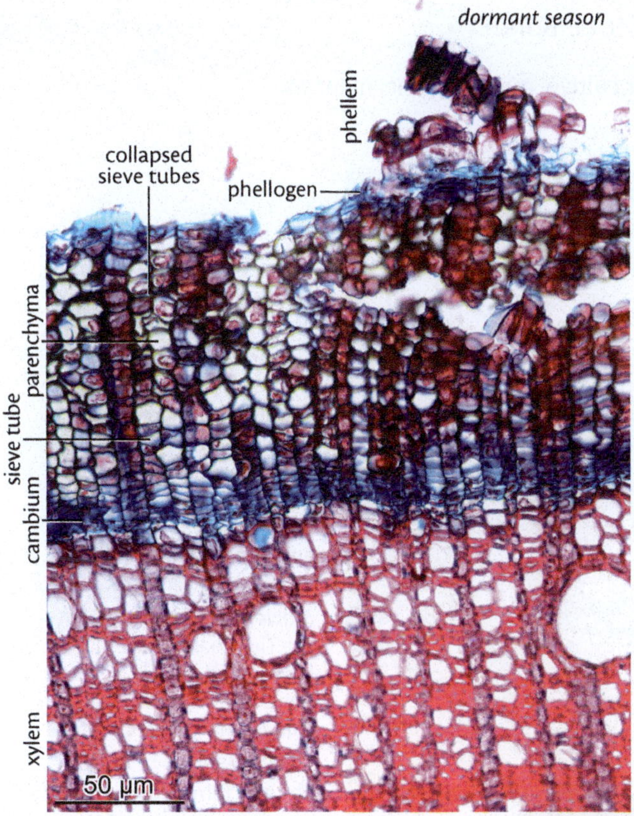

Bark anatomy: Tangentially layered structure, layers scaling off easily. Sieve tubes angular, parenchyma cells round, irregularly mixed. Sieve tubes collapsing within a short time after formation. Fibers absent. Rays hardly recognizable in transmitting light, but in polarized light clearly indicated by many crystal druses.

Shed rhytidome consisting of a layer of living, unlignified, square cells.

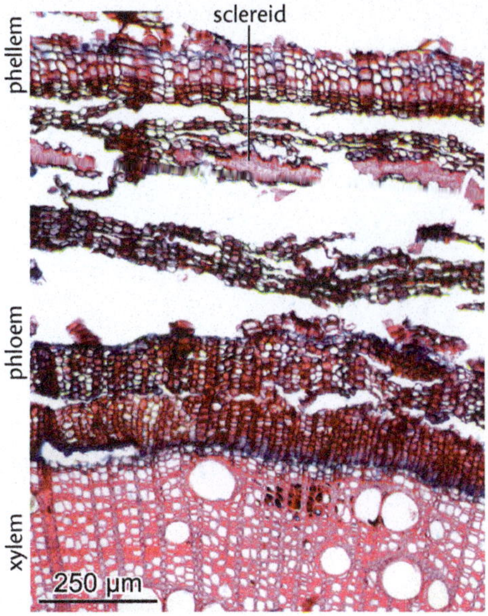

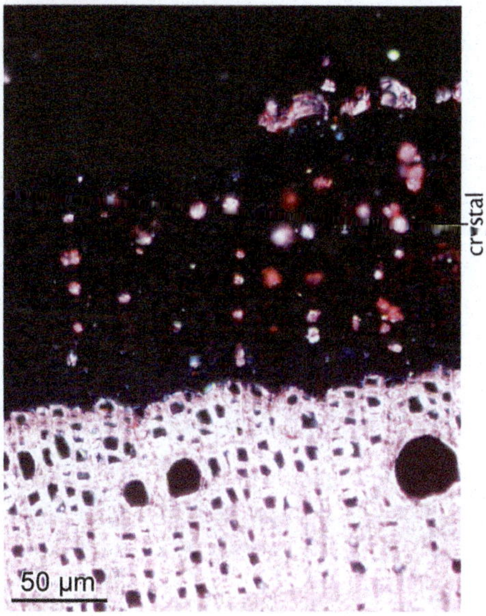

Malus x domestica Borkh. Rosaceae

apple, Kulturapfel, pommier cultivé, melo comune, manzano

Form: Upright tree or shrub, cultivars of the Central Asian *Malus sieversii*, to 10 m tall.

Features: Leaves deciduous, alternate, ovate, underside usually hairy, margin regularly toothed or entire, to 6 cm long. Flowers pinkish white, five-parted, to 5 cm across. Fruit a globular, yellow or red apple, to 15 cm wide.

Habitat: The wild form *Malus sieversii* exclusively in the Tien Shan Mountains of Kirghistan, Kazakhstan and Uzbekistan; hundreds of cultivars have been selected and planted.

The wild form on dry, stony, mostly calcareous soils, in warm, sunny positions, to 2200 m.

Bark: Young glossy and smooth, gray brown, with age fissured and cracking into small, square plates.

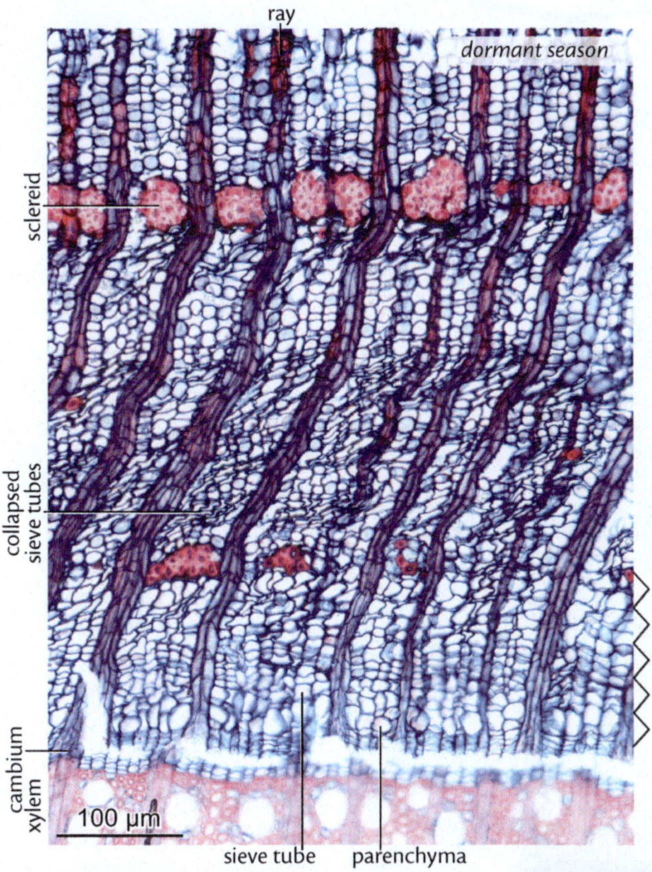

Bark anatomy: Tangentially layered structure, with sieve tubes and parenchyma cells indistinctly tangentially arranged. Sieve tubes collapsing within a short time after formation. Fibers absent. Sclereids in roundish groups in tangential layers, some solitary. Rays slightly wavy. Dilatations rare. Many prismatic crystals, mainly in axial parenchyma cells.

Rectangular phellem cells alternating with small, thick-walled cells and larger, thin-walled cells.

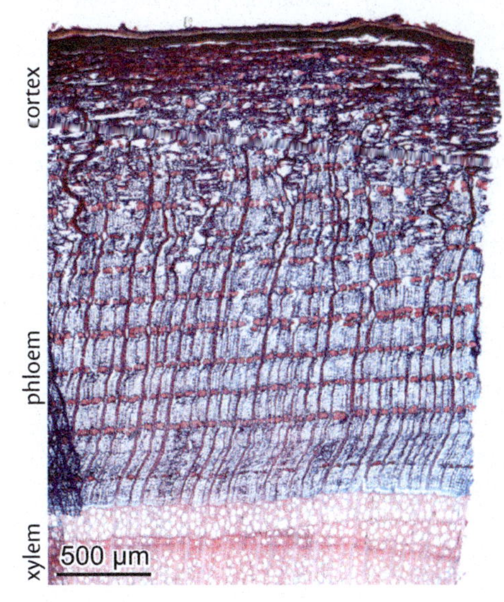

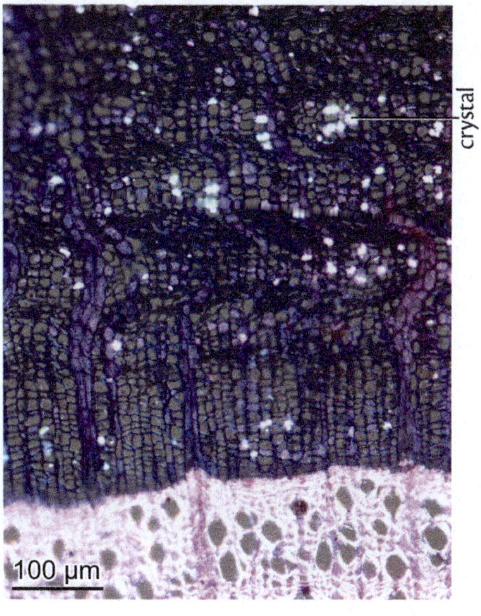

Ostrya carpinifolia Scop. Betulaceae

hop hornbeam, Hopfenbuche, charme houblon, carpino nero, carpe negro

Form: Upright tree, crown irregular shaped, to 25 m tall.

Features: Leaves deciduous, alternate, long ovate, margin double-toothed, tip pointed, to 10 cm long. Flowers monoecious, female catkins hanging, small, yellow green, male catkins hanging, yellow, to 6 cm long. Fruit an ovate nut, embedded in a papery, light brown bract, grouped in dense, hanging racemes, resembling hop fruits.

Habitat: Southern Europe, from Southern France northeast to the Southern Alps, the Balkans and the Caucasus, southeast to Lebanon.

On dry to moist soils, often calcareous, preferring warm locations, to 1300 m.

Bark: Young smooth, gray, with age gray to black brown, fissured and cracking into small, square plates.

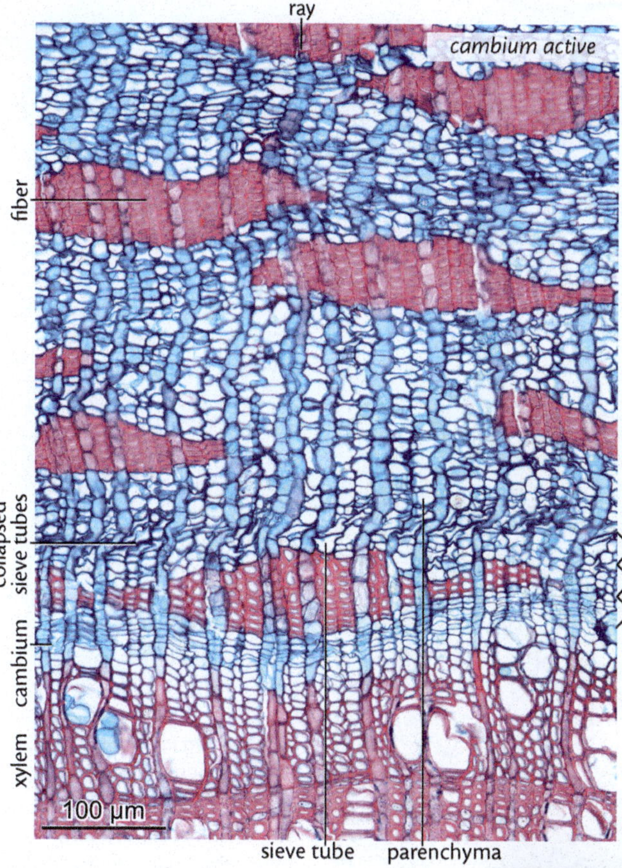

Bark anatomy: Tangentially structured tissue with many large, oval groups of fibers.

Sieve tubes angular, parenchyma cells roundish. Sieve tubes collapsing within a short time after formation, indicated by irregular, dark lines. Fibers in radial lines in tangentially oval groups. Sclereids absent. Rays slightly wavy, dilatations rare. Prismatic crystals large, in rays, especially within fiber groups. Crystal druses mostly in axial parenchyma cells, rarely in association with fibers.

Phellem not observed.

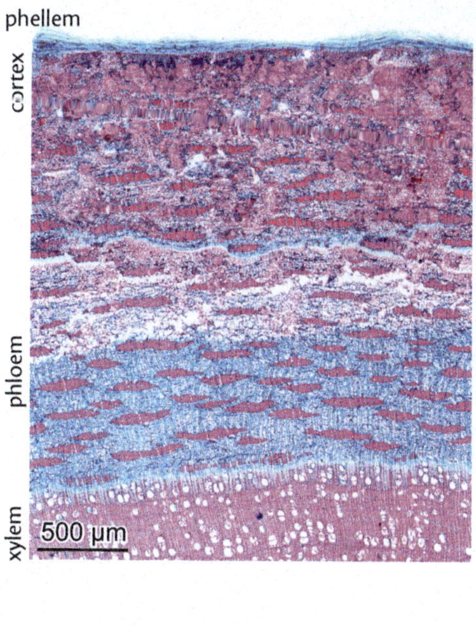

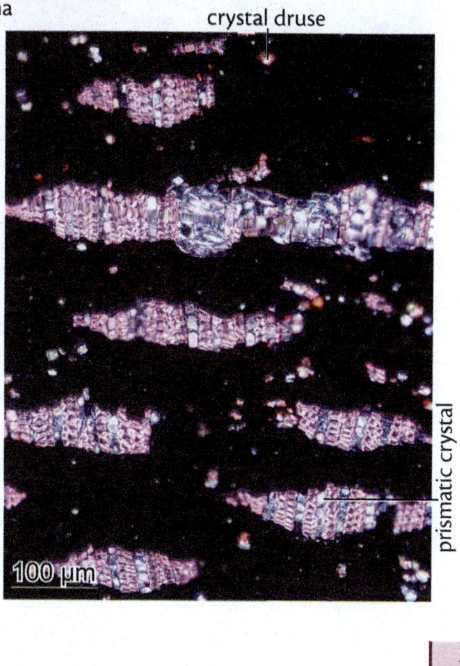

Prunus laurocerasus L.

Rosaceae

cherry laurel, Kirschlorbeer, laurier-cerise sauvage, lauroceraso, laurel cerezo

Form: Shrub or small tree, crown large, spreading, to 6 m tall.

Features: Leaves evergreen, leathery, alternate, glossy green above, long ovate, margin entire, to 15 cm long. Flowers white, five-parted, to 6 mm long, in dense, upright, scented racemes. Fruit a glossy black, berry-like drupe, to 1 cm wide.

Habitat: Western Asia, from northern Turkey to the Caucasus and Northern Iran, isolated in the Bulgarian Strandja Mountains; widely planted and often escaped.

On moist, deep and loamy, often calcareous soils, to 1500 m.

Bark: Dull and smooth, light gray.

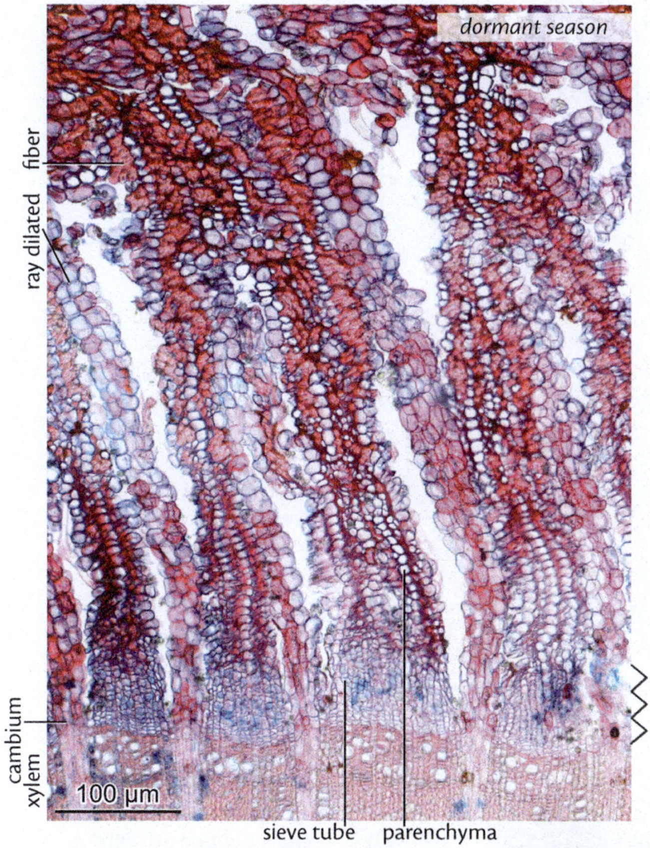

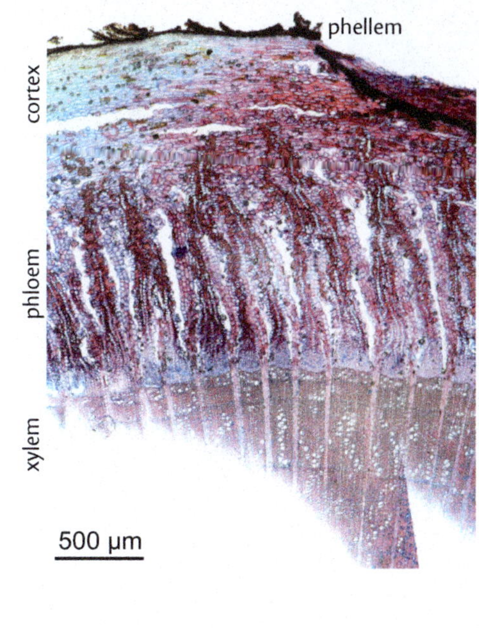

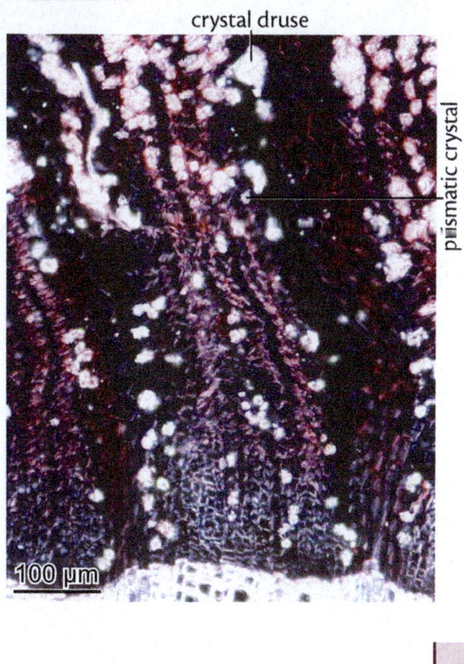

Bark anatomy: Radial structure with large intercellulars.

Sieve tubes, parenchyma cells and sclereids in intensively wavy radial strips between large, bent rays. Rays dilated. Fibers in radial rows. Structure chaotic in external parts. Many crystal druses and prismatic crystals, mainly in axial parenchyma cells.

Phellem consisting of a large band of unlignified, thin-walled, rectangular cells.

Prunus mahaleb L. Rosaceae

St. Lucie cherry, Felsenkirsche, faux merisier, ciliegio di Santa Lucia, cerezo de Santa Lucía

Form: Dense, broad-crowned shrub, rarely a small tree, spineless, to 6 m tall.

Features: Leaves deciduous, alternate, round ovate, margin regularly toothed, to 4 cm long. Flowers white, five-parted, to 1 cm across. Fruit a globular, glossy black drupe, to 1 cm long.

Habitat: Southern Europe and Southwest Asia, from Southern France north to Belgium, Austria, the Balkans and the Caucasus, south to Syria.

On dry, rocky, mostly calcareous soils, in warm, sunny positions, to 2000 m.

Bark: Glossy and smooth, reddish gray brown, with short, horizontal strips.

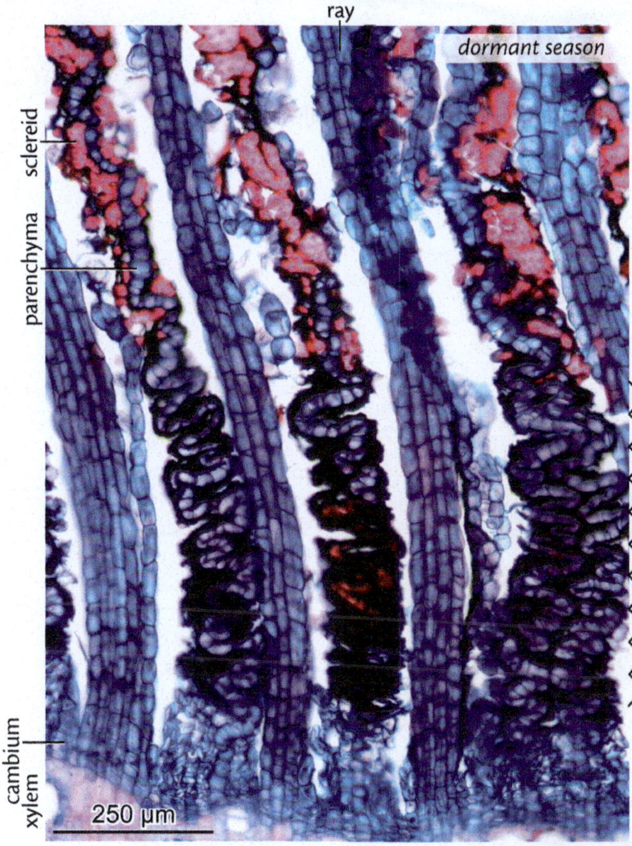

Bark anatomy: Radial structure with large intercellulars.

Sieve tubes, parenchyma cells and sclereids in intensively wavy radial strips between large, bent rays. Fibers absent. Dilatations not observed. Structure chaotic in external parts. Many crystal druses and prismatic crystals, mainly in axial parenchyma cells.

Phellem consisting of a large band of unlignified, thin-walled, rectangular cells.

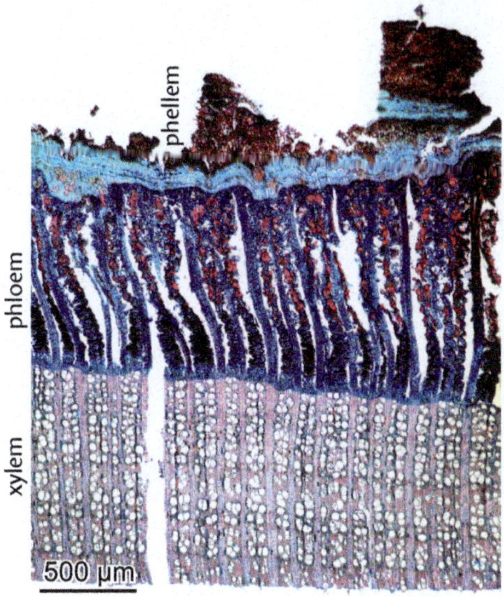

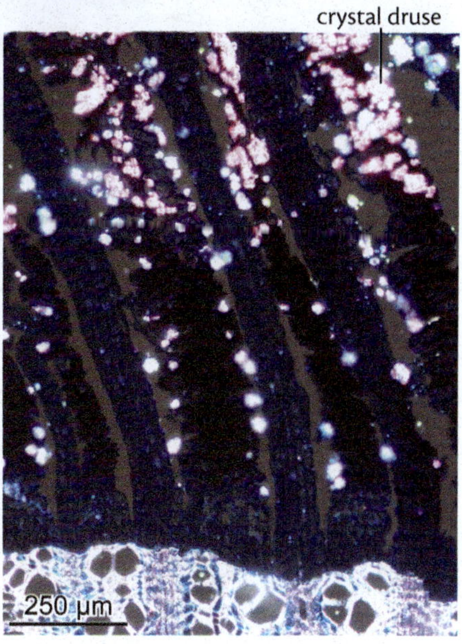

Quercus cerris L.

Fagaceae

Turkey oak, Zerreiche, chêne chevelu, quercia cerro, marojo

Form: Upright tree, crown large, irregular shaped, to 35 m tall.

Features: Leaves deciduous, alternate, long ovate, margin regularly sharply lobed, to 15 cm long. Flowers monoecious, female flower small, short, reddish, male flowers in long, thin, yellow green catkins, to 12 cm long. Fruit an ovate, red brown nut, sitting in a cup with spreading scales.

Habitat: Southeastern Europe, from the Western Alps and southern Switzerland to the Vienna Basin and Slovakia, southeast to Turkey and Crete.

On dry, stony, often acidic soils, to 1000 m.

Bark: Young smooth, gray brown, with age deeply fissured.

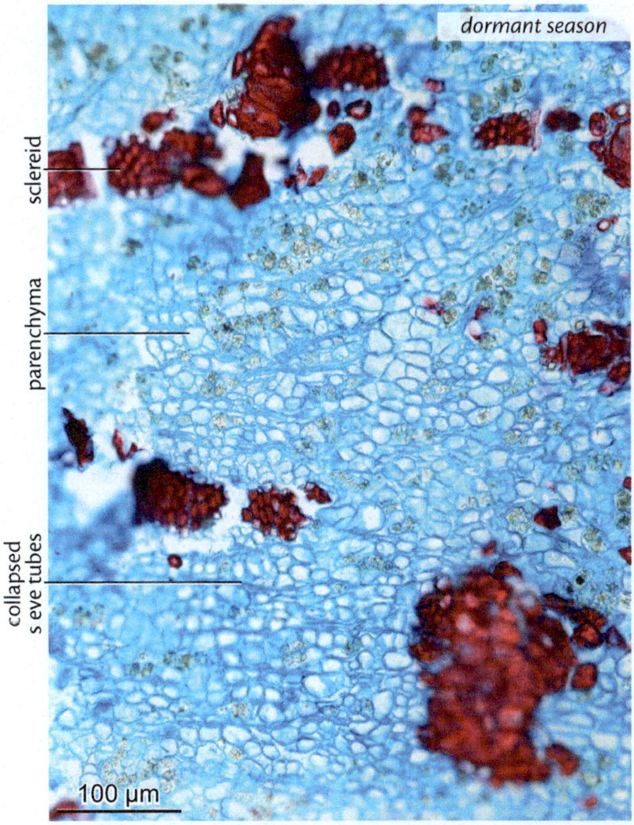

Bark anatomy: Structure indistinctly tangentially layered.

Sieve tubes angular, parenchyma cells round, radially arranged. Sieve tubes collapsing within a short time after formation. Groups of fibers alternating with dominant zones of sieve tubes and parenchyma. Sclereids in irregular groups. Rays hardly recognizable. Crystal druses in axial parenchyma cells, surrounding fiber bands and within sclereids.

Phellem layered, cells rectangular, thin- to thick-walled.

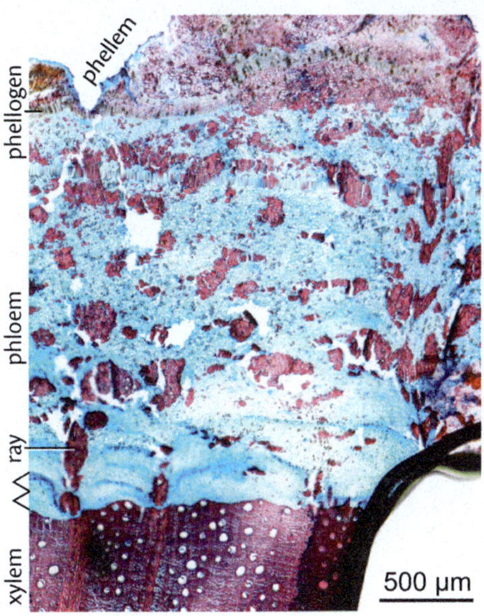

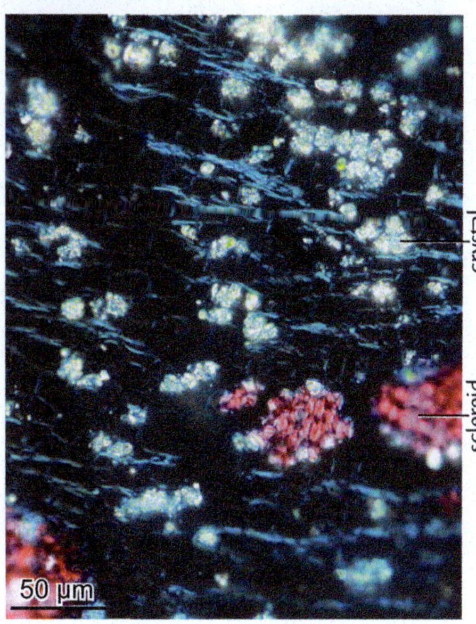

Quercus pubescens Willd. Fagaceae

downy oak, Flaumeiche, chêne pubescent, roverella, roble pubescente

Form: Upright tree, crown irregular shaped, to 20 m tall.

Features: Leaves deciduous, alternate, long ovate, often densely hairy below, margin regularly lobed, to 15 cm long. Flowers monoecious, female flowers small, short, reddish, male flowers in long, thin, yellow green catkins, to 8 cm long. Fruit an ovate, dull brown nut, sitting in a cup.

Habitat: Southern Europe, from the Ebro Valley northeast to Belgium, Slovakia, Crimea and the Caucasus, south to Greece and Sicily.

On dry, rocky, calcareous or acidic soils, to 1500 m.

Bark: Young smooth, gray brown, with age dark brown gray and deeply fissured.

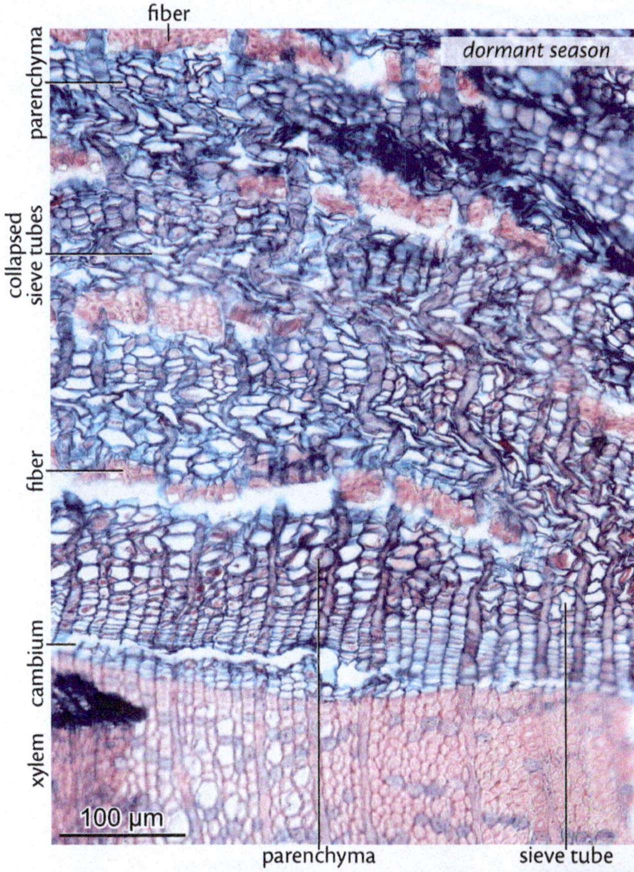

Bark anatomy: Tangentially layered structure.

Sieve tubes angular, parenchyma cells round, radially arranged. Sieve tubes collapsing within a short time after formation. Fiber bands alternating with zones of sieve tubes and parenchyma. Sclereids in round groups, dominating large rays. Rays large and straight or uniseriate and wavy. Some large rays dilating. Prismatic crystals surrounding fiber bands, crystal druses in parenchymatic zones. Many small crystal druses in axial parenchyma cells.

Phellem formed by layers of rectangular, thin-walled cork cells.

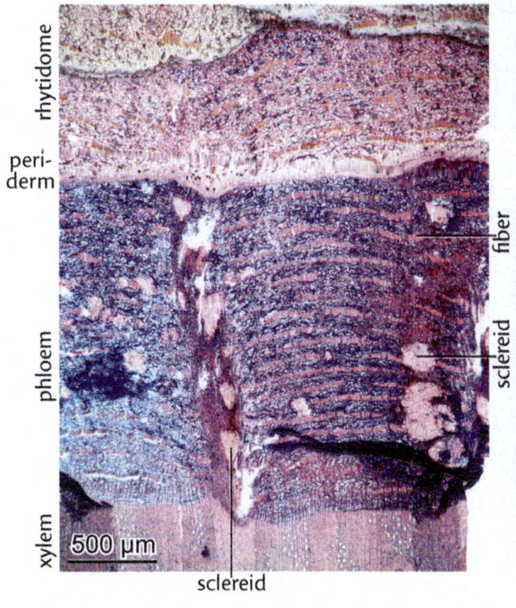

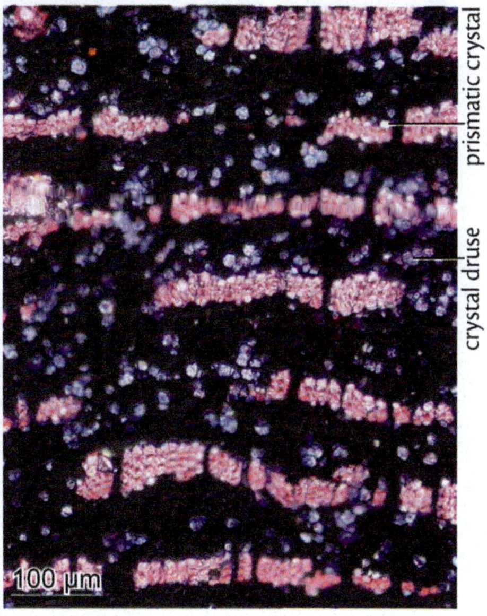

Sorbus domestica L. Rosaceae

service tree, Speierling, sorbier domestique, sorbo domestico, serbal común

Form: Upright tree, crown regular shaped, to 20 m tall.

Features: Leaves deciduous, alternate, pinnate with 9–19 leaflets, hairy below, margin regularly toothed, to 25 cm long. Flowers white, five-parted, to 5 mm long, in pseudo-umbels. Fruit a yellow to red, ovate or globular pome, to 3 cm wide.

Habitat: Mediterranean, northeast to Central Germany and the Vienna Basin, east to the Caucasus south to Algeria, Sicily and the Sierra Nevada.

On dry, stony, calcareous soils, to 1700 m.

Bark: Young smooth, reddish gray, with conspicuous horizontal diamonds, with age dull gray and fissured.

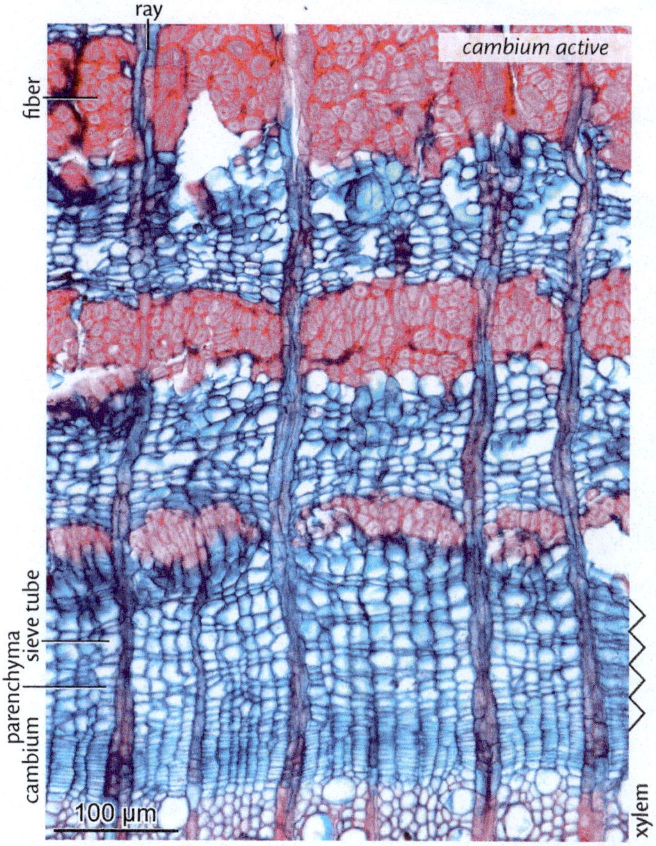

Bark anatomy: Regular, tangentially layered structure.

Sieve tubes collapsing before beginning of sclereid formation. Fibers and sclereids in tangential bands. Rays straight in the phloem. Dilatations not observed. Many small and large prismatic crystals in axial parenchyma cells.

Phellem layered, cells rectangular, thick-walled, separating living from dead bark (rhytidome).

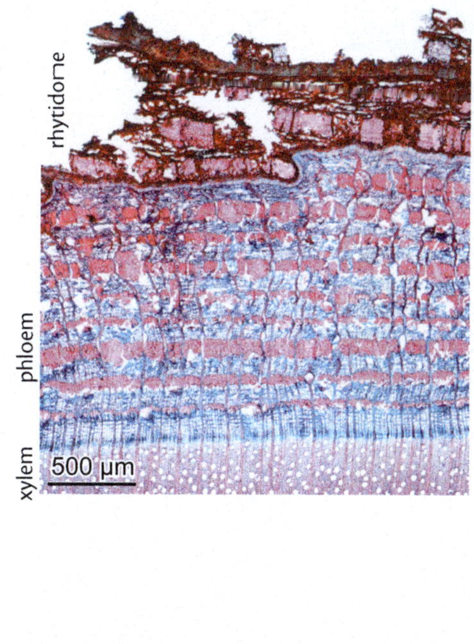

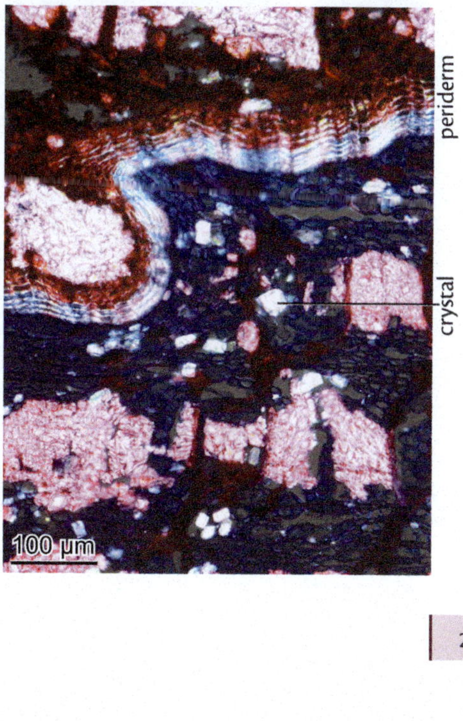

Staphylea pinnata L. Staphyleaceae

bladdernut, Pimpernuss, staphylier penné, bossolo, estafilario

Form: Dense, upright shrub, to 5 m tall.

Features: Leaves deciduous, opposite, pinnate with three to seven leaflets, margin finely toothed, to 10 cm long. Flowers white, five-parted, to 1 cm long, in hanging panicles. Fruit an ovate, inflated, light brown capsule, two- or three-parted, with glossy brown nuts, to 4 cm long.

Habitat: Southern Europe and Southwest Asia, from the French Jura and the Piedmont northeast to the Danube River, southern Poland and Ukraine, the Caucasus and Armenia, south to Syria, Macedonia and Calabria.

On moist, calcareous soils, in warm positions, to 1800 m.

Bark: Twigs dark gray, green or red brown, the white lenticels soon combining into vertical white lines, stem bark light greenish gray, mottled with white lenticels, with age slightly fissured.

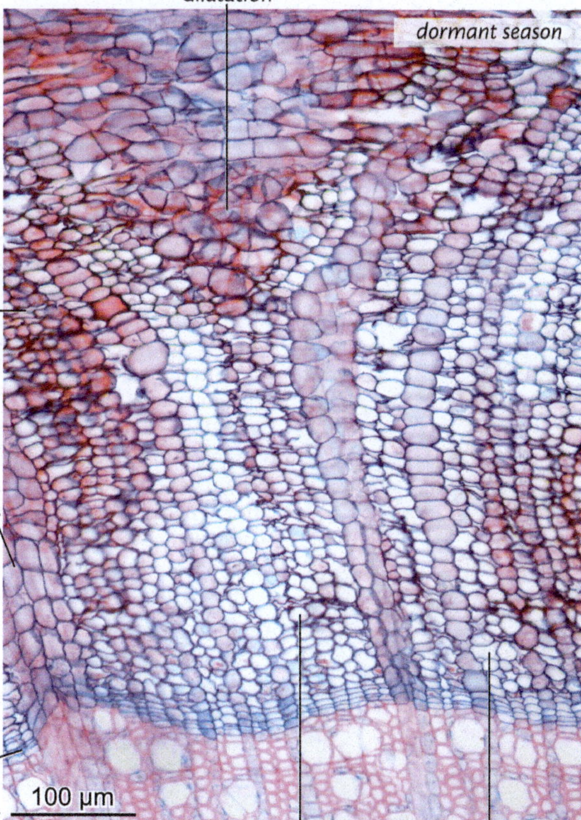

Bark anatomy: Soft tissue with few groups of sclereids. Sieve tubes angular, parenchyma cells round. Sieve tubes collapsing within a short time after formation, collapsed cells indicated by dark tangential and radial lines. Few fibers in a line between phloem and cortex. Sclereids often absent or in small tangential groups. Rays straight or bent. Axial parenchyma cells forming large dilatations. Crystal druses mainly in dilatations.

Phellem layered, cells rectangular or square, thin-walled.

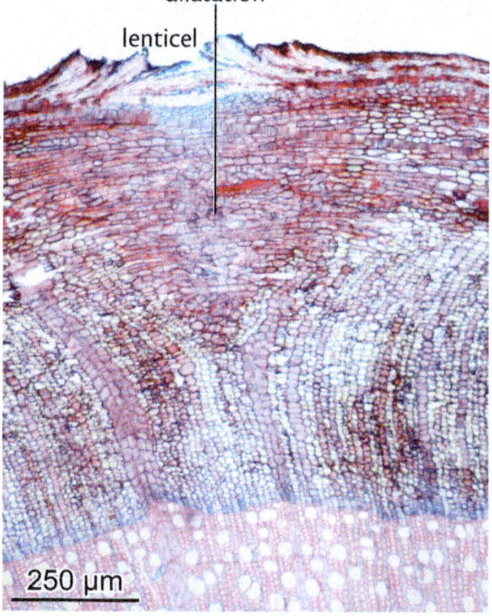

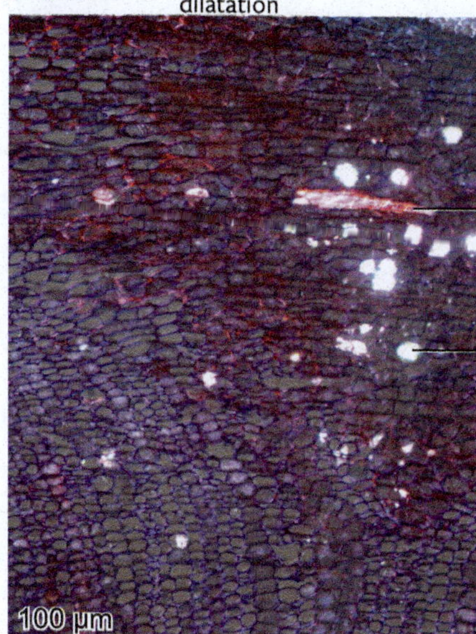

Syringa vulgaris L.

Oleaceae

lilac, Flieder, lilas, serenella, lilo

Form: Upright, densely branched shrub, to 6 m tall.

Features: Leaves deciduous, opposite, long ovate to cordate, margin entire, to 12 cm long. Flowers lilac, scented, four-parted, to 12 mm wide, in dense, upright panicles. Fruit a woody, long ovate capsule, splitting into two parts, to 3 cm long.

Habitat: Southeastern Europe, the Balkans, north to the Banat and Serbia, Central Germany, southwestern Ukraine and the Caucasus, south to Bulgaria and Northern Greece; widely planted and often escaped.

On dry, rocky, calcareous soils, to 1800 m.

Bark: Twigs olive green to yellow brown, stem bark light gray, soon fissured and shedding in papery flakes.

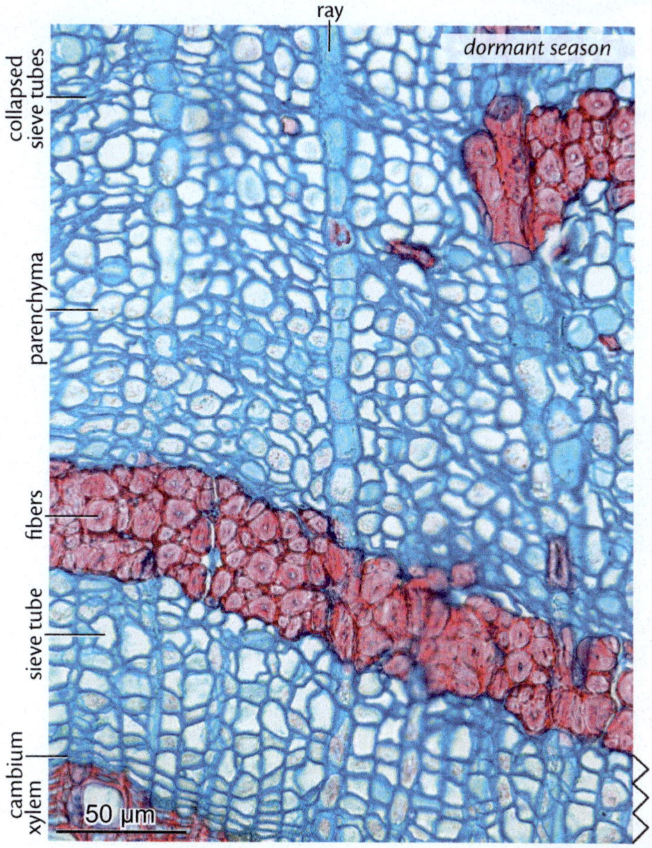

Bark anatomy: Regular, tangentially layered structure. Sieve tubes angular, parenchyma cells round. Fibers in tangential layers. Sclereids absent. Rays straight in the phloem. Dilatations absent. Many small acicular crystals, mainly in rays.

Phellem layered, cells square or irregular shaped.

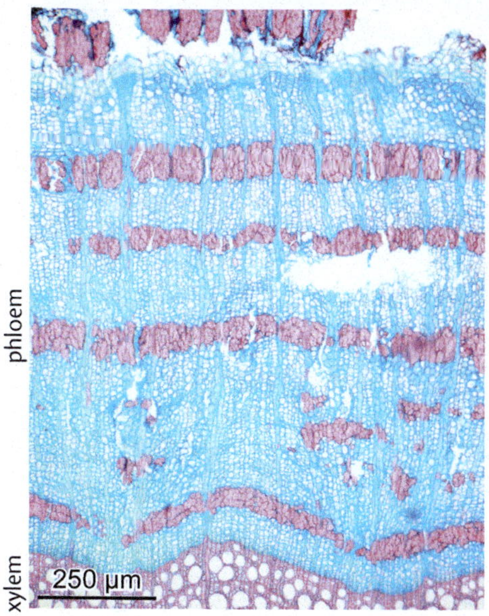

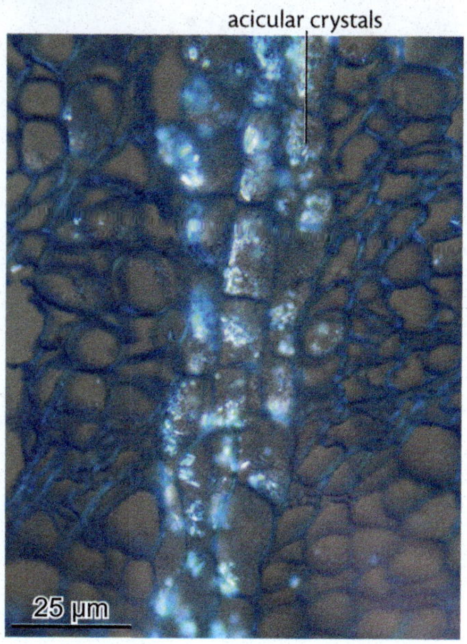

Vitis vinifera L.

Vitaceae

grape vine, Weinstock, vigne, vite, vid

Form: Vine with winding tendrils, to 30 m tall.

Features: Leaves deciduous, alternate, three- to five-lobed, margin irregularly toothed, to 15 cm long. Flowers greenish, inconspicuous, to 3 mm long, in long panicles. Fruit a globular, blue black, edible berry, to 2 cm long, in hanging panicles.

Habitat: The wild form *Vitis sylvestris* in Southern Europe and Western Asia, northeast to the Rhine and Danube rivers, Central Russia and Central Asia; widely planted as various cultivars.

The wild form strictly on moist to wet, riverine soils, to 1200 m.

Bark: Yellow to red brown, with age scaly fissured.

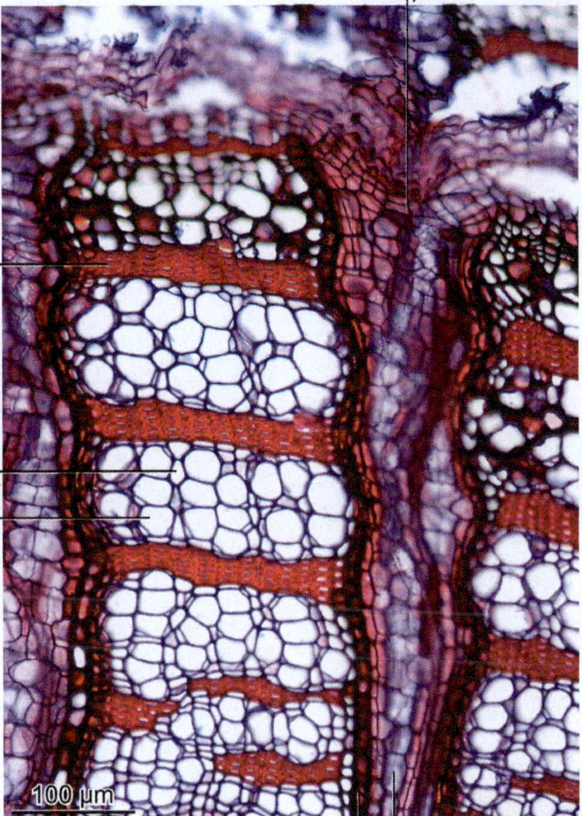

Bark anatomy: Regular, tangentially layered structure. Sieve tubes large, angular, companion cells small, parenchyma cells small, all arranged in groups between fiber bands. Fiber bands of different width in tangential layers. Rays straight, most distinctly dilating, composed of thin-walled central and thin- to thick-walled lateral cells. Prismatic crystals in lateral ray cells. Very few very large crystals in large parenchyma cells and cortex. Few cells with raphids.

Phellem layered, cells rectangular, thick-walled.

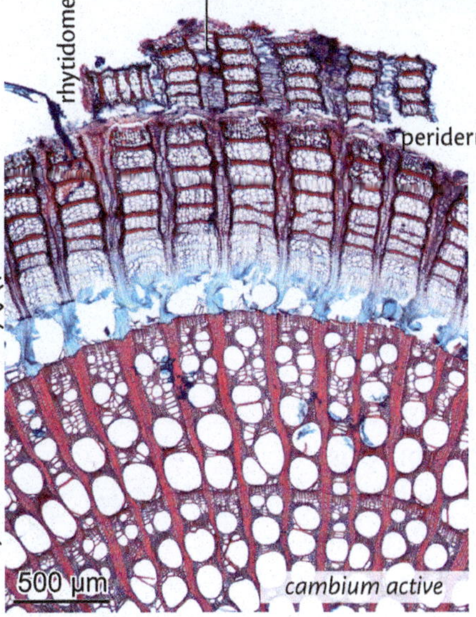

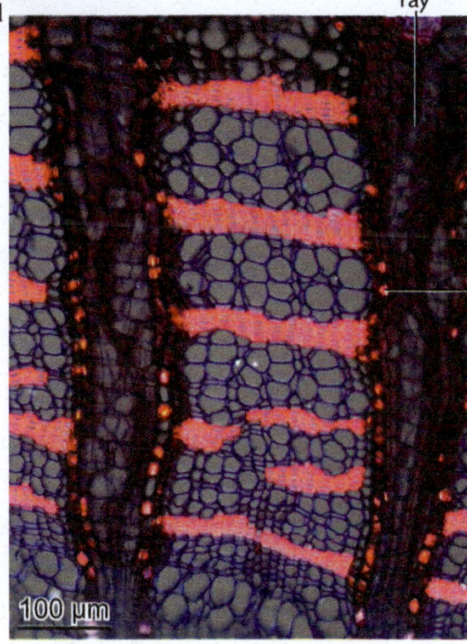

The **deciduous forest of Eastern Asia** with its enormous biodiversity and subtropical influences reaching far north, is represented in this book only by a small selection of typical species, with the purpose to include some of the plant families centered in Eastern Asia, and also some species with a worldwide distribution and familiarity as planted trees and shrubs.

F. H. Schweingruber et al., *Bark Anatomy of Trees and Shrubs in the Temperate Northern Hemisphere*, https://doi.org/10.1007/978-3-030-14056-4_7

7 Deciduous forest of Eastern Asia

Cryptomeria japonica (Thunb. ex L.f.) D.Don Cupressaceae

Japanese sugi pine, Sicheltanne, cèdre du Japon, cedro rosso del Giappone, cedro japonés

Form: Upright conifer, crown straight, conical, to 60 m tall.

Features: Leaves evergreen, needle-like, spirally arranged and crescent, to 2 cm long. Flowers monoecious, yellowish, spherical, in both sexes to 5 mm across. Fruit a globular cone with many spiny scales, to 2.5 cm long.

Habitat: Native only in scattered localities on Japans main island of Honshu; widely planted also in China.

On moist, deep soils, to 1000 m.

Bark: Red brown, peeling in thin, vertical strips.

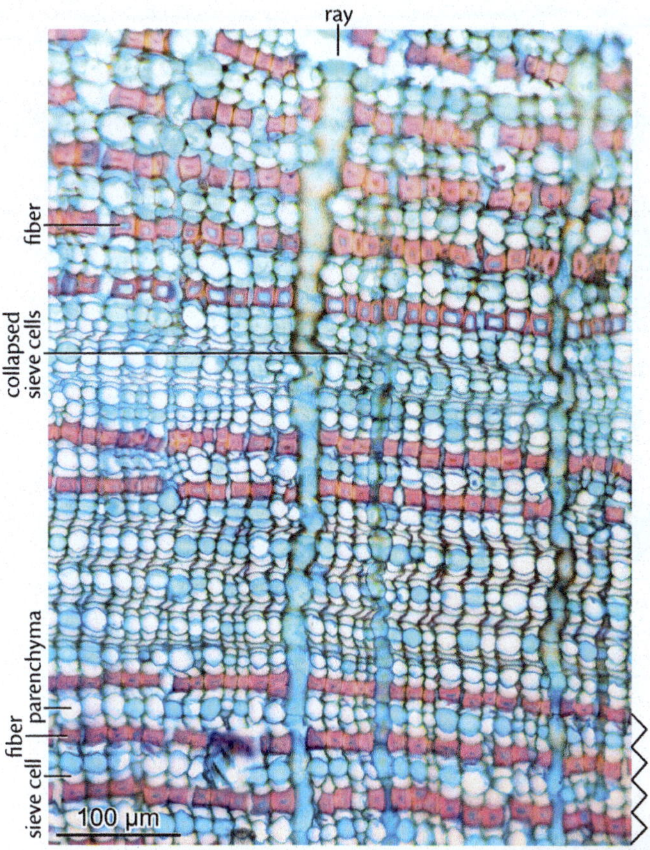

Bark anatomy: Regular tangential layers of sieve cells, fibers and parenchyma cells, characteristic for Cupressaceae.

Sieve cells thin-walled, rectangular. Parenchyma cells round to oval. Fibers thick-walled, square. Rays remaining straight, dilating in older stages. Crystal sand in ray cells and radial intracellular layers.

Annual rings mostly indistinct. Ring boundaries of fast-growing trees indicated by thick-walled fibers in early phloem. Phellem cells thin-walled, irregular shaped.

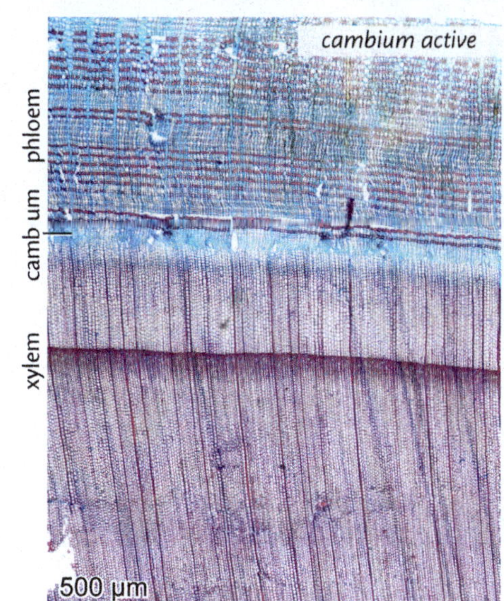

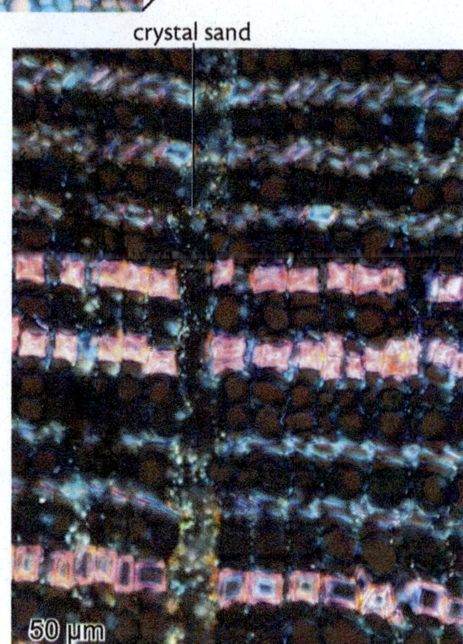

Ginkgo biloba L. Ginkgoaceae

maidenhair tree, Ginkgo, arbre aux quarante écus, albero di capelvenere, árbol de los cuarenta escudos

Form: Upright tree, crown cylindrical, to 40 m tall.

Features: Leaves deciduous, leathery, alternate, fan-shaped, with parallel veins, to 10 cm long. Flowers dioecious, female flower inconspicuous, green, hanging, to 3 mm long, male catkins greenish, hanging, to 5 cm long. Fruit a globular to ovate, orange yellow, fetid drupe, with an oily, roasted edible nut, to 3 cm long.

Habitat: Native only in the Tianmushan Mountains in Zhejiang in East China; widely planted.

On moist, deep soils, to 800 m.

Bark: Young gray brown, later dark gray, fissured with wide furrows.

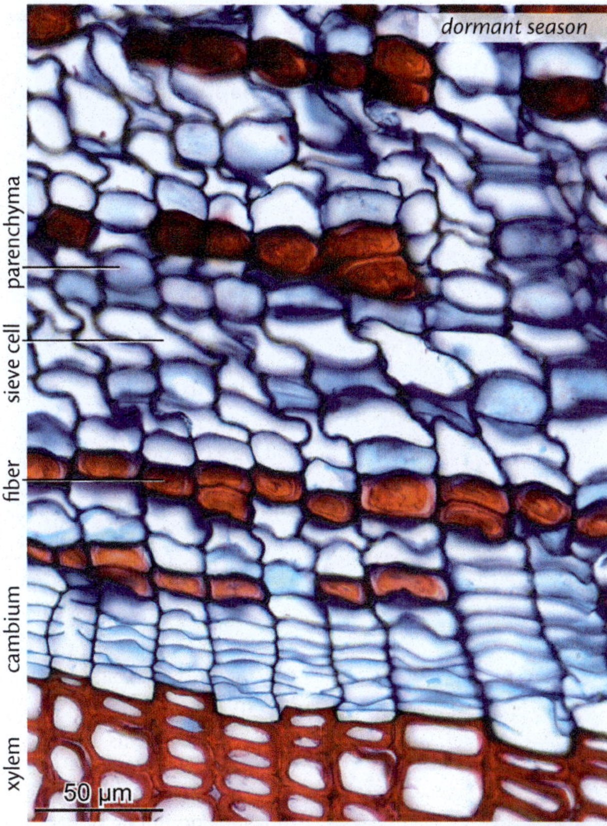

Bark anatomy: Sieve cells thin-walled, rectangular, collapsing in later stages. Parenchyma cells round to oval. Fibers very thick-walled, rectangular. Rays remaining straight, dilating in older stages. With a few large crystal druses outside of rays.

Annual rings indistinct. Phellem layered, early phellem consisting of several rows of thin-walled, regularly formed, square cells and one row of latewood phellem cells.

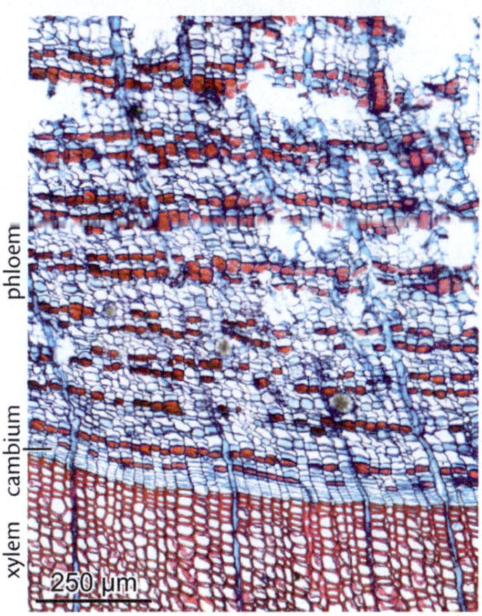

Larix kaempferi (Lamb.) Carrière

Pinaceae

Japanese larch, Japanische Lärche, mélèze du Japon, alerce del Japón, larice del Giappone

Form: Deciduous conifer, crown conical, later broadly columnar, with horizontal branches, to 45 m tall.

Features: Needles deciduous, soft, to 4 cm long, in star-shaped bundles of 20–40 on short shoots. Flowers monoecious, female cone upright, red, 1–2 cm long, male cone hanging, yellow, globular, to 2 cm long. Seed cones soft, globular, red to gray brown, with scales bent upwards, to 3.5 cm long.

Habitat: Mountains of Central Honshu, Japan. Fresh to moist, deep soils on subalpine mountain slopes, to 2400 m.

Bark: Thin and flaky, gray to gray brown, slightly corky with longitudinal furrows revealing a red brown inner bark.

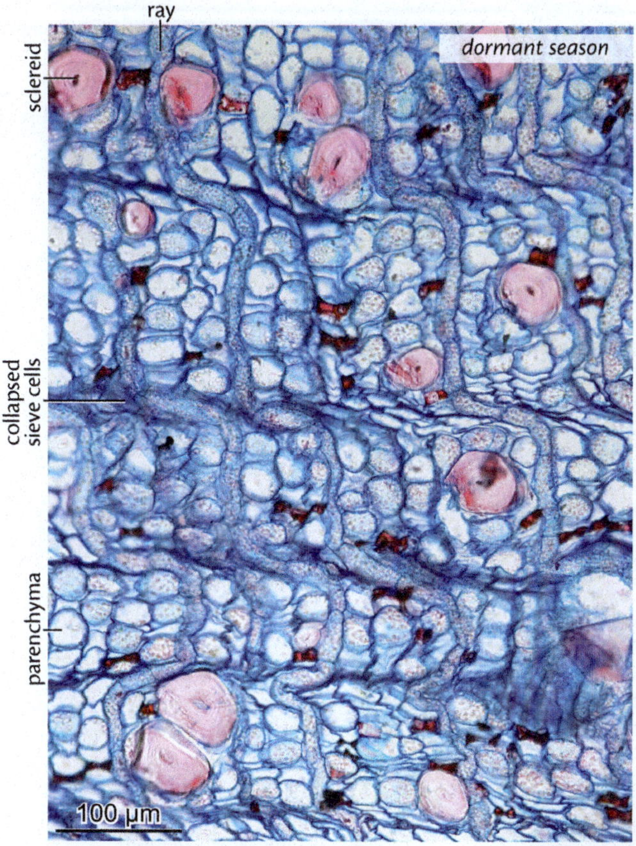

Bark anatomy: Regular tangential layers in the phloem. Characteristic presence of isolated sclereids.

Sieve cells rectangular, arranged in 3–5-seriate tangential rows, collapsing within a few years. Parenchyma cells round in the juvenile and enlarged in the adult phloem. Fibers absent. Rays radially wavy. Crystal sand in all parenchymatic cells.

Annual rings distinct for a few years near the cambium. Rings indistinct in the adult phloem due to sieve-cell collapse.

Periderm active for several years. Phellem cells rectangular, thin-walled.

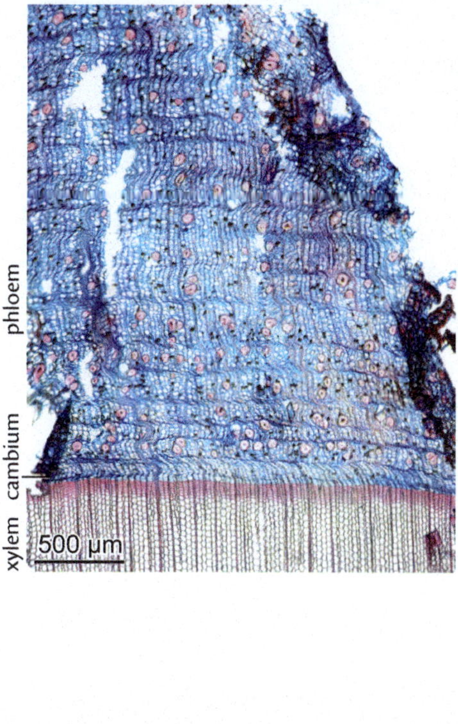

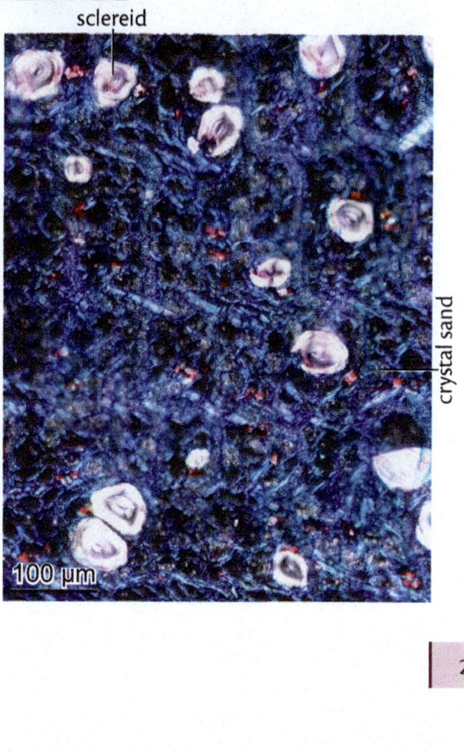

Metasequoia glyptostroboides Hu & W.C.Cheng

Cupressaceae

dawn redwood, Urweltmammutbaum, metasequoia, abete d'acqua, leño rojo oriental

Form: Deciduous conifer, crown lax, conical, to 50 m tall.

Features: Leaves deciduous, soft, needle-like, nearly opposite in two rows, light green, to 2 cm long. Flowers monoecious, female flower yellowish, spherical, to 6 mm across, male catkins yellow brown, hanging, to 10 cm long. Fruit a globular, dull brown cone, to 3 cm long, hanging on a long stalk.

Habitat: Native only in scattered relictual populations in eastern Szechuan, western Hubei and northwestern Hunan in China; widely planted.

On moist, deep humus and nutrient-rich soils, to 1500 m.

Bark: Young red brown, later gray brown, peeling in thin, vertical strips.

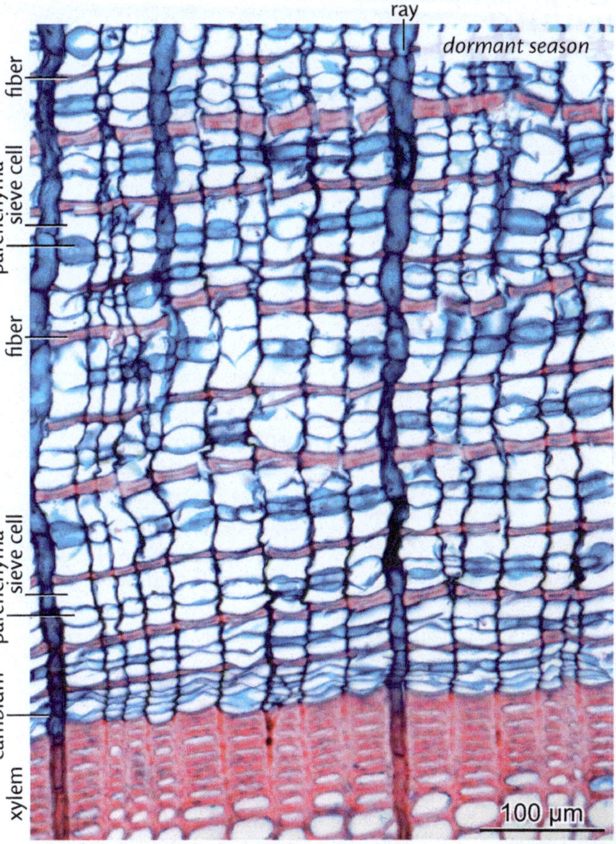

Bark anatomy: Regular tangential layers of sieve cells, fibers and parenchyma cells, characteristic for Cupressaceae.

Sieve cells thin-walled, rectangular. Parenchyma cells round to oval. Fibers thick-walled, rectangular. Sieve cells tending to collapse in later stages. Rays uniseriate, remaining straight, dilating in older stages. Ducts absent. Crystal sand in ray cells and radial intracellular layers.

Annual rings absent. Phellem consisting of a few layers of thin-walled, rectangular cells.

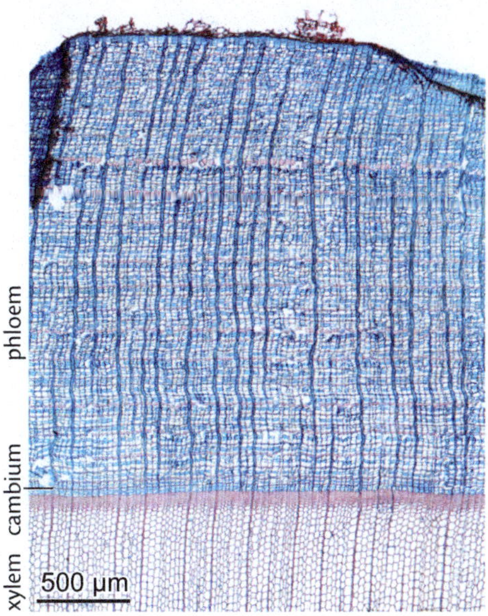

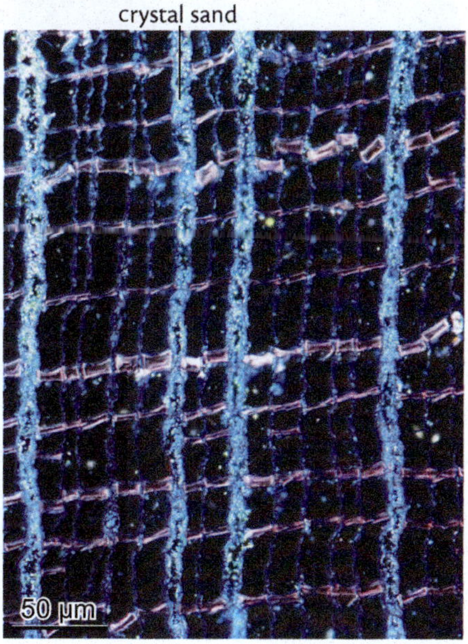

Sciadopitys verticillata (Thunb.) Siebold & Zucc.

Sciadopityaceae

Japanese umbrella pine, Japanische Schirmtanne, pin parasol du Japon, sciadopitys, pino sombrilla del Japon

Form: Upright conifer, crown straight, conical, to 45 m tall.

Features: Function of leaves performed by evergreen, needle-like cladodes, arranged in pseudo-whorls, to 15 cm long. Flowers monoecious, yellowish, spherical, in both sexes to 5 mm across. Fruit a long ovate, scaly cone, to 12 cm long.

Habitat: Southern Japan on southern Honshu, Kyushu and Shikoku.

On moist, deep, loamy and mildly acidic soils, in humid locations, to 1000 m.

Bark: Young red brown, smooth or peeling in small flakes, with age brown gray, with long, peeling fissures, revealing the deep red inner bark.

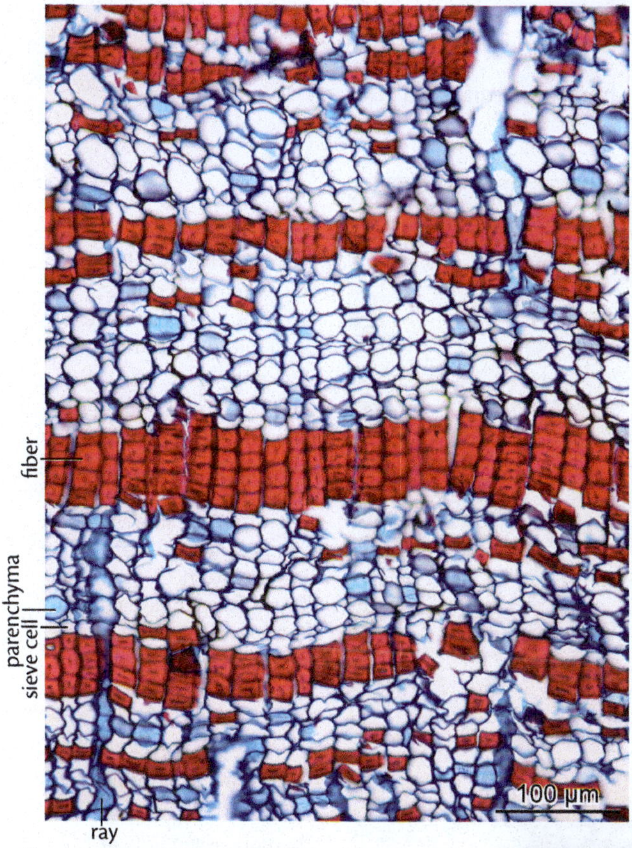

Bark anatomy: Regular and irregular tangential layers of sieve cells, fibers and parenchyma cells.

Sieve cells thin-walled, rectangular. Parenchyma cells round to oval. Fibers very thick-walled, square to rectangular. Sieve cells tending to collapse. Rays uniseriate, mostly remaining straight, dilating in later stages. Ducts absent. Crystal sand in ray cells and radial intracellular layers.

Annual rings absent. Phellem consisting of a few layers of thin-walled, rectangular cells.

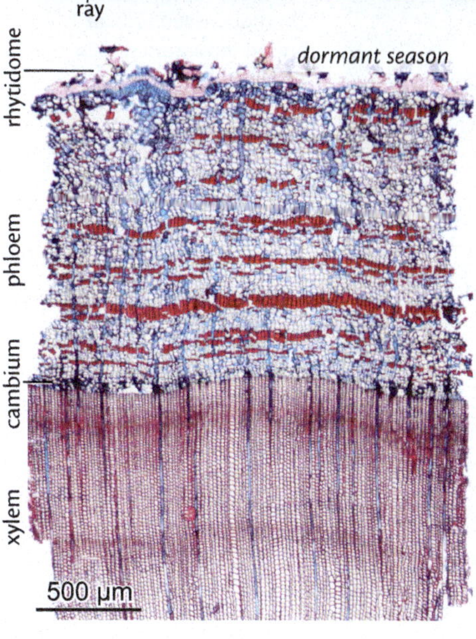

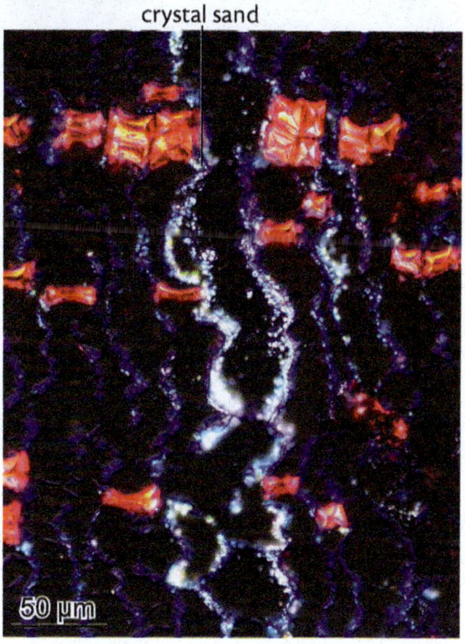

Thujopsis dolabrata (L.f.) Siebold & Zucc. Cupressaceae

hiba cedar, Hibalebensbaum, hiba, falsa tuia, hiba

Form: Upright, single- or multi-stemmed conifer, crown straight, conical, to 30 m tall, in cultivation often only shrubby.

Features: Leaves evergreen, flat scales, fan-like spreading, glossy green, with two silver white patches below, to 8 mm wide. Flowers monoecious, globular, male flower reddish, to 4 mm wide, female flower bluish green, to 10 mm wide. Fruit a globular to ovate, nearly upright cone with thickened, pointy scales, to 1.5 cm long.

Habitat: Japan, mainly in northernmost Honshu; widely planted.

On moist, deep soils, at humid mountain slopes, to 1400 m.

Bark: Red brown, peeling in thin, vertical strips and flakes.

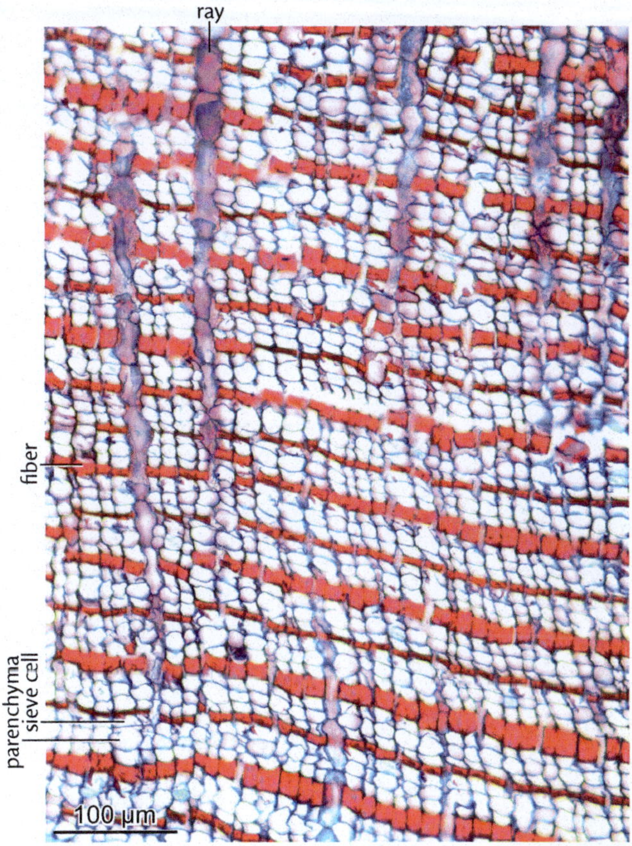

Bark anatomy: Regular tangential layers of sieve cells, fibers and parenchyma cells.

Sieve cells thin-walled, rectangular. Parenchyma cells round to oval. Fibers thick-walled, square to rectangular. Rays remaining straight, dilating in older stages. Rarely with large ducts. Crystal sand in ray cells and radial intracellular layers.

Annual rings absent. Phellem cells thin-walled, regularly formed.

dormant season

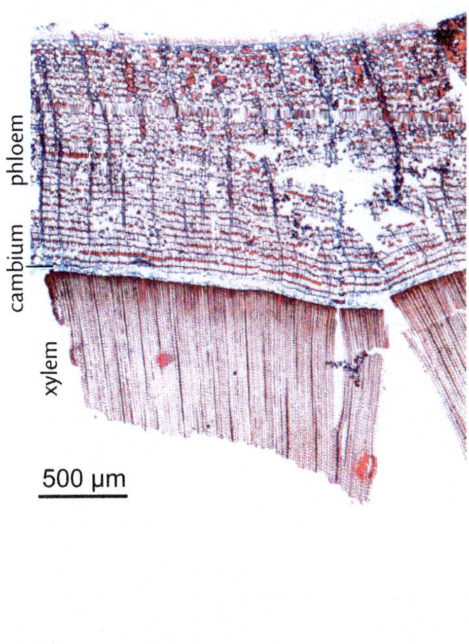

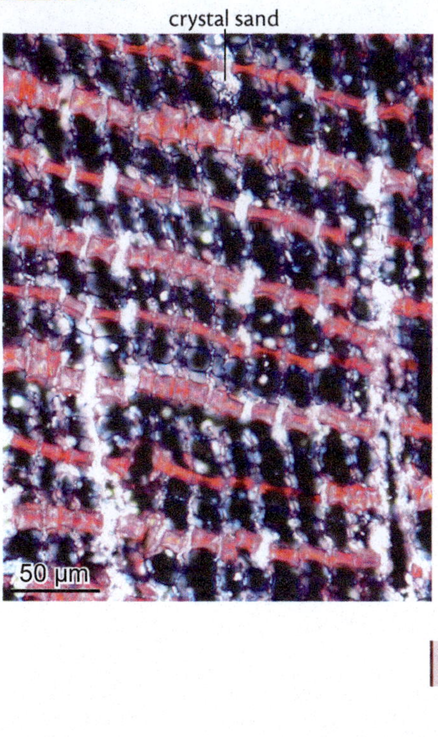

Ailanthus altissima (Mill.) Swingle Simaroubaceae

tree of heaven, Götterbaum, ailante, ailanto, árbol del cielo

Form: Upright tree, crown broad, irregular shaped, to 25 m tall.

Features: Leaves deciduous, alternate, pinnate with 13–43 long ovate, fetid leaflets, to 90 cm long. Flowers dioecious, both sexes similar, yellow white, with 5–6 petals, to 4 mm long, in upright panicles. Fruit a long-winged, yellowish or reddish nut, to 6 cm long, in dense, hanging panicles.

Habitat: Eastern China, from Jilin southwest to Guizhou, Yunnan and northern Vietnam; widely planted and often invasive.

Mostly on dry, stony soils in warm situations, to 2500 m.

Bark: Twigs olive green to red brown, stem bark dull light gray, young with diamonds, later shortly fissured.

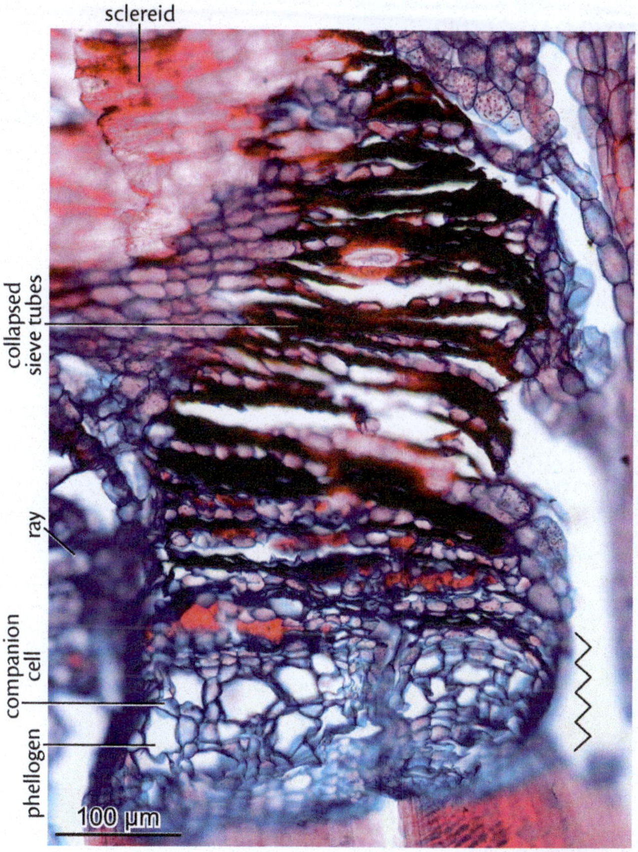

Bark anatomy: Dense structure with distinct ray dilatations and large tangential layers of compressed sieve tubes and parenchyma cells.

Sieve tubes large, irregular shaped, collapsing within a short time after formation, forming dense layers between large rays. Parenchyma cells roundish. Tangential small fibers in the early phloem. Sclereids in tangential layers and in large irregular groups. With distinct ray dilatations. With a few prismatic crystals.

Periderm consisting of many layers of rectangular, thick-walled cells.

dormant season

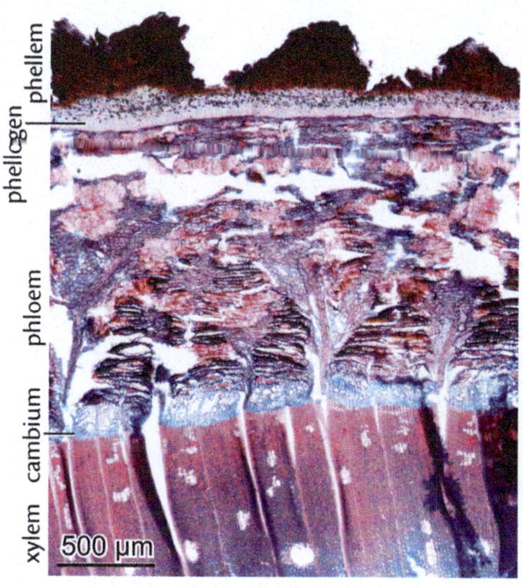

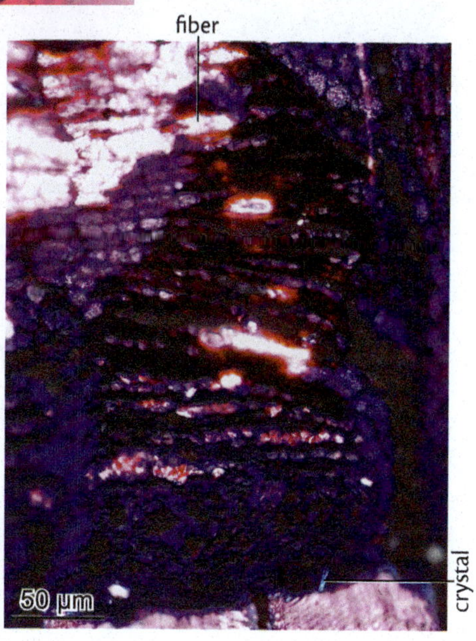

Cercidiphyllum japonicum Siebold & Zucc. ex J.J.Hoffm. & J.H.Schult.bis

Cercidiphyllaceae

katsura, Kuchenbaum, arbre au caramel, katsura, arbol de katsura

Form: Upright, often multi-stemmed tree, crown round, to 40 m tall.

Features: Leaves deciduous, opposite, round ovate with a heart-shaped base, margin toothed, to 8 cm long. Flowers dioecious, female flower yellowish, inconspicuous, male flower scarlet, to 8 mm long. Fruit a cluster of bent follicles with a tip, to 1.5 cm long.

Habitat: Western to Central China and Japan.

On deep, moist soils, to 2700 m.

Bark: Young smooth, light gray brown, later fissured or cracking into small plates.

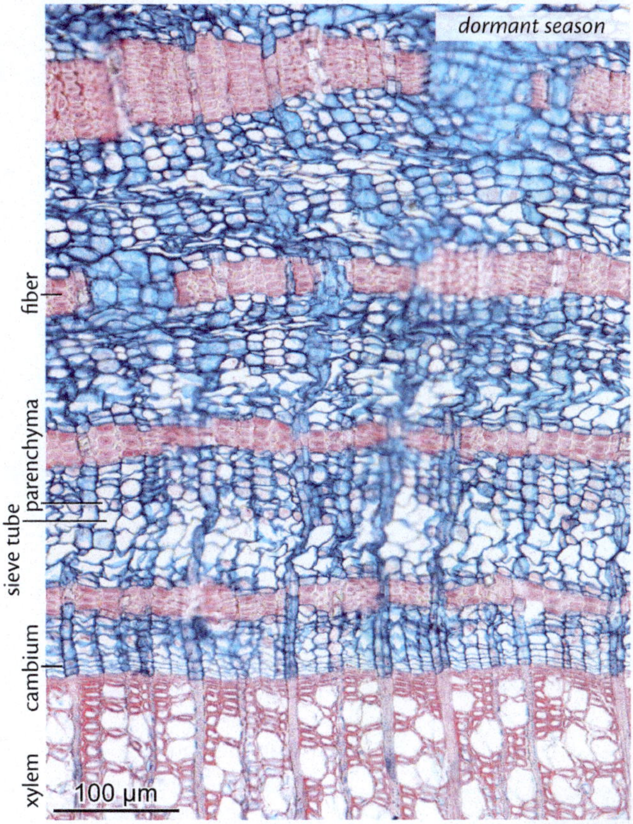

Bark anatomy: Regular tangential layers of fiber bands and zones of sieve tubes and parenchyma.

Sieve tubes angular, parenchyma cells roundish in early phloem. Sieve tubes collapsing within a short time after formation. Fibers in 2–4-layered tangential bands, laterally interrupted by dilated rays. Sclereids absent. Ray structure remaining for many years. Prismatic crystals mainly within rays.

Annual ring boundaries usually corresponding with fiber bands. Phellem consisting of a few layers of rectangular cells.

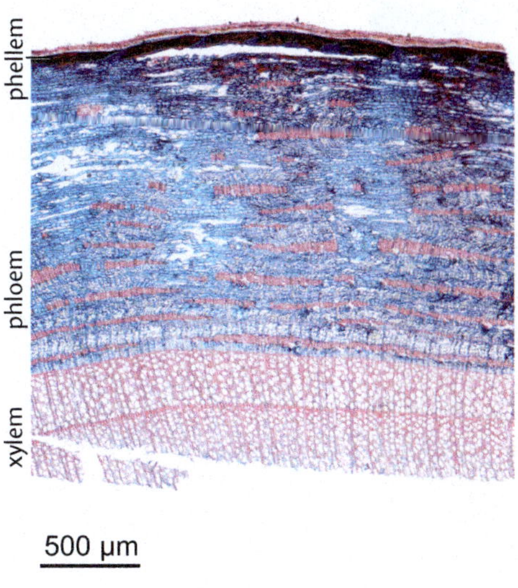

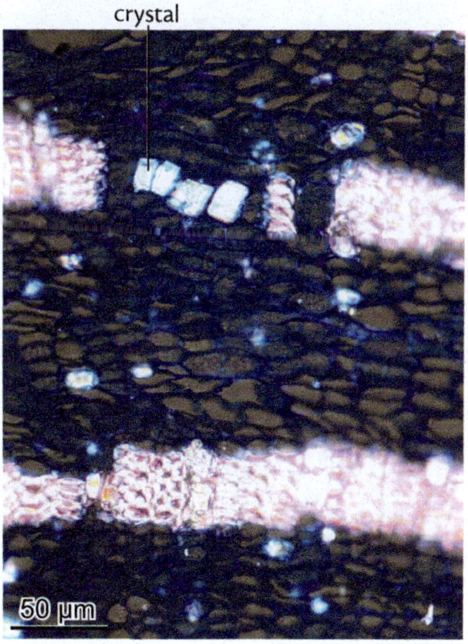

Decaisnea insignis
(Griff.) Hook.f. & Thomson

Lardizabalaceae

blue bean, Blauschote, arbre aux haricots bleus, fagiolo azzurro, judia azul

Form: Shrub, stiff upright, to 5 m tall.

Features: Leaves deciduous, opposite, pinnate with 13–25 long ovate leaflets, to 14 cm long. Flowers with six long, pale yellow sepals, without petals, hanging, to 5 cm long. Fruit a cylindrical, fleshy cobalt blue berry, with glossy black seeds embedded in white flesh, to 10 cm long.

Habitat: The Himalayas from northwestern India northeast to Central China as far as Henan and Anhui.

On moist soils, often on steep slopes and in ravines, to 3600 m.

Bark: Olive green to brown, with a conspicuous, bluish white coat.

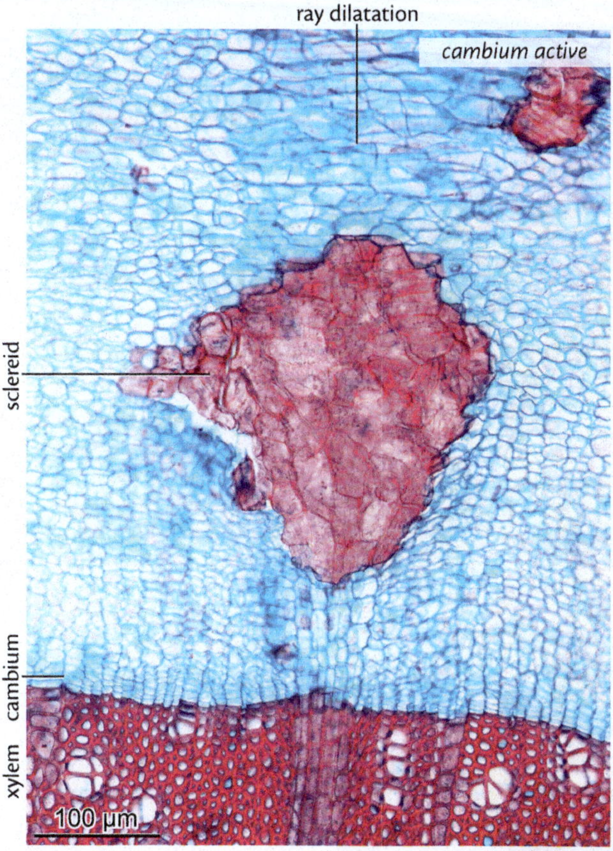

Bark anatomy: Irregular, unlignified tissue with a few large, round groups of sclereids.

Sieve tubes irregular shaped, parenchyma roundish. Fibers absent. Sclereids in round groups. Rays distinctly dilating. Crystals absent.

Annual ring boundaries absent. Phellem layered, consisting of 2–4 rows of thin-walled, square cells and one row of rectangular cells.

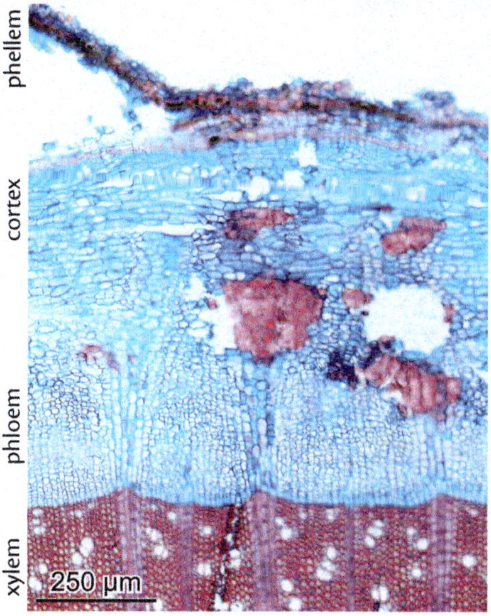

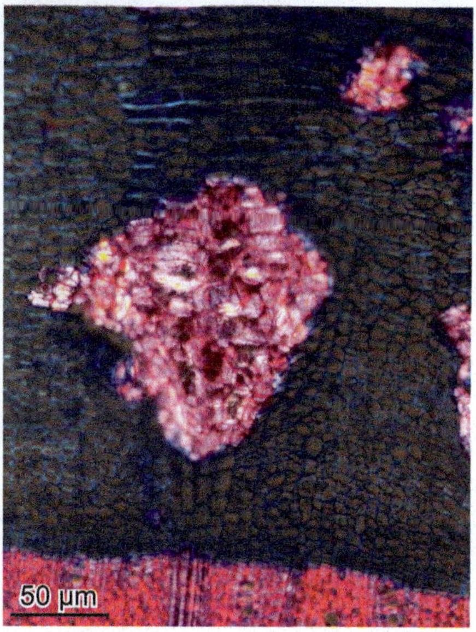

Diospyros lotus L. Ebenaceae

date plum, Lotuspflaume, plaqueminier lotier, albero di Sant'Andrea, lodoñero

Form: Upright, round-crowned tree, to 30 m tall.

Features: Leaves deciduous, alternate, long ovate, margin entire or slightly wavy, to 13 cm long. Flowers dioecious, in both sexes yellow red, to 8 mm large. Fruit an elliptical berry, yellow at first, turning blue black, to 2 cm long.

Habitat: Western and Northern China, widely planted and often naturalized.

On dry to moist soils on slopes and in gorges, to 2500 m.

Bark: Brown black to gray brown, soon finely fissured and cracking into small squares.

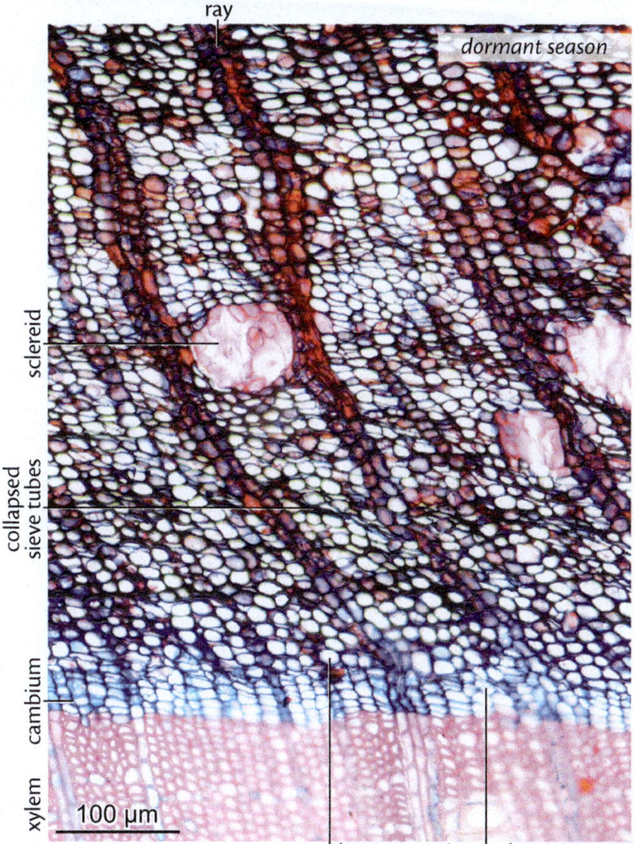

Bark anatomy: Irregular, unlignified tissue with a few round groups of sclereids.

Sieve tubes rectangular, collapsing within a short time after formation. Parenchyma roundish. Fibers absent. Sclereids in round groups. Rays slightly bent. Dilatations at the outside of the stem. Many prismatic crystals in axial parenchyma cells.

Annual ring boundaries absent. Phellem layered, consisting of 2–4 rows of thin-walled, square cells and one row of rectangular cells.

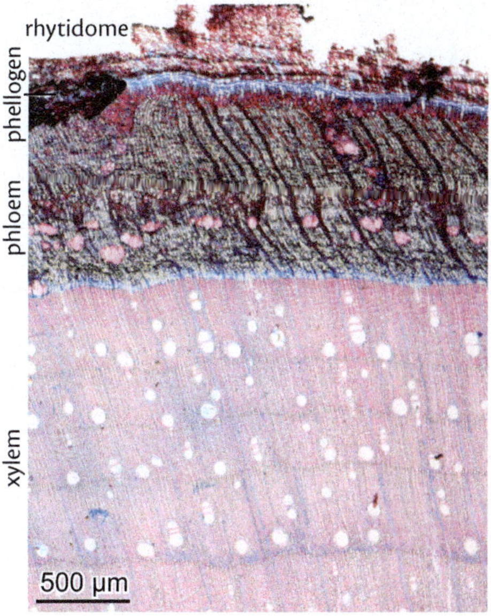

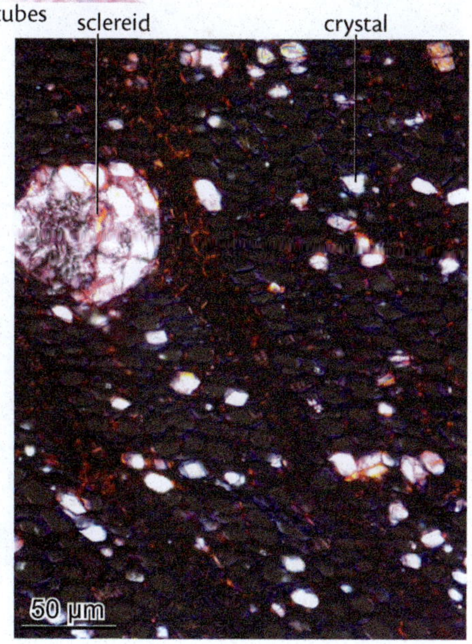

Edgeworthia chrysantha Lindl. Thymelaeaceae

paperbush, Papierstrauch, arbre à papier, arbusto della carta, arbusto de papel

Form: Shrub, many-stemmed, crown broad, flat, to 2 m tall.

Features: Leaves deciduous, alternate, crowded at the branch ends, long ovate, margin entire, to 10 cm long. Flowers only sepals, with a long silver white tube and yellow, intensely fragrant interior, in dense, hanging clusters, to 3 cm long, emerging before the leaves. Fruit a dry drupe, to 2 cm long.

Habitat: Burma and China east to Henan and Zhejiang; often planted and used for paper in Japan.

On moist, deep and humus-rich soils, to 2200 m.

Bark: Light brown to brown gray, shedding papery.

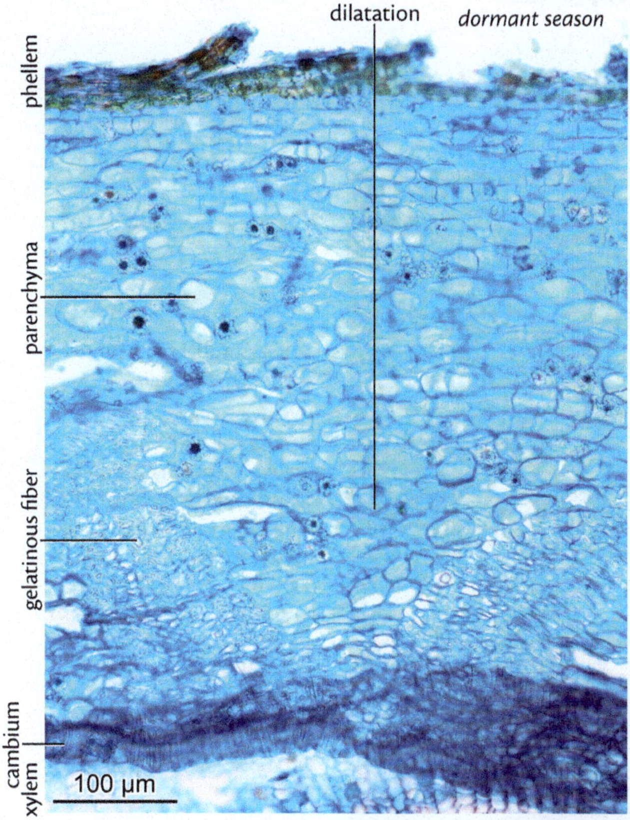

Bark anatomy: Irregular, unlignified tissue with gelatinous fibers and laticifers. Tissue intensively distorted due to tension from the phellem.

Sieve tubes rectangular, collapsing within a short time after formation. Parenchyma roundish. Sclereids absent. Rays dilating. Many large crystal druses in parenchyma cells.

Annual ring boundaries absent. Phellem layered, consisting of 2–4 rows of thin-walled, square cells and one row of rectangular cells.

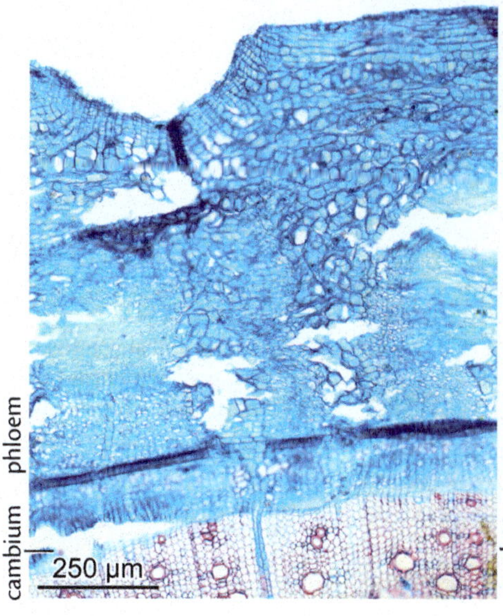

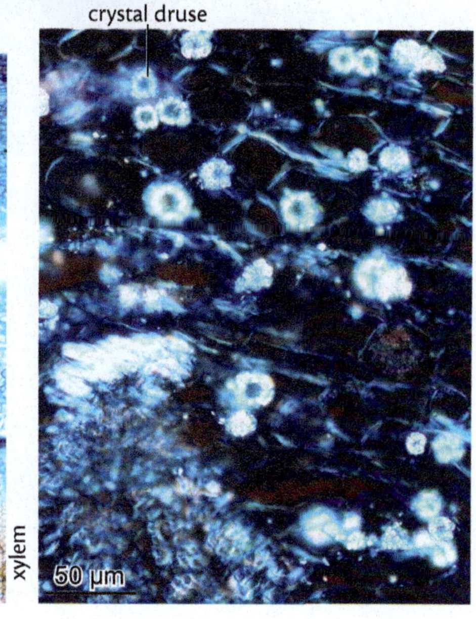

Hamamelis mollis

Oliv. ex F.B.Forbes & Hemsl.

Hamamelidaceae

witch hazel, Zaubernuss, hamamélis, hamamelis, avellano de bruja

Form: Shrub, broad, funnel-shaped, to 5 m tall.

Features: Leaves deciduous, alternate, broad ovate, margin irregularly toothed, to 12 cm long. Flowers with four long, thin, ribbon-like, deep yellow, well-scented petals, to 4 cm long, emerging in late winter. Fruit an ovate, woody capsule, to 3 cm long.

Habitat: East China from Guizhou and Hubei to Zhejiang.

On moist, deep soils, to 800 m.

Bark: Young dull gray brown, later light gray, sometimes with fine, irregular fissures.

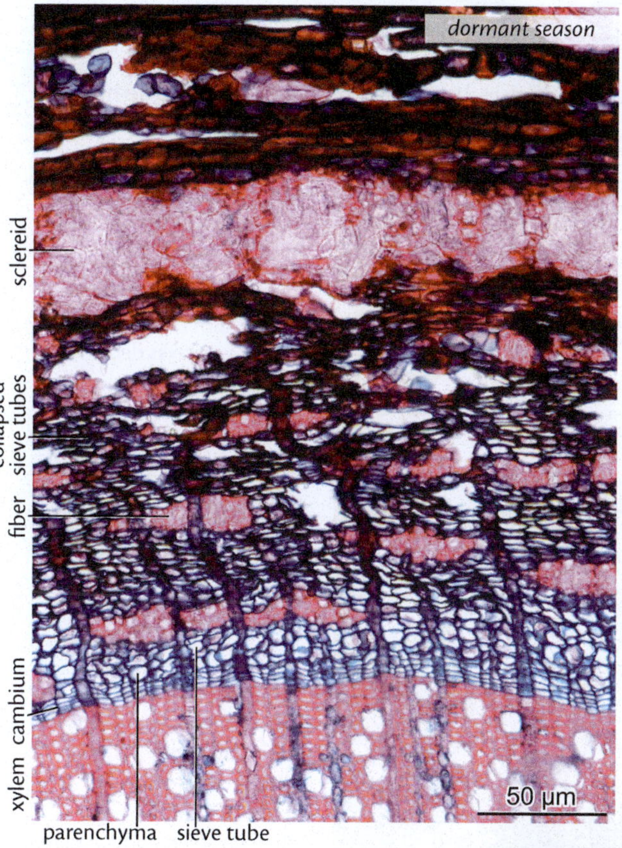

Bark anatomy: Almost uniform structure of sieve tubes and parenchyma.

Sieve tubes irregular shaped, parenchyma cells round to oval. Fibers in tangential groups. Sclereids in large groups, mainly at the outside of the phloem. Rays straight. Many irregular shaped crystals in axial parenchyma cells. Phellem consisting of layers of large, rectangular, thin-walled cells and small, thick-walled cells.

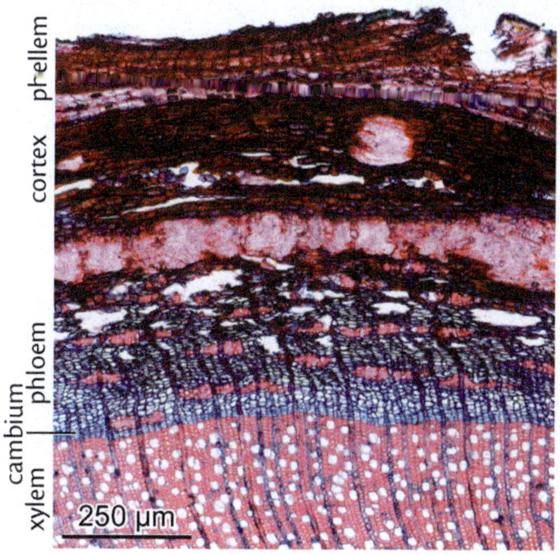

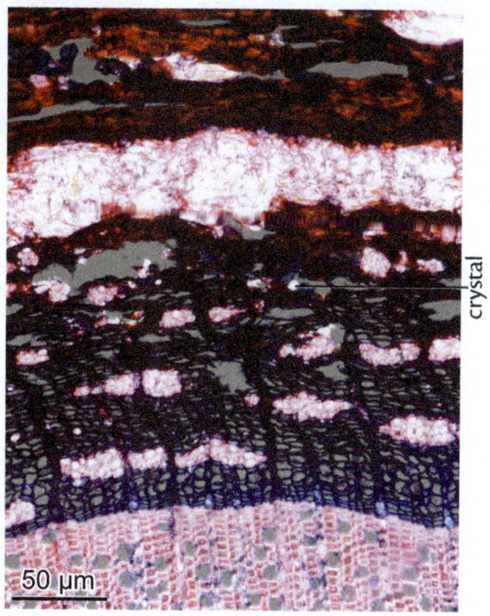

Hydrangea aspera W. Bartram

Hydrangeaceae

rough-leaved hydrangea, Raue Hortensie, hortensia rugueux, hortensia de hojas ásperas, ortensie aspera

Form: Stiff upright shrub, sparsely branched, to 3 m tall.

Features: Leaves deciduous, opposite, long ovate, serrated, densely felty hairy below, to 23 cm long. Flowers in large pseudo-umbels, fertile flowers small and pink to lilac, sterile flowers at the rim of the umbel white to pink, to 6 cm wide. Fruits in small brown capsules, to 1 cm long.

Habitat: The Himalayas from northwestern India and Myanmar northeast to Central China as far as Shaanxi and Henan.

On fresh to moist soils, often on steep slopes and in ravines, to 4000 m.

Bark: Papery scaly, light gray to cinnamon brown.

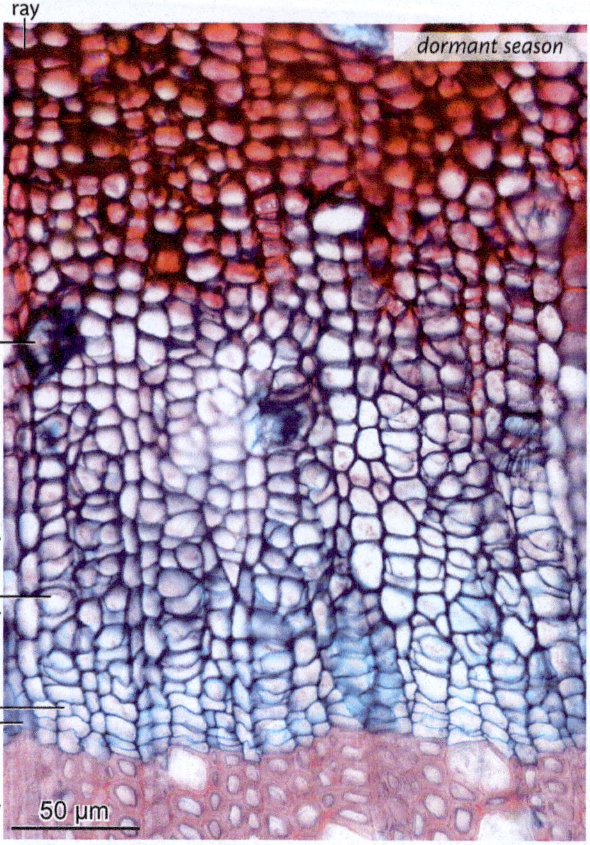

Bark anatomy: Uniform structure of sieve tubes and parenchyma with crystal chambers.

Sieve tubes irregular shaped, parenchyma cells round to oval. Fibers and sclereids absent. Rays dilating. Large chambers with bundles of raphids.

Phellem consisting of rectangular cells, flaking off after the first year.

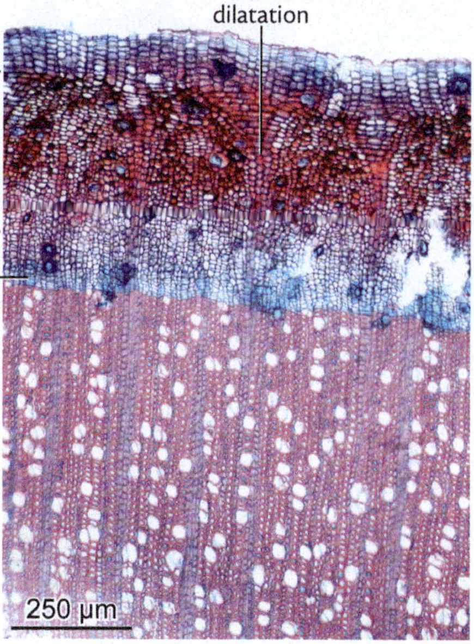

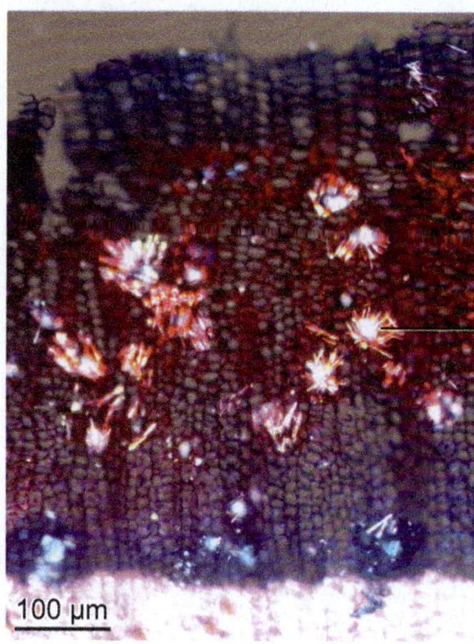

Koelreuteria paniculata Laxm. Sapindaceae

goldenrain tree, Blasenbaum, savonnier, albero delle lanterne cinesi, jabonero de la China

Form: Upright, short-stemmed tree, crown globular, with age umbellate, to 15 m tall.

Features: Leaves deciduous, alternate, double pinnate with 7–15 long ovate, irregularly toothed leaflets, to 35 cm long. Flowers yellow, scented, to 1 cm wide, in showy, upright panicles, to 35 cm long. Fruit a long ovate, woody capsule, to 5 cm long.

Habitat: East Asia, from Szechuan northeast to Hebei, Korea and Japan; widely planted.

Mostly on dry soils in warm locations, to 1800 m.

Bark: Young gray brown with corky lenticels, later narrowly fissured, often revealing an orange inner bark.

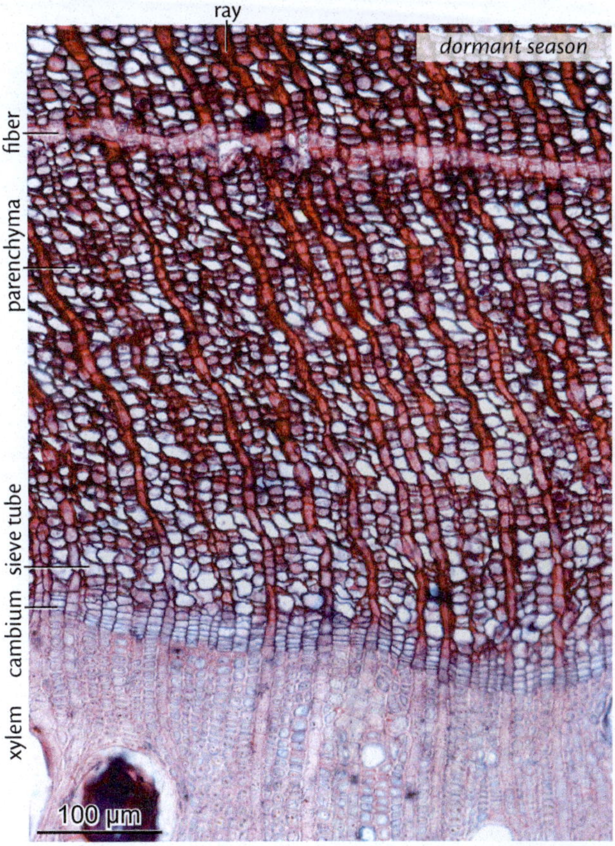

Bark anatomy: Almost uniform structure of sieve tubes and parenchyma.

Sieve tubes irregular shaped, parenchyma cells round to oval. Fibers in bands, mainly within the phloem. Rays slightly wavy. Many irregular shaped crystal druses in axial parenchyma cells.

Phellem consisting of layers of large, rectangular, thin-walled cells and small, thick-walled cells.

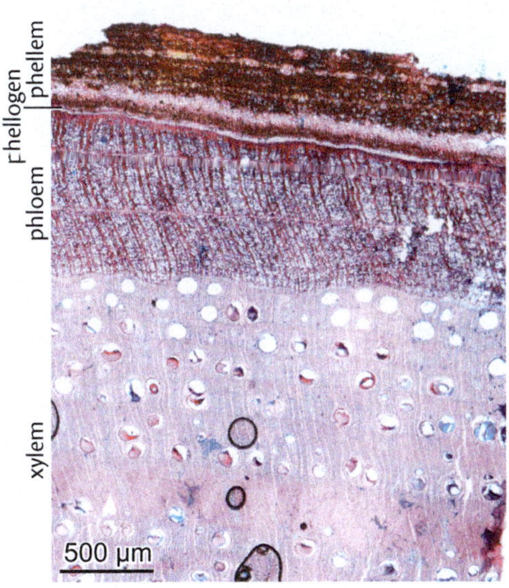

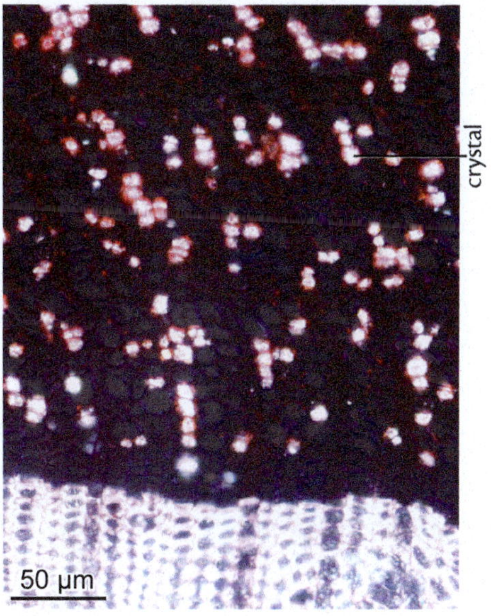

Morus alba L.

Moraceae

white mulberry, Weiße Maulbeere, mûrier blanc, gelso bianco, morera blanca

Form: Upright, short-stemmed tree, crown flat, with white latex, to 15 m tall.

Features: Leaves deciduous, alternate, heart-shaped or irregularly lobed, margin toothed, to 19 cm long. Flowers mono- or dioecious, both sexes inconspicuous, light green, in hanging catkins, to 1.5 cm long. Fruit an ovate to cylindrical spurious fruit, resembling a white-colored blackberry, to 2.5 cm long, edible.

Habitat: Originally endemic to Central and Northern China; since antiquity widely planted, also as forage plant for silkworm.

Mostly on dry soils in warm locations, to 2500 m.

Bark: Young smooth, gray brown or light brown, with widely spaced lenticels, later irregularly, roughly fissured.

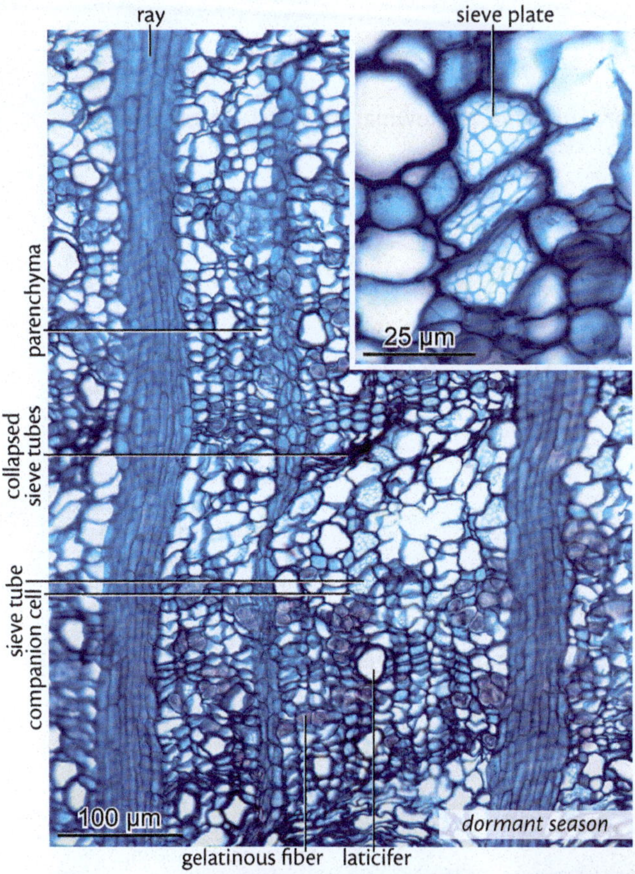

Bark anatomy: Tangentially structured tissue of sieve tubes and parenchyma without lignified elements in early formed parts (annual rings?).

Sieve tubes angular, accompanied by companion cells and round to oval parenchyma cells. Laticifers embedded in parenchymatic zones. Fibers unlignified, gelatinous. Large irregular groups of sclereids in external parts. Dilatations absent. Large prismatic crystals in sclereids.

Rhytidome not observed.

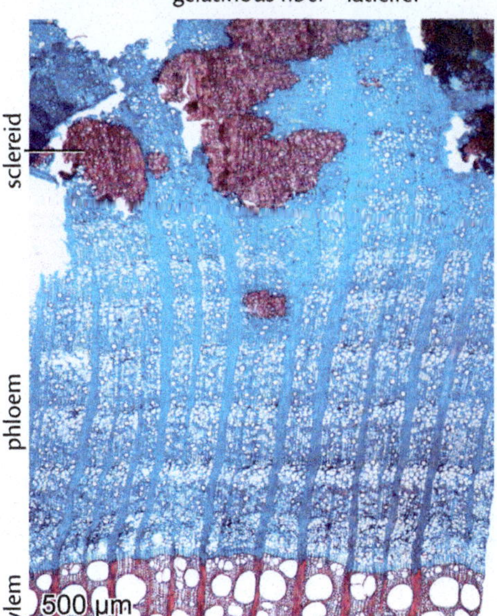

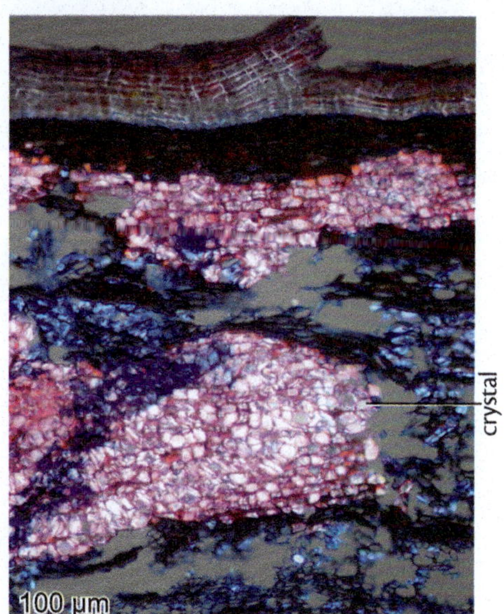

Paulownia tomentosa Steud. Paulowniaceae

paulownia, Blauglockenbaum, paulownia, paulownia, arbol de la emperatriz

Form: Upright tree, crown large and light, to 30 m tall.

Features: Leaves deciduous, opposite, heart-shaped or broad ovate, margin entire, to 30 cm long. Flowers lilac blue, inside yellow, sweetly scented, two-lipped and funnel-shaped, to 6 cm long, in showy, upright panicles. Fruit an ovate, woody capsule, splitting in two halves, to 4.5 cm long.

Habitat: China from Hebei south to Guangdong and Yunnan; widely planted.

Mostly on dry soils in hot locations, to 1500 m.

Bark: Twigs olive brown to red or gray brown, stem bark first smooth, light gray, later finely fissured.

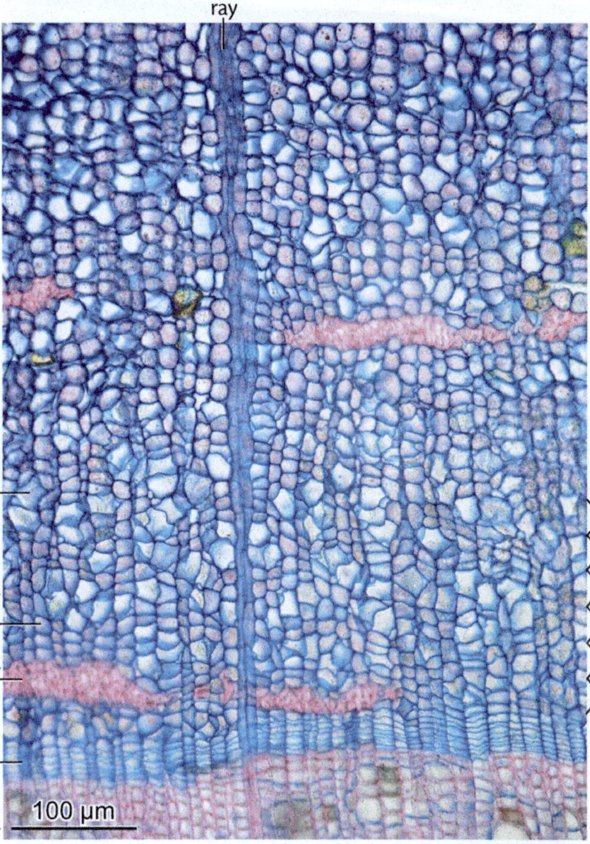

Bark anatomy: Regular tangential layers consisting of small bands of fibers and large zones of sieve tubes and parenchyma.

Early phloem formed by irregular shaped sieve tubes and roundish parenchyma cells. Sieve tubes collapsing within a short time after formation, not disturbing the overall structure. Fibers grouped in tangential bands 2–4 cells wide. Irregular groups of sclereids at the periphery of the bark. Ray structure remaining for many years. Crystals sand in most parenchyma cells. Crystals absent in rays.

Periderm consisting of alternating layers of rectangular, thick-walled cells and thin-walled cells.

dormant season

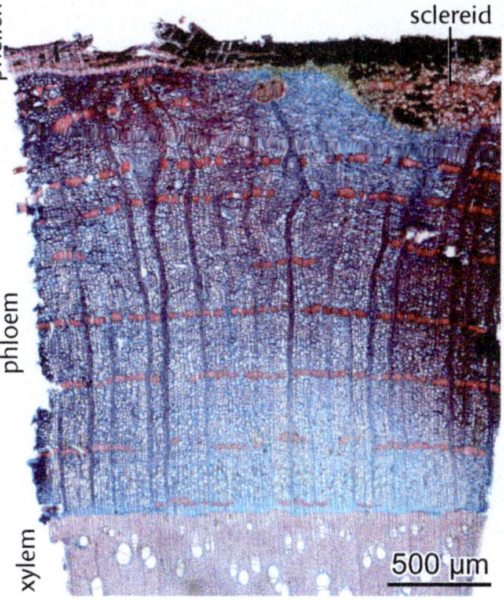

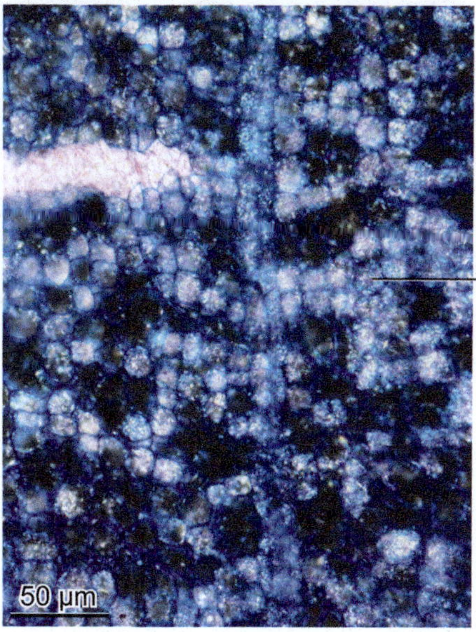

Stachyurus praecox Siebold & Zucc. Stachyuraceae

early stachyurus, Japanische Schweifähre, stachyurus, stachyurus, stachyurus

Form: Shrub, broad, upright, to 4 m tall.

Features: Leaves deciduous, alternate, long ovate, margin toothed, to 14 cm long. Flowers yellowish-white, bell-shaped, to 8 mm long, in hanging racemes, to 8 cm long, emerging in early spring. Fruit a globular, reddish green berry, to 8 mm long.

Habitat: Japan from Honshu south to the Ryukyu and Ogasawara Islands.

On dry to moist, deep soils, to 1000 m.

Bark: Young wine red, with numerous, whitish lenticels, later reddish gray brown.

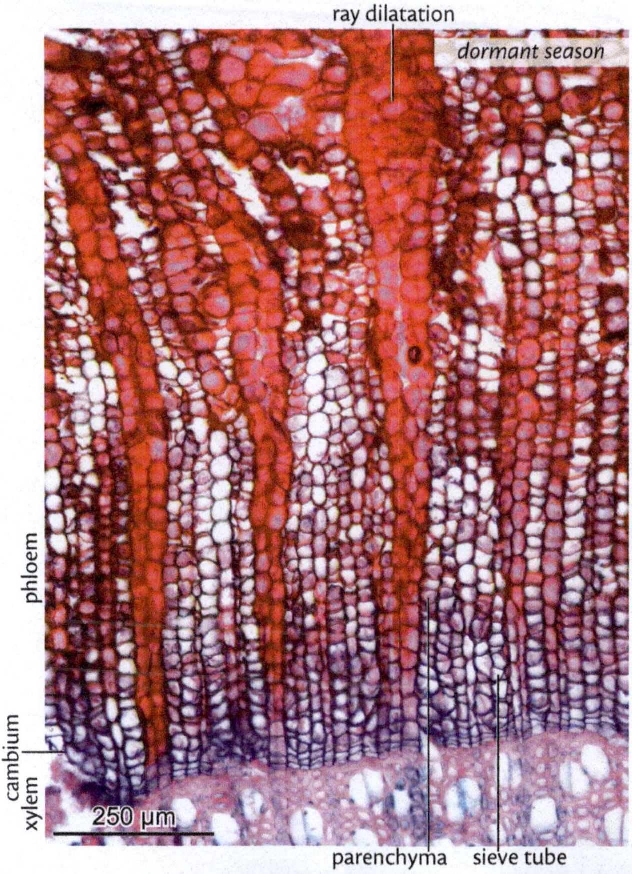

Bark anatomy: Soft tissue with a few groups of sclereids.

Sieve tubes angular, parenchyma cells round, collapsing within a short time after formation. Few fibers in a line between phloem and cortex. Sclereids often absent or in small tangential groups. Rays straight or bent. Axial parenchyma cells forming large dilatations. Crystal druses mainly in dilatations.

Phellem consisting of layers of thin-walled, rectangular cells and square cells.

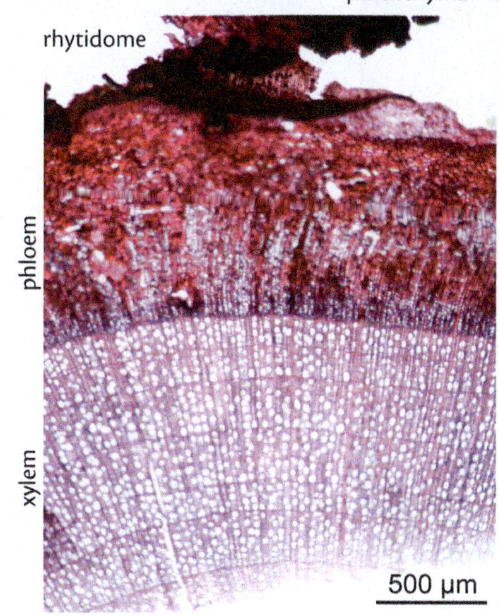

Trochodendron aralioides Siebold & Zucc.

Trochodendraceae

Japanese wheel tree, Radbaum, arbre aux roues, albero della ruota, árbol de la rueda

Form: Upright tree, to 20 m tall.

Features: Leaves evergreen, leathery, alternate and crowded at the twig ends, long ovate, margin entire, to 12 cm long. Flowers yellowish green stamens, spreading wheel-like from a green disk without sepals and petals, to 18 mm large, in dense, upright panicles, to 13 cm long. Fruit a star-shaped, 4–11-parted follicle, to 3 cm long.

Habitat: Taiwan, the Ryukyu Islands and South Korea.

On moist, deep soils, to 2700 m.

Bark: Young smooth, yellow green, with widely spaced lenticels, with age light brown gray, often finely fissured.

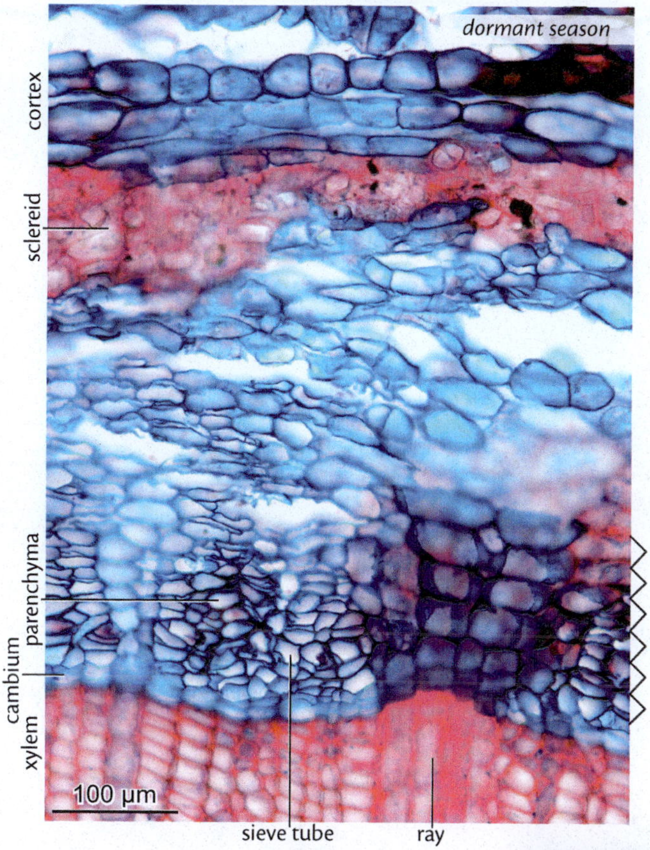

Bark anatomy: Structure with irregular distribution of sclereids, sieve tubes and parenchyma zones.

Early phloem formed by irregular shaped sieve tubes and roundish parenchyma cells. Sclereids in irregular, tangentially enlarged groups. Rays large, intensively dilating, structure remaining for a few years. Few prismatic crystals in the cortex.

Annual rings absent. Phellem consisting of layers of thin-walled, rectangular cells.

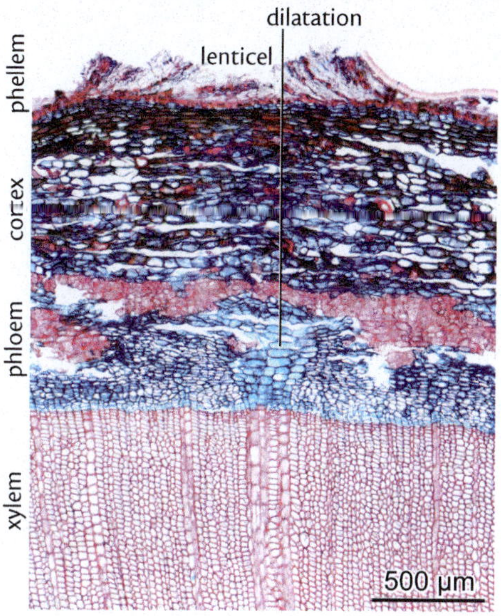

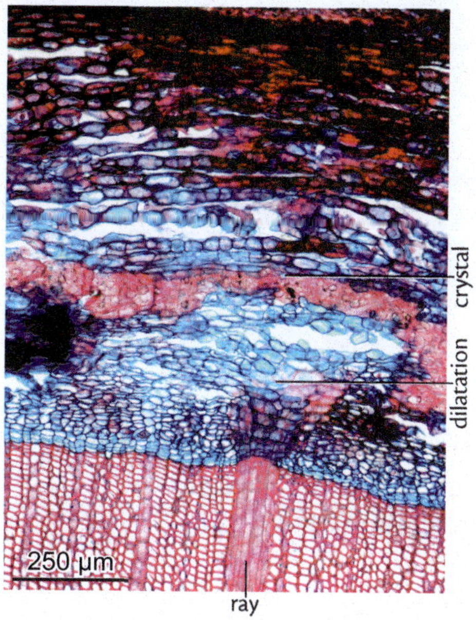

A limited representation is given to the **deciduous forest of Eastern America**, although its level of biodiversity and species richness lies somewhere in between Eastern Asia and Europe.

F. H. Schweingruber et al., *Bark Anatomy of Trees and Shrubs in the Temperate Northern Hemisphere*, https://doi.org/10.1007/978-3-030-14056-4_8

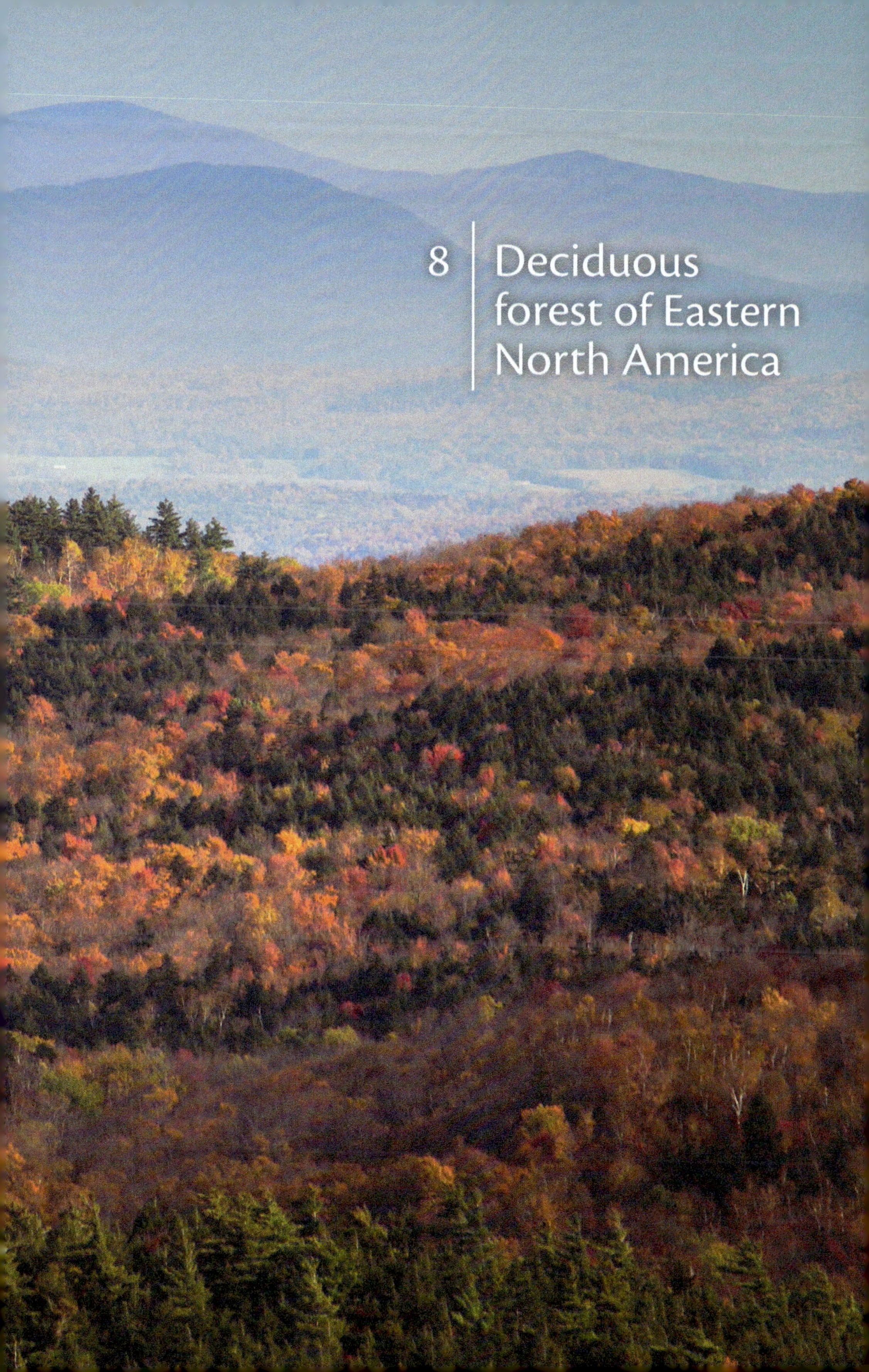

8 Deciduous forest of Eastern North America

Pinus strobus L.

Pinaceae

Weymouth pine, Weymouthskiefer, pin blanc, pino strobo, pino de Weymouth

Form: Upright conifer, crown straight, conical, with age turning umbellate, to 60 m tall.

Features: Needles evergreen, thin, soft, in bundles of five, to 15 cm long. Flowers monoecious, both sexes a globular cone, to 8 mm long, male cone light brown, female cone pink to scarlet. Seed cones long, cylindrical, flexible, hanging, to 15 cm long.

Habitat: Eastern North America from Newfoundland south to North Georgia and west to Manitoba and Illinois; widely planted.

On moist, deep soils, to 1200 m.

Variation: Also variety *chiapensis* in southeastern Mexico and Guatemala.

Bark: Young smooth, gray brown, later with small scales and characteristic rings around the stem, marking former branches.

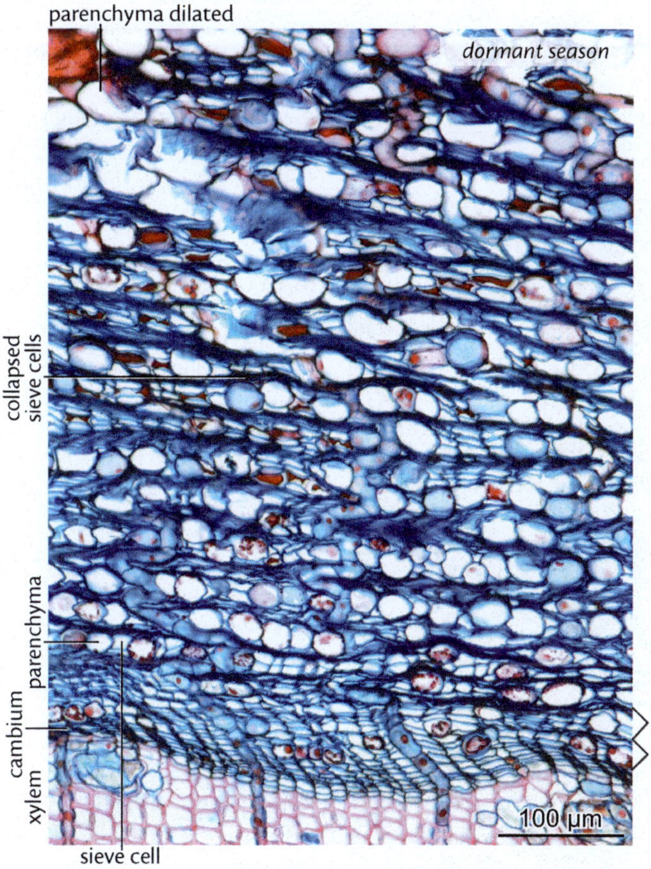

Bark anatomy: Sieve cells rectangular, arranged in 3–5-seriate tangential rows, collapsing within a few years after formation. Parenchyma cells small and oval in the juvenile, enlarged in the adult phloem. Sclereids and fibers absent. Rays radially wavy. Crystals absent in the phloem. Large resin ducts in the older phloem. Many small prismatic and acicular crystals in the periderm.

Annual rings mostly indistinct due to the irregular occurrence of parenchyma cells.

Phellem irregularly layered, consisting of rectangular, thick-walled cells (sclereids) and thin-walled cells.

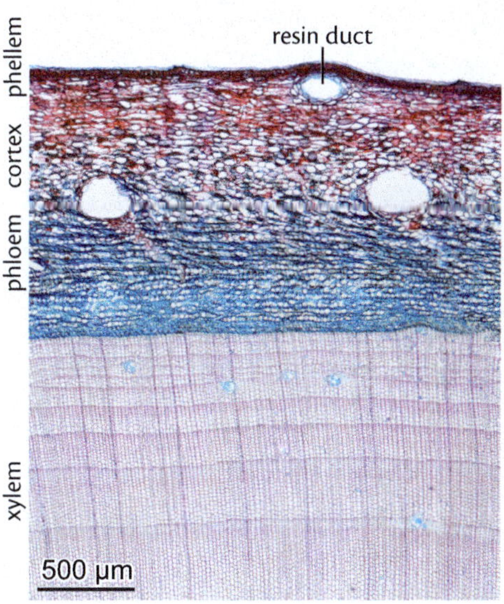

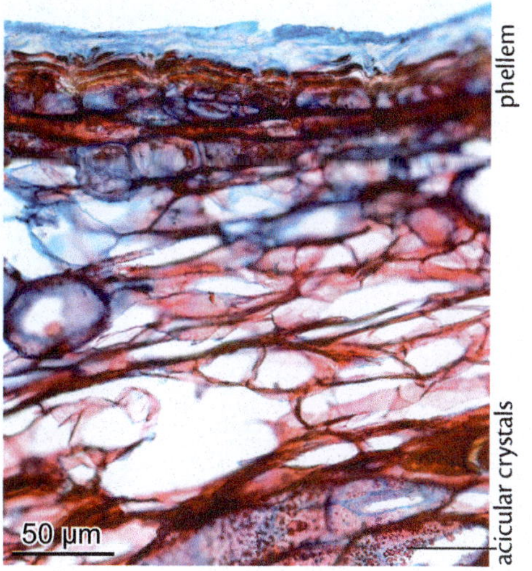

Tsuga canadensis (L.) Carrière — Pinaceae

eastern hemlock, Kanadische Hemlocktanne, pruche de l'est, tsuga orientale, tsuga oriental

Form: Upright conifer, crown broad, lax, conical crown, branches horizontal, the top usually hanging over, to 50 m tall.

Features: Needles evergreen, thin, linear, in two flat rows, to 2 cm long. Flowers monoecious, both sexes a globular cone, to 8 mm long, male cone light brown, female cone reddish. Seed cone ovate, soft, hanging, to 3 cm long.

Habitat: Eastern North America from Nova Scotia and Ontario south to North Georgia and west to Minnesota and North Alabama.

On moist, deep, loamy soils, preferring shadowy slopes, to 1700 m.

Bark: Young smooth, silvery gray brown to red brown, later cracking into small, square scales.

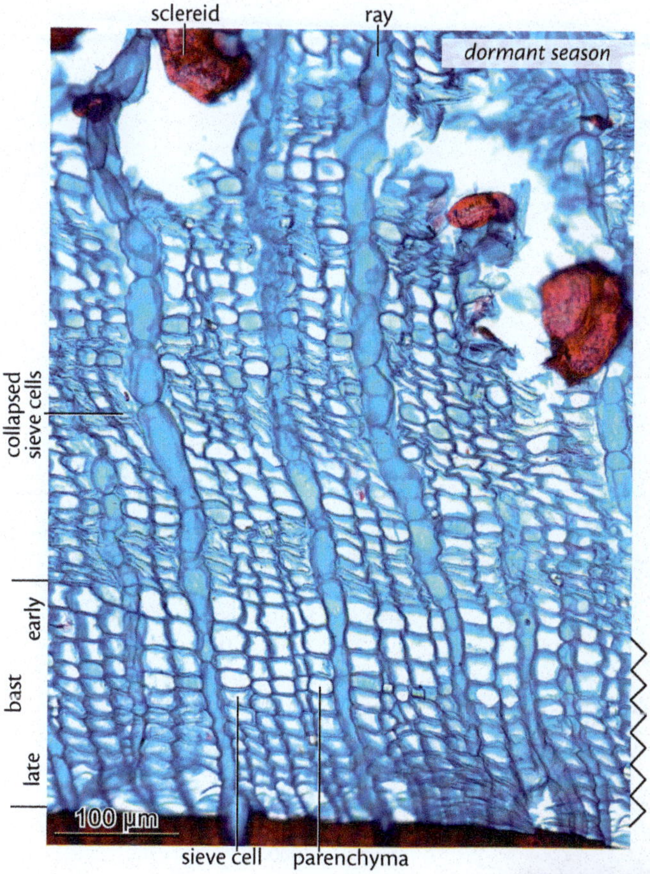

Bark anatomy: Regular tangential layers in young phloem, adult phloem with many irregular shaped groups of sclereids.

Sieve cells rectangular, collapsing within a few years. Parenchyma cells small and oval in the juvenile and enlarged in the adult phloem. Fibers absent. Groups of sclereids frequent in the adult phloem. Rays radially wavy in the older phloem. Prismatic crystals outside of sclereids. Resin ducts absent.

Annual rings distinct for a few years near the cambium. Rings indistinct in the sclerotic part of the bark. Periderm small. Phellem cells rectangular, thin-walled.

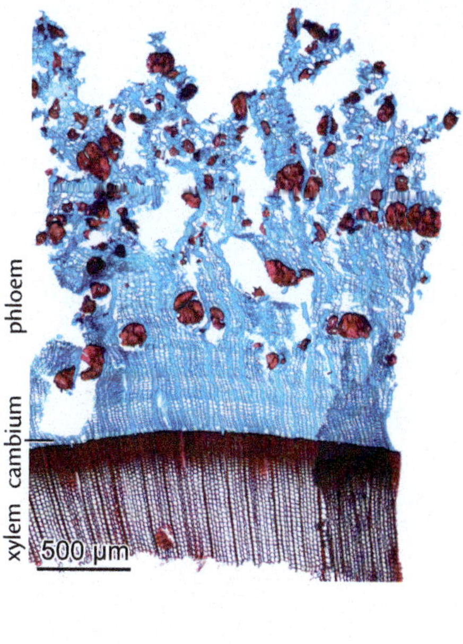

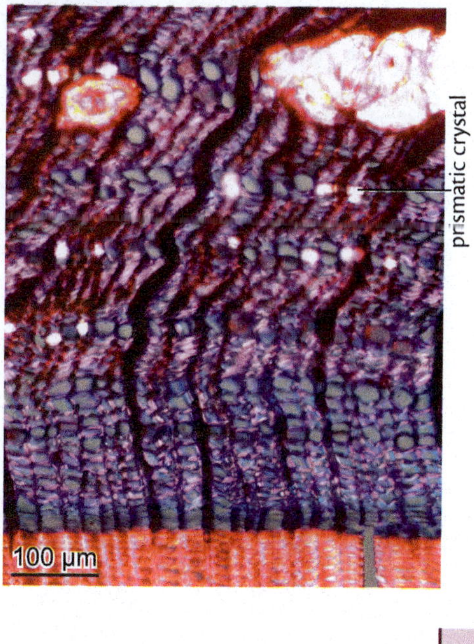

Acer rubrum L. Sapindaceae

red maple, Rotahorn, érable rouge, acero rosso, arce rojo

Form: Upright tree, crown pyramidal to round, flattened, to 40 m tall.

Features: Leaves deciduous, opposite, three- or five-lobed, lobes ending in short tips, to 10 cm long, with deep red autumn color. Flowers dark red, in side-standing panicles. Fruits splitting, with two wings at right angles.

Habitat: Eastern North America from Nova Scotia west to Manitoba and Texas, south to Louisiana and Florida.

On deep, moist soils, to 1800 m.

Bark: Young smooth, light red brown or gray brown, later silver gray, with shedding plates.

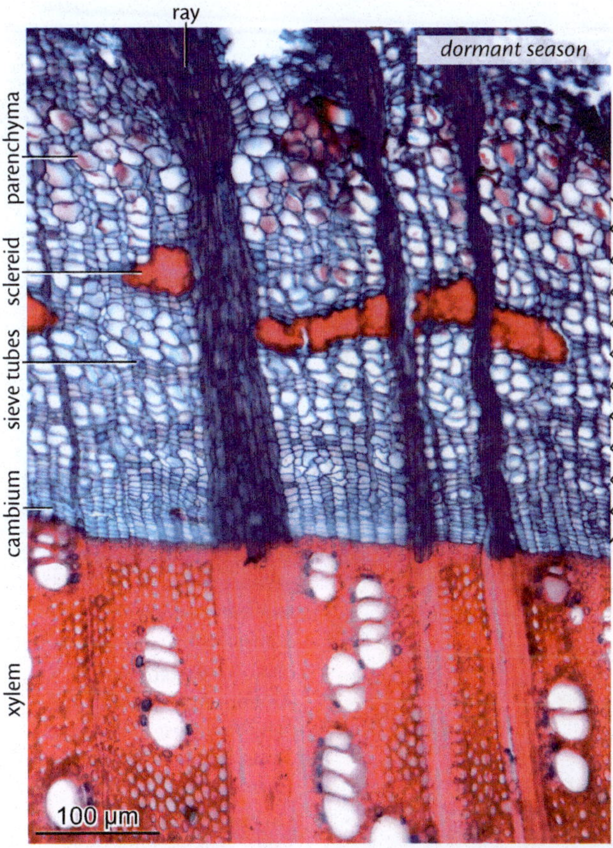

Bark anatomy: Regular tangential layers consisting of bands of sclereids, sieve tubes and parenchyma zones.

Early phloem formed by irregular shaped sieve tubes and roundish parenchyma zones. A few thin- to thick-walled fibers between large bands of very thick-walled sclereids. Rays dilating after a few years. Prismatic crystals in parenchymatic cells and sclereids.

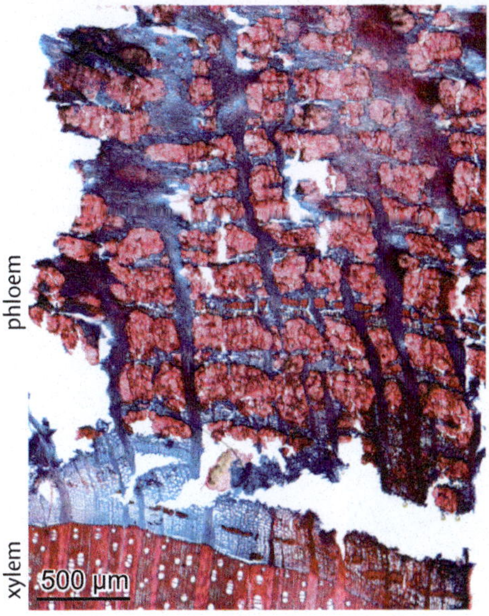

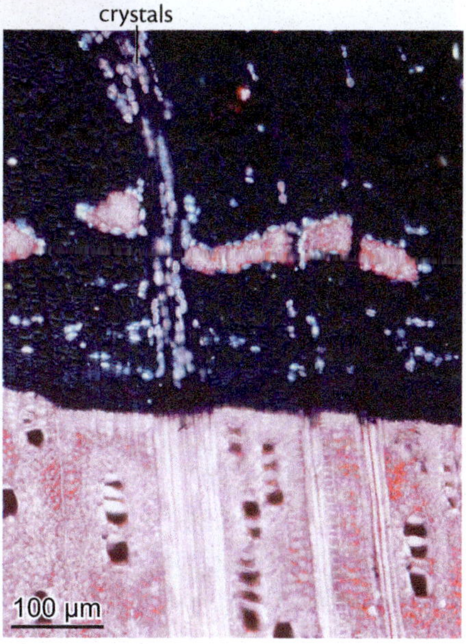

Asimina triloba (L.) Dunal

Annonaceae

paw paw, Dreilappige Papau, asiminier trilobé, asimina, pawpaw

Form: Upright tree, crown broad to round, to 12 m tall.

Features: Leaves deciduous, alternate, long ovate, margin entire, to 25 cm long. Flowers dark brown red, with six petals, nodding, to 6 cm long. Fruit a banana-like, smooth, yellow green berry, with black seeds embedded in white, edible flesh, to 15 cm long.

Habitat: Eastern North America from Ontario and New York west to Iowa and Texas, south to Louisiana and Florida.

On deep, moist, slightly acidic soils, to 1200 m.

Bark: Young smooth, light gray, later brown gray, with small plates.

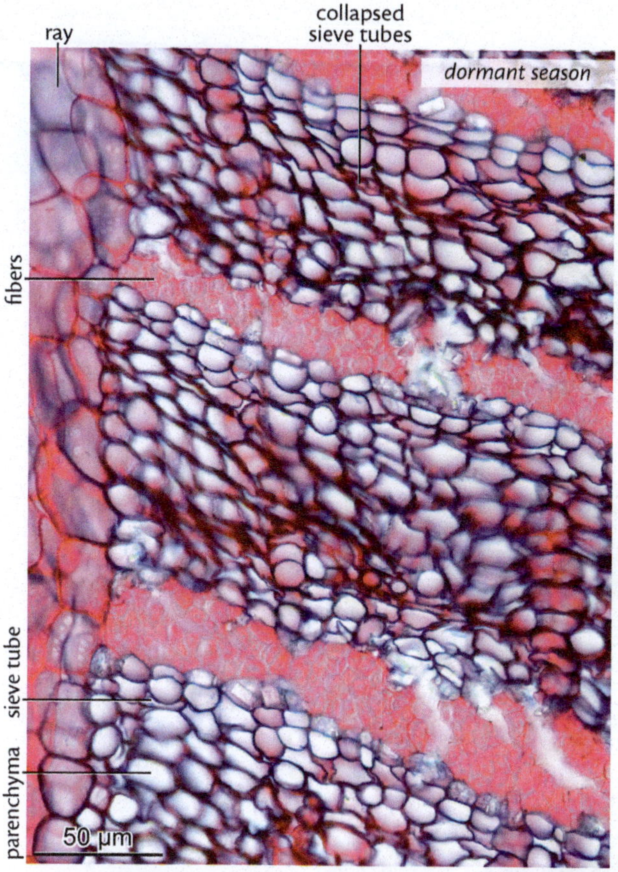

Bark anatomy: Regular tangential layers consisting of bands of fibers and zones of sieve tubes and parenchyma.

Early phloem formed by irregular shaped sieve tubes and roundish parenchyma zones. Fibers in 2–4-layered tangential bands. Rays dilating. Prismatic crystals around the fiber bands.

Annual ring boundaries usually corresponding with fiber bands.

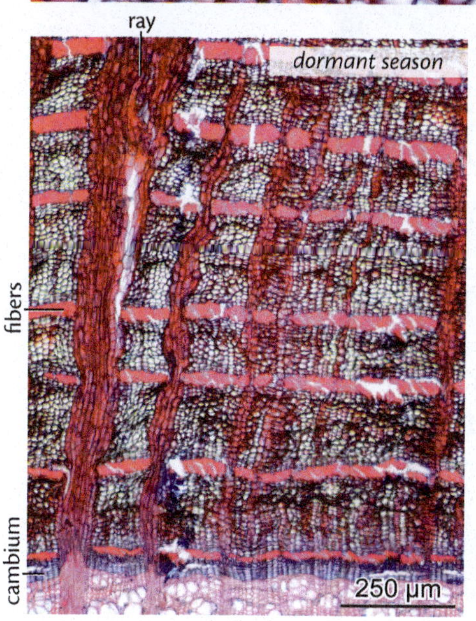

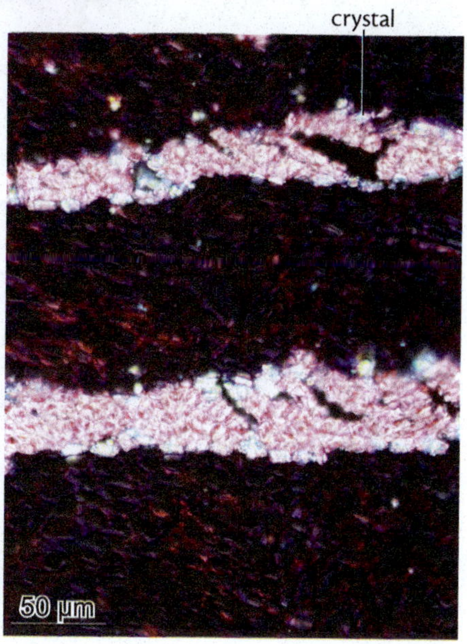

Calycanthus floridus L. Calycanthaceae

sweetshrub, Gewürzstrauch, arbuste aux anémones, calicanto, calicanto

Form: Shrub, twigs upright, to 3 m tall.

Features: Leaves deciduous, opposite, long ovate, margin entire, to 12 cm long, with spicy aromatic smell when crushed. Flowers dark brown red, cider scented, with 10–20 petals, upright, to 5 cm long. Fruit a fig-like, leathery brown, upright aggregate fruit, to 7 cm long.

Habitat: Eastern North America from New York west to Missouri and Texas, south to Louisiana and Florida.

On moist soils, to 1500 m.

Bark: Young gray brown to cinnamon brown, with longitudinal lenticels, later brown gray, spicily scented.

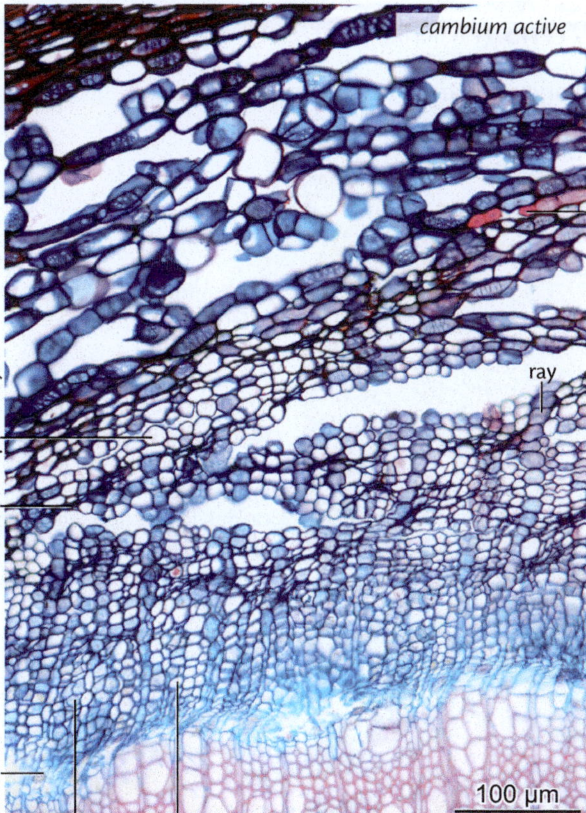

Bark anatomy: Irregularly arranged structure of sieve tubes and parenchyma with a small layer of fibers and sclereids in the cortex.

Sieve tubes angular, irregular shaped, parenchyma cells oval, forming a homogeneous tissue. Small bands of fibers, interrupted by sclereids, just outside of the phloem. Rays wavy, not dilating, disappearing soon. Crystals almost absent.

Phellem consisting of a few layers of rectangular, dark-stained cells.

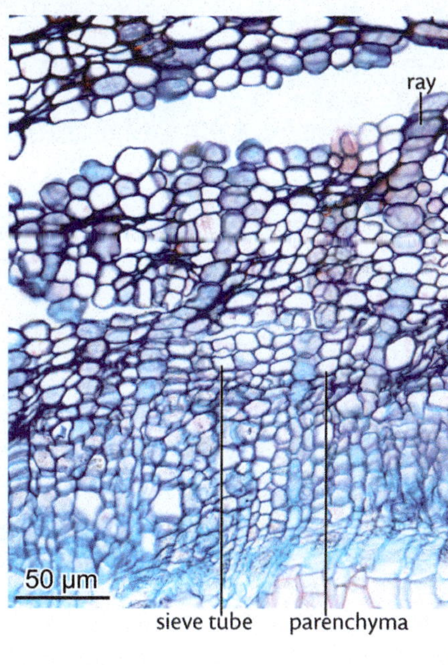

Carya myristiciformis (F.Michx.) Nutt. ex Elliott Juglandaceae

nutmeg hickory, Muskat-Hickorynuss, caryer, carya, carya

Form: Upright tree, with a pyramidal crown, to 35 m tall.

Features: Leaves deciduous, alternate, pinnate, with 5–11 toothed leaflets, to 40 cm long. Flowers monoecious, male in greenish, hanging catkins, female in greenish, upright spikes, to 11 cm long. Fruit an ellipsoid nut, smooth, with four seams, to 4 cm long.

Habitat: Southeastern North America from South Carolina west to Arkansas, south to Texas, Nuevo Léon and Tamaulipas.

On fresh to moist soils, often on limestone-rich riverine soils, to 1800 m.

Bark: First smooth and gray brown, later shedding in long, vertically strips and plates, curling on both ends.

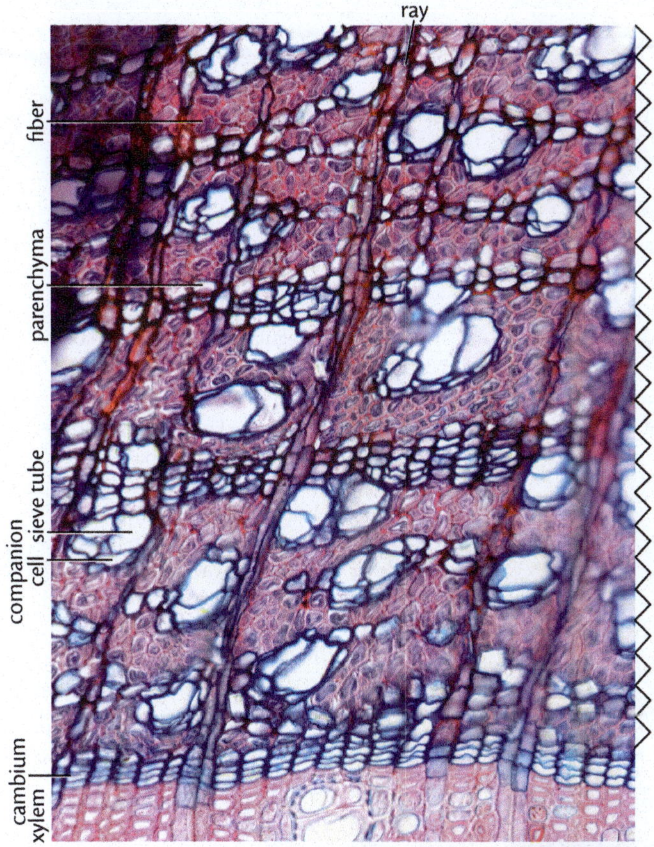

Bark anatomy: Dense structure with distinct ray dilatations and tangential layers of compressed sieve tubes and parenchyma cells.

Sieve tubes large, irregular shaped, with distinct companion cells, collapsing very soon after formation, forming dense layers between the large rays. Sieve tubes and layers of rectangular parenchyma cells embedded in tangential bands of thick-walled fibers. Sclereids absent. With a few distinct dilatations. Many prismatic crystals in the living phloem. Crystal druses of different size in the dead phloem.

Periderm consisting of many layers of rectangular cells.

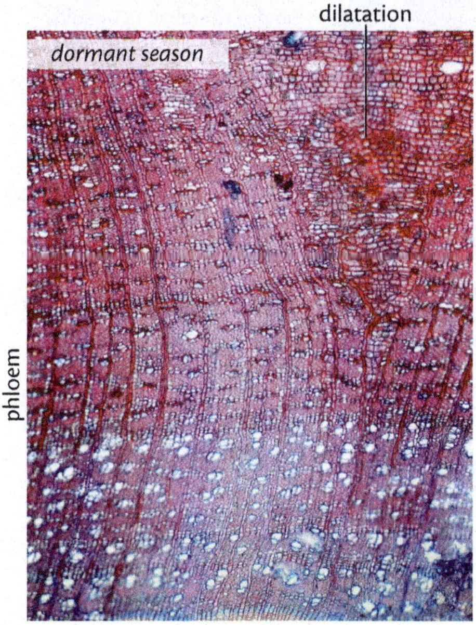

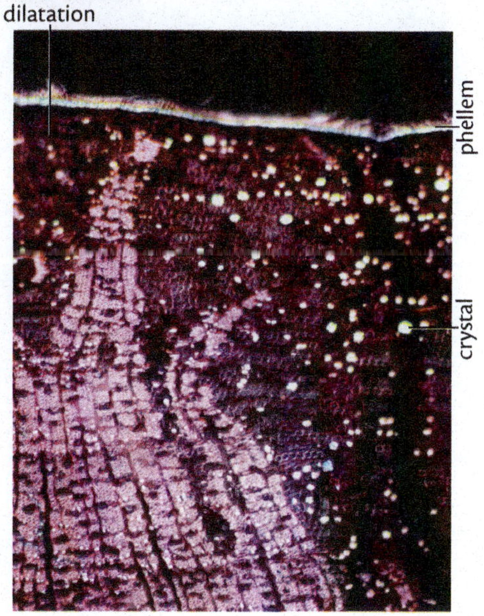

Gleditsia triacanthos L. Fabaceae

honey locust, Amerikanische Gleditschie, févier d'Amérique, spino di Giuda, gleditsia de tres espinas

Form: Upright tree, crown irregular shaped, broad, lax, pyramidal, twigs thorny, to 45 m tall.

Features: Leaves deciduous, alternate, double pinnate with 14–32 slightly toothed leaflets, to 25 cm long. Flowers yellow green pea flowers, in racemes, to 13 cm long. Fruit a long, brown pod, curved and twisted longitudinally, to 50 cm long.

Habitat: Eastern Central North America from Pennsylvania and Ontario west to South Dakota and Texas, south to Louisiana and Alabama.

On dry to moist, light soils, to 1200 m.

Bark: Young gray brown, soon with clusters of long, ramified thorns. Later peeling and shedding in long, vertical strips, revealing the reddish inner bark.

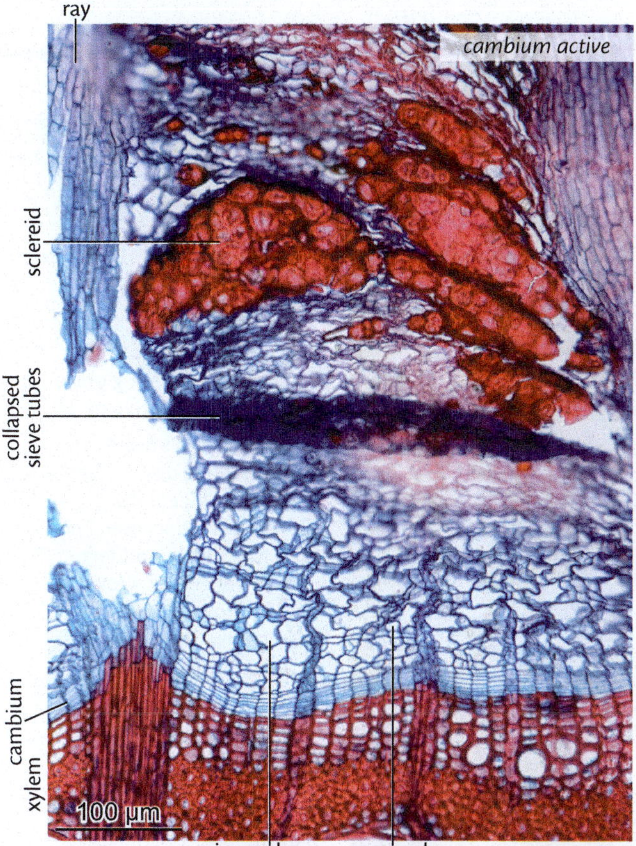

Bark anatomy: Tangentially layered structure.

Sieve tubes angular, companion cells small, parenchyma cells round, all distinctly visible in the phloem close to the cambium. Sieve tubes collapsing within a short time after formation. Fibers absent. With many very dense, irregular groups of sclereids. Rays large, straight, dilating after a few years. Many small crystal druses in unlignified parenchyma cells and large prismatic crystals in the sclereids.

Phellem formed by strictly radially arranged, rectangular, thin-walled cork cells.

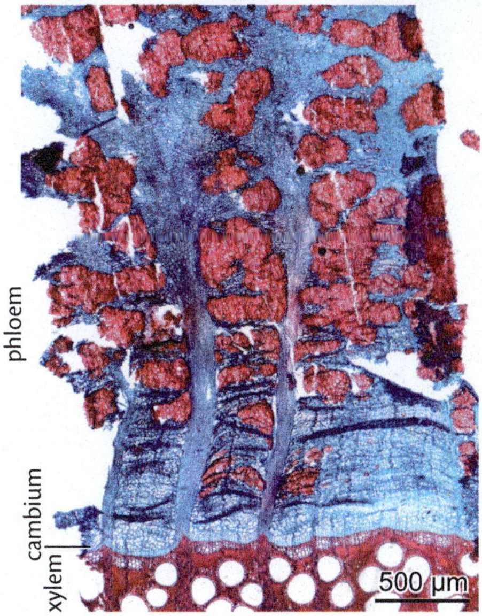

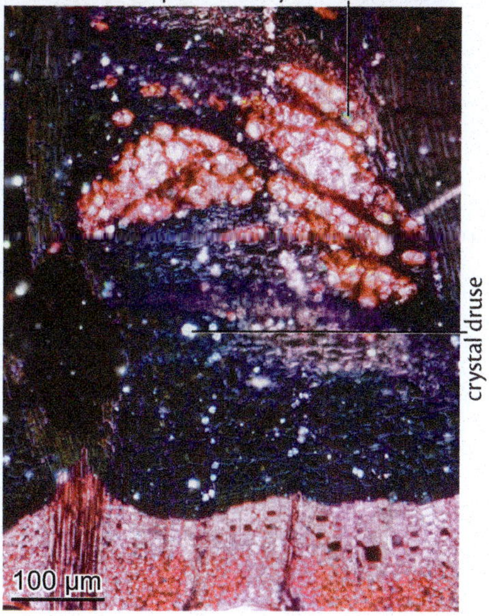

Halesia carolina L. Styracaceae

Carolina silverbell, Schneeglöckchenbaum, halésie, halesia, alesia

Form: Shrub or small tree, upright, crown semi-globular or funnel-shaped, to 6 m tall.

Features: Leaves deciduous, alternate, long ovate, margin toothed, to 12 cm long. Flower a white, scented bell with four petals, nodding, to 2 cm long. Fruit a four-ribbed, brown drupe, to 3.5 cm long.

Habitat: Southeastern North America from West Virginia west to Arkansas, south to Mississippi and Florida. On deep, moist and humus-rich, slightly acidic soils, to 1500 m.

Bark: Young gray, finely fissured, later brown gray, deeper fissured with darker ribs.

dormant season

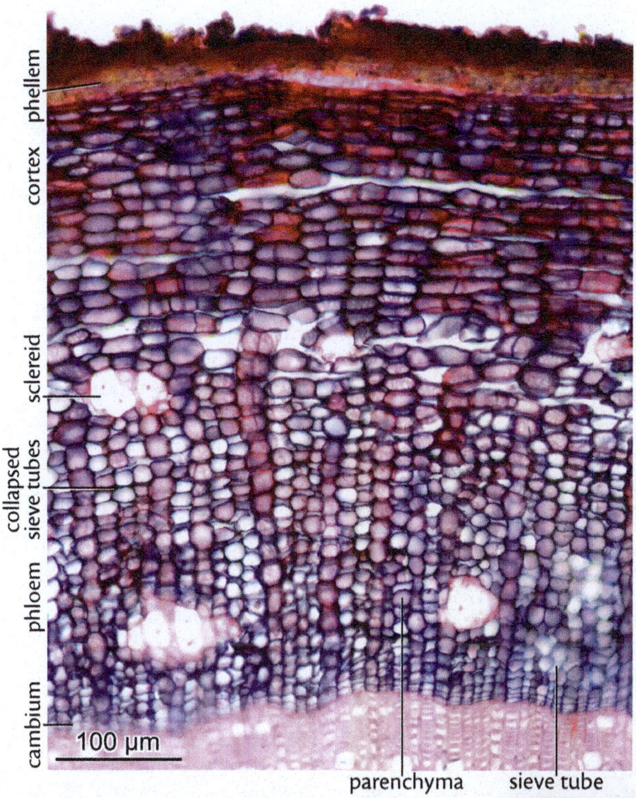

Bark anatomy: Phloem poorly structured. Sieve tubes angular, collapsing within a short time after formation. Parenchyma cells round. Solitary fibers absent. Sclereids in small groups. Rays straight and slightly dilating. Prismatic crystals in axial parenchyma cells.

Phellem thin, consisting of layers of thin-walled, rectangular cells and square cells.

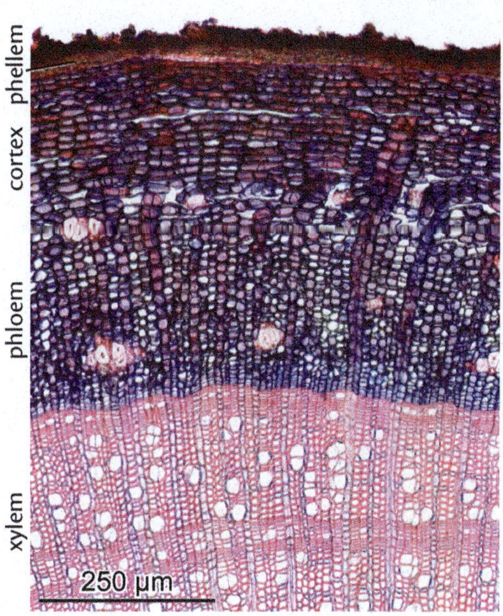

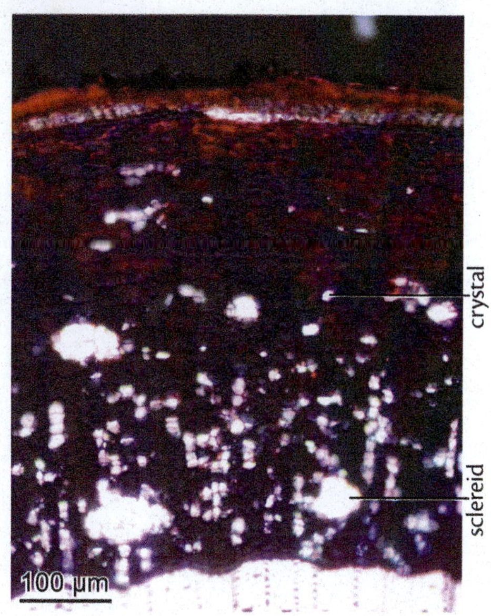

Hamamelis virginiana L. Hamamelidaceae

American witch-hazel, Virginische Zaubernuss, noisetier des sorcières, amamelide americana, avellano de bruja americana

Form: Shrub, broad, funnel-shaped, to 7 m tall.

Features: Leaves deciduous, alternate, broad ovate, margin irregularly toothed, to 15 cm long. Flowers with four long, thin, ribbon-like, deep yellow, scented petals, to 3 cm long, emerging in autumn. Fruit an ovate, woody capsule, to 3 cm long.

Habitat: Eastern North America from Nova Scotia and Quebec west to Minnesota and Texas, south to Florida and Tamaulipas.

On deep, moist and humus-rich, slightly acidic soils, to 1600 m.

Bark: Young dull, gray brown, later light gray, sometimes with fine, irregular fissures.

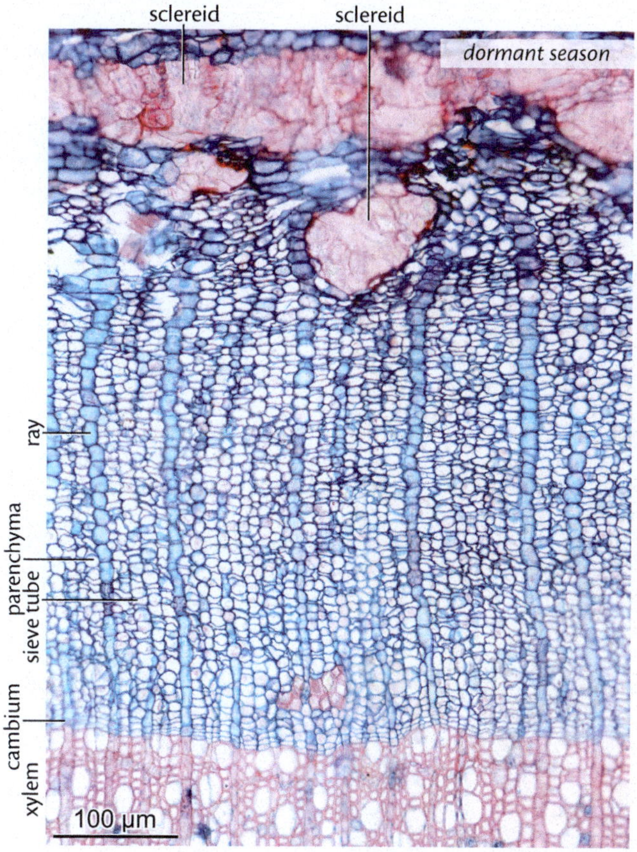

Bark anatomy: Uniform structure of sieve tubes and parenchyma.

Sieve tubes irregular shaped, parenchyma cells round to oval. Large groups of sclereids mainly at the outside of the phloem. Rays straight. Many irregular shaped crystals in axial parenchyma cells.

Phellem consisting of layers of large, rectangular, thin-walled cells and small, thick-walled cells.

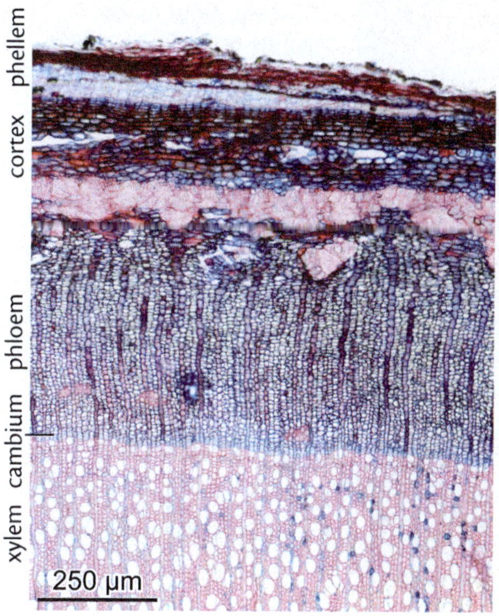

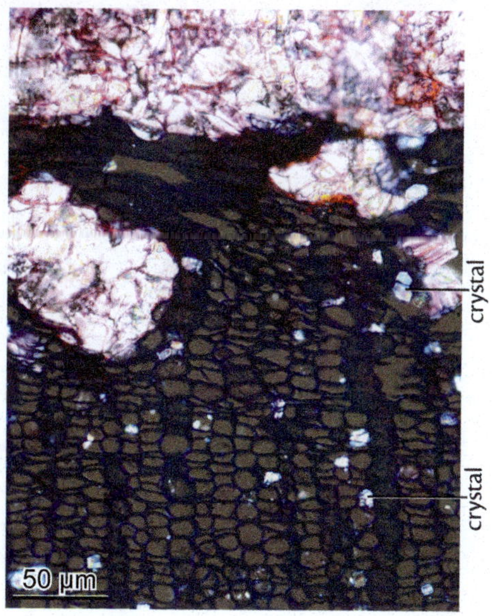

Liquidambar styraciflua L. Hamamelidaceae

sweetgum, Amerikanischer Amberbaum, copalme d'Amérique, storace americano, ocozote

Form: Upright, straight tree, crown small, pyramidal at first, later ovate, to 45 m tall.

Features: Leaves deciduous, alternate, maple-like lobed with five to seven pointy lobes, margin finely toothed, to 18 cm long, with deep red autumn color. Flowers monoecious, both sexes reddish green, globular, without petals, male upright, female hanging, to 3 cm wide. Fruit a globular, hooked capsule, to 3 cm wide, hanging on a long stalk.

Habitat: Eastern North America from New York west to Oklahoma and Texas, south to Florida, eastern Mexico and Nicaragua.

On deep, moist to wet, humus-rich soils, to 2100 m.

Bark: Young dull, gray brown, with corky lenticels, later deeply fissured with long, often corky ribs, revealing the cinnamon-colored inner bark.

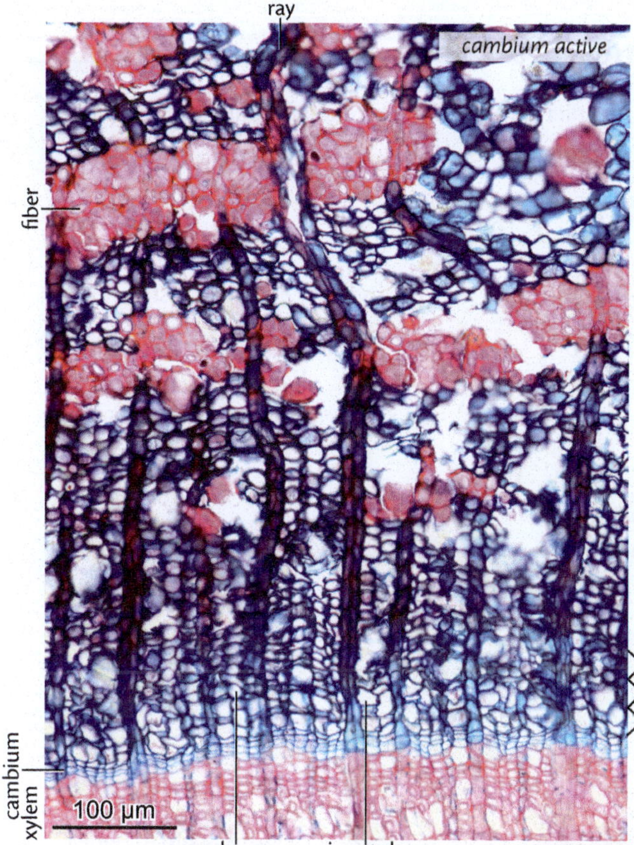

Bark anatomy: Regular tangential layers of fiber bands, sieve tubes, and parenchyma zones.

Early phloem formed by irregular shaped sieve cells, roundish parenchyma cells and small zones of fibers. Fibers and sclerenchyma in 2–4-seriate, tangential, laterally interrupted bands. Rays very distinct. With crystal druses.

Annual ring boundaries usually corresponding with fiber bands. Phellem layered.

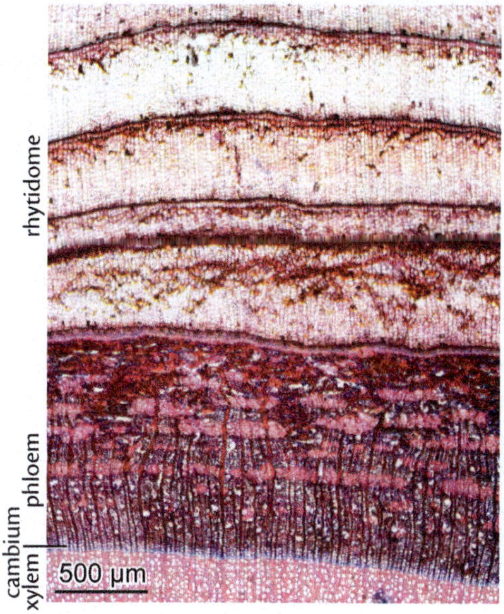

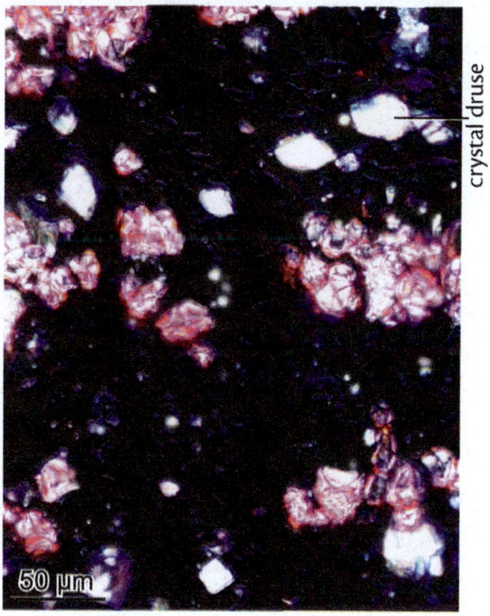

Liriodendron tulipifera L. Magnoliaceae

yellow-poplar, Tulpenbaum, tulipier de Virginie, albero dei tulipani, tulípero de Virginia

Form: Upright, straight tree, crown pyramidal, to 50 m tall.

Features: Leaves deciduous, alternate, nearly quadrangular, lobed with four to six pointed lobes, margin entire, to 15 cm long, with deep red autumn color. Flowers showy, light yellow with an orange yellow center, scented, with six upright, tulip-like petals, to 6 cm long. Fruit a cone-like, cylindrical, upright aggregate fruit, to 8 cm long.

Habitat: Eastern North America from Vermont and Ontario west to Missouri and Louisiana, south to Florida.

On deep, moist, humus-rich soils, to 1400 m.

Bark: Young dull, gray brown, slightly fissured, later deeply fissured, often in a regular, net-shaped pattern.

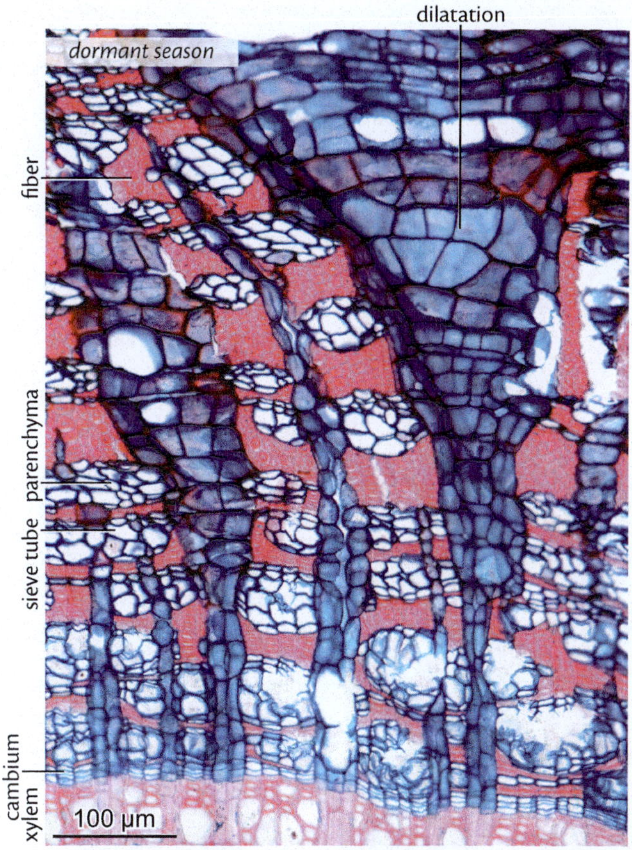

Bark anatomy: Regular tangential layers of fiber bands, sclerenchyma, sieve tubes and parenchyma zones.

Early phloem formed by irregular shaped sieve tubes and roundish parenchyma cells. Fibers and sclerenchyma in 2–4-seriate, tangential, laterally interrupted bands. Prismatic crystals large, mainly between fiber-sclerenchyma bands. Crystals absent in rays.

Annual ring boundaries usually corresponding with fiber bands.

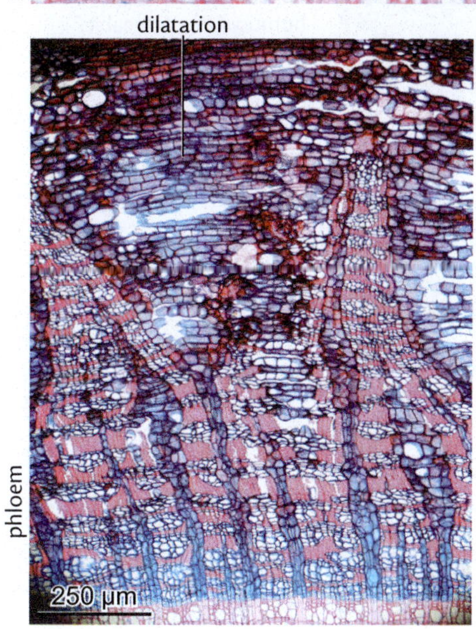

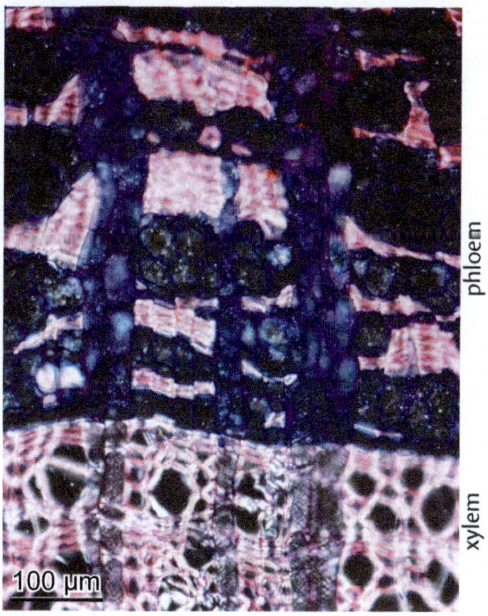

Magnolia grandiflora L. Magnoliaceae

southern magnolia, Immergrüne Magnolie, magnolia à grandes fleurs, magnolia a fiori grandi, magnolia común

Form: Upright tree, crown regular shaped, pyramidal, to 25 m tall.

Features: Leaves evergreen, alternate, long ovate, glossy leathery with a rusty underside, margin entire, to 20 cm long. Flowers showy, white, sweetly scented, upright, shell-like, with 9–12 tepals, to 20 cm wide. Fruit an ovate, felty brown, upright aggregate fruit, with red arilled seeds, to 10 cm long.

Habitat: Southeastern North America from Georgia and Florida west to Mississippi and Texas.

On deep, moist to wet, humus-rich, permeable, acidic soils, to 150 m.

Bark: Young dull, gray brown, later scaly shedding, revealing the cinnamon red inner bark, with many lenticels.

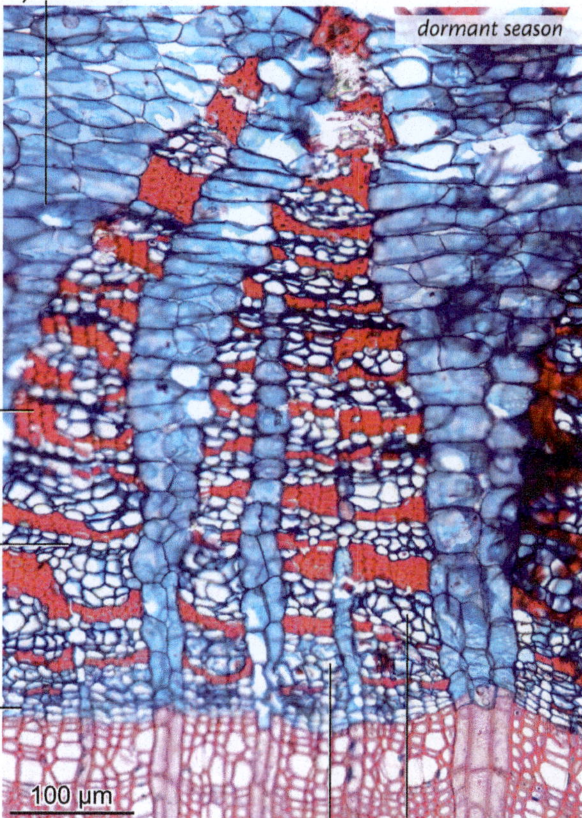

Bark anatomy: Tangentially layered structure of sieve tubes and parenchyma cells.

Sieve tubes collapsing within a short time after formation. Fibers rectangular, in distinct tangential bands. Sclereids mostly absent in the phloem, frequent in the cortex. Rays distinctly dilating. Many prismatic crystals, mainly in axial parenchyma cells.

Phellem cells rectangular, mostly filled with dark-staining substances.

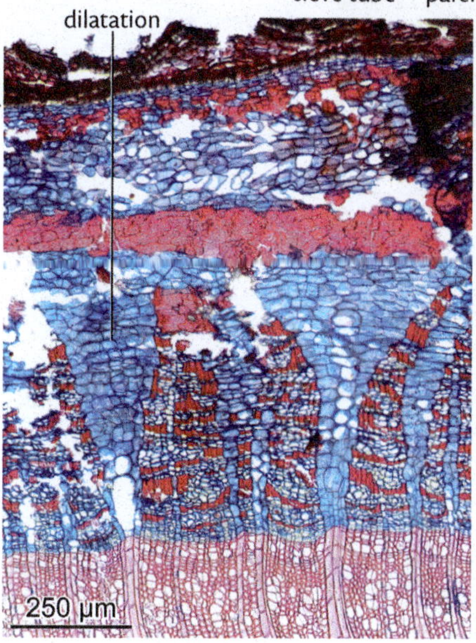

Nyssa sylvatica Marshall

Cornaceae

black tupelo, Wald-Tupelobaum, tupélo noir, tupelo nero, tupelo negro

Form: Upright tree, crown regular shaped, pyramidal, with age more rounded, to 30 m tall.

Features: Leaves evergreen, alternate, long ovate, glossy green with a bluish underside, margin entire, to 15 cm long. Flowers dioecious, inconspicuous and greenish white in both sexes, to 4 mm across. Fruit an ovate, fleshy, blue black, one-seeded drupe, to 1 cm wide.

Habitat: Eastern North America from Maine and Ontario west to Texas, south to Florida and Mississippi, isolated in Veracruz and Chiapas.

On deep, moist, slightly acidic soils, to 1700 m.

Bark: Young dull, gray brown, later finely fissured.

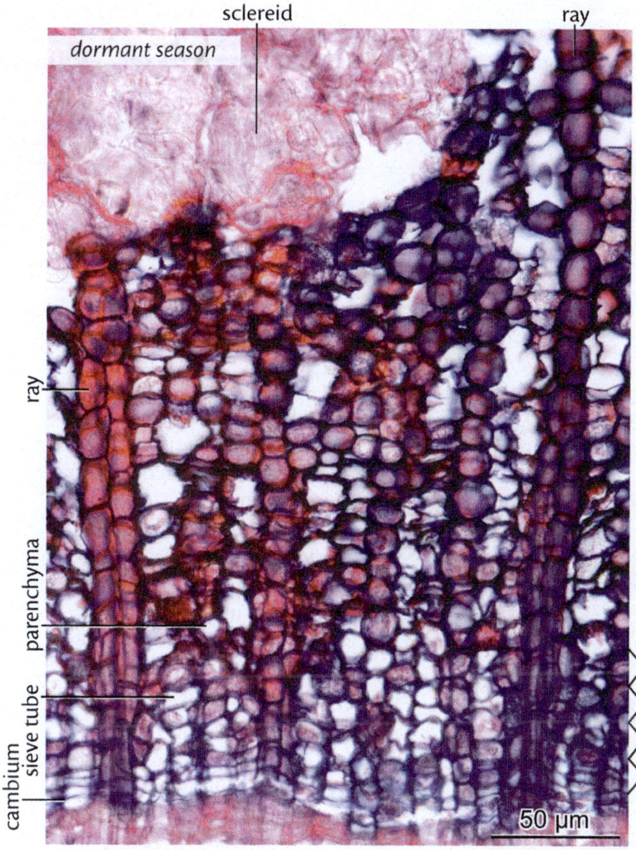

Bark anatomy: Irregularly structured tissue of sieve tubes and parenchyma with some round groups of sclereids.

Sieve tubes angular, parenchyma cells round to oval. Sclerenchyma in few, small groups in the phloem. Dilatations absent. Prismatic crystals numerous, small, rectangular, in axial parenchyma cells and rays.

Periderm consisting of many square, unlignified, rectangular cells.

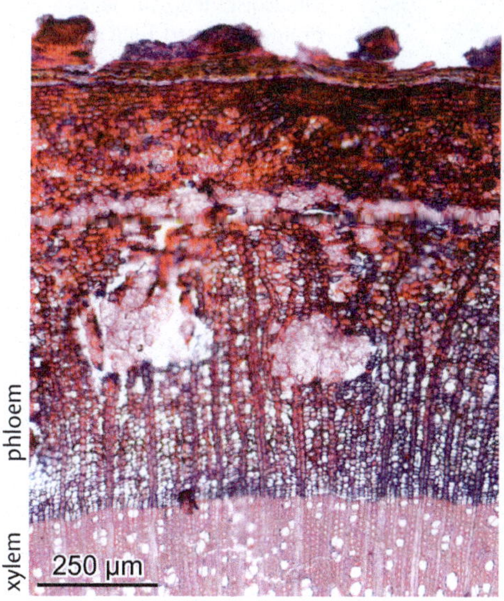

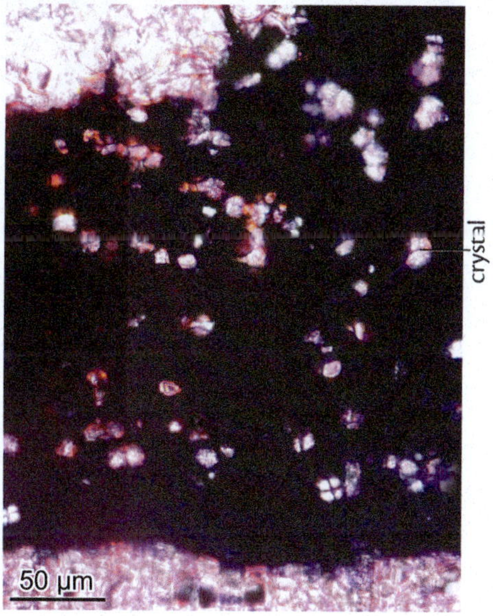

Quercus rubra L.

Fagaceae

northern red oak, Roteiche, chêne rouge d'Amérique, quercia rossa, roble rojo americano

Form: Upright tree, crown large, funnel-shaped, to 50 m tall.

Features: Leaves deciduous, alternate, sessile, acutely lobed with four to six toothed and aristate lobes, to 20 cm long, with deep scarlet autumn colors. Flowers monoecious, female flower very small, reddish, grouped in clusters, hanging, male flowers in hanging, green yellow catkins, to 5 cm long. Fruit an ovate nut, to 3 cm long, sitting in a scaly cup.

Habitat: Eastern North America from Nova Scotia and Quebec west to Nebraska and Oklahoma, south to Arkansas and Alabama.

On moist, deep and well-drained soils, to 1600 m.

Bark: Young smooth, brown gray, with age longitudinally fissured.

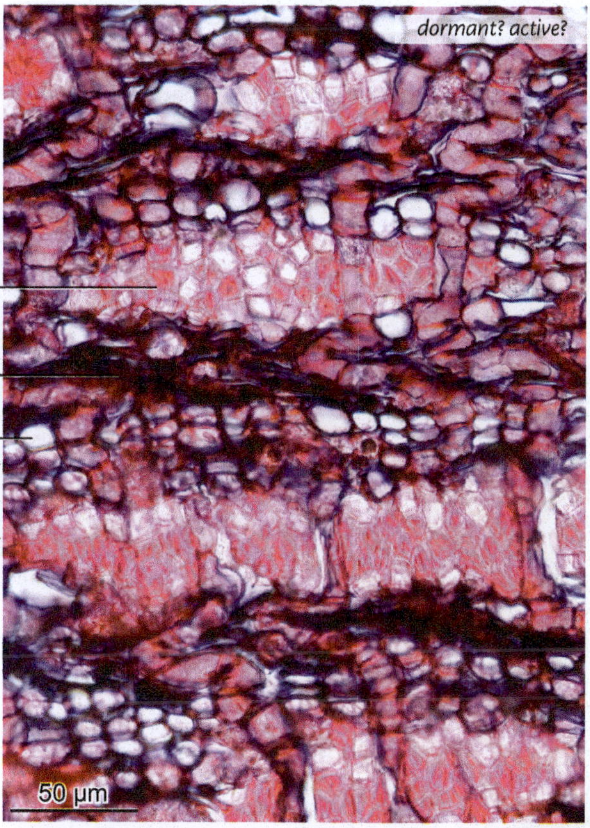

Bark anatomy: Tangentially layered structure. Sieve tubes angular, parenchyma cells round, radially arranged. Sieve tubes collapsing within a short time after formation. Bands of sclereids alternating with zones of sieve tubes and parenchyma. Fibers absent. Sclereids in round groups, dominating large rays. Large rays straight, uniseriate rays wavy. Prismatic crystals within groups of sclereids and surrounding them. Crystal druses in parenchymatic zones and the cambium zone. Many small crystal druses in axial parenchyma cells.

Phellem formed by layers of rectangular, thin-walled cork cells.

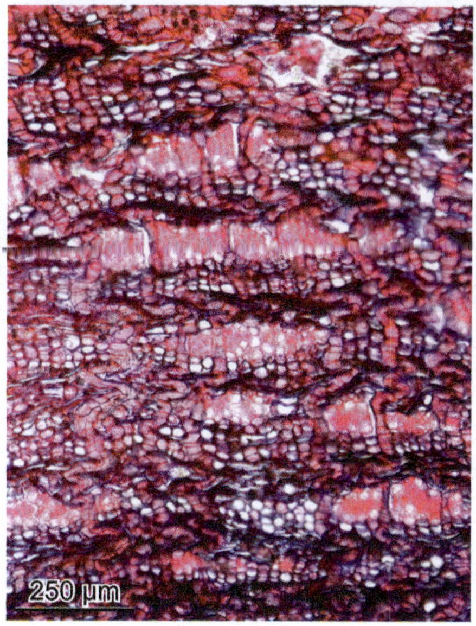

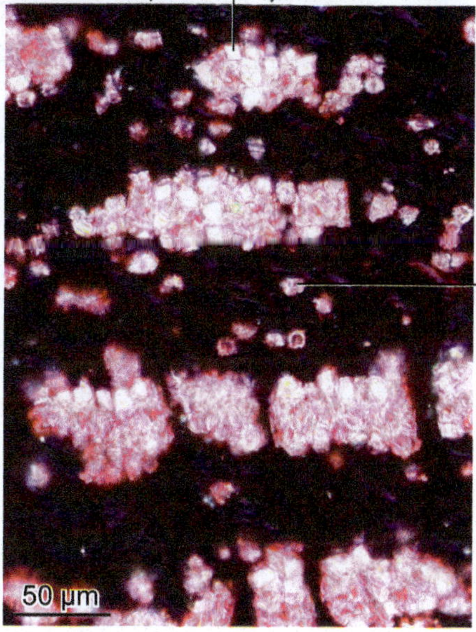

Rhus typhina L. Anacardiaceae

staghorn sumac, Essigbaum, vinaigrier, sommaco americano, zumaque de Virginia

Form: Shrub or small tree, scarcely branched, crown umbellate, to 10 m tall. Often forming dense colonies by root-sprouting.

Features: Leaves deciduous, alternate, pinnate with 11–31 toothed leaflets, to 80 cm long, with deep orange red autumn colors. Flowers monoecious or dioecious, male flowers in greenish, loose panicles, female flowers in dense, hairy, creamy white, upright spikes, to 15 cm long. Fruit a globular, scarlet drupe, to 5 mm wide, in dense, pyramidal, hirsute panicle.

Habitat: Northeastern North America from Nova Scotia and Ontario west to Minnesota, south to Tennessee and South Carolina; widely planted and often invasive.

On dry to moist soils, often as a pioneer, to 1200 m.

Bark: Young smooth, gray brown, with lenticels, later shedding in irregular shaped scales and diamonds, often revealing the orange brown inner bark.

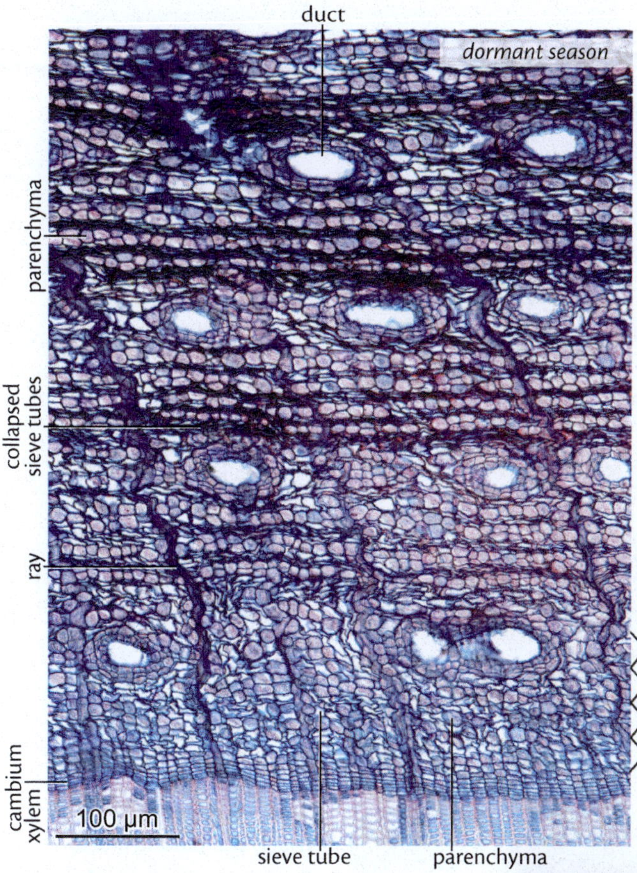

Bark anatomy: Irregular, unlignified tissue with many oval groups of fibers.

Sieve tubes angular, parenchyma cells roundish. Sieve tubes collapsing within a short time after formation, forming small, tangential bands. Fibers absent. Sclereids rare, occurring in groups towards the periderm. Rays straight, dilating. Ducts frequent, tangentially arranged. Prismatic crystals surrounding fiber groups. Few large crystal druses occupying axial parenchyma in young phloem. Druses very frequent near the periderm.

Phellem layered, consisting of several rows of square cells and one row of tangential cells.

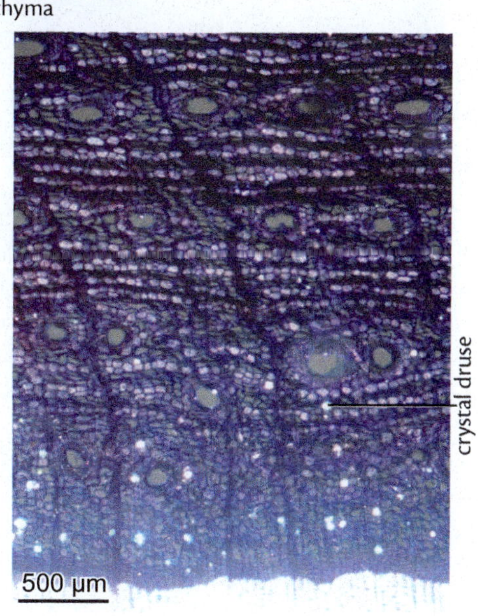

Robinia pseudoacacia L. Fabaceae

black locust, Robinie, robinier faux-acacia, robinia, falsa acacia

Form: Upright tree, crown irregular shaped, pyramidal, later umbellate, twigs thorny, to 25 m tall.

Features: Leaves deciduous, alternate, pinnate with 7–21 ovate leaflets, margin entire, to 30 cm long. Flowers white pea flowers, intensely scented, in hanging racemes, to 20 cm long. Fruit a brown, long ovate, smooth pod, to 10 cm long.

Habitat: Eastern North America in two disjunct areas: more eastern from Pennsylvania to Georgia, and more western from Missouri to Arkansas and Oklahoma. Widely planted and invasive.

On dry to moist, light, often sandy soils, to 1600 m.

Bark: Young dull, gray brown, with age cracking into rough, long, vertical strips, often shedding.

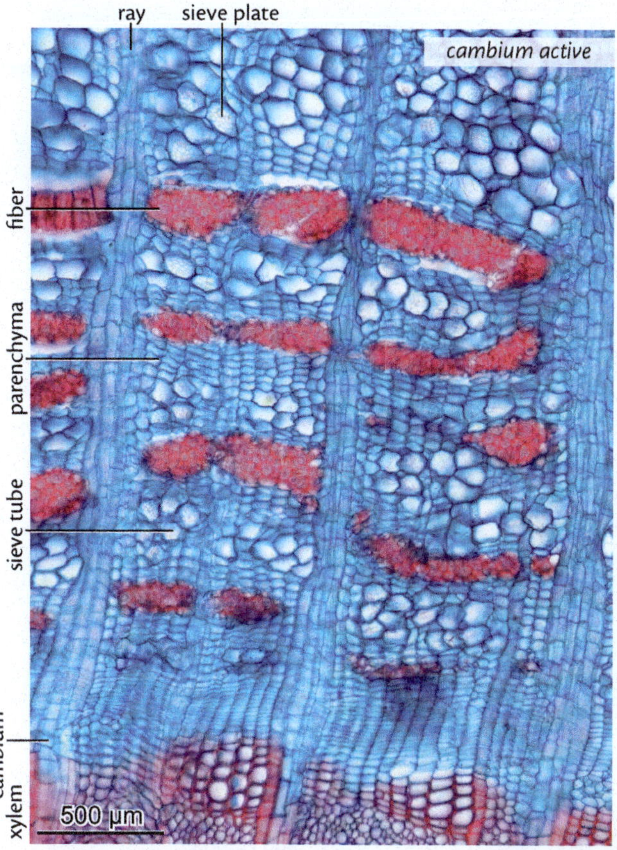

Bark anatomy: Tangentially structured phloem and periderm with groups of sclerenchyma.

Sieve tubes large, arranged tangentially in the early bast. In the adjacent tissue, parenchyma alternating in pure layers with bands of sieve tubes and tangentially enlarged fiber groups. Dilatations absent. Isodiametric crystals mainly located around sclerenchymatic groups.

Phellem consisting of many layers of large, square cells and small, rectangular cells.

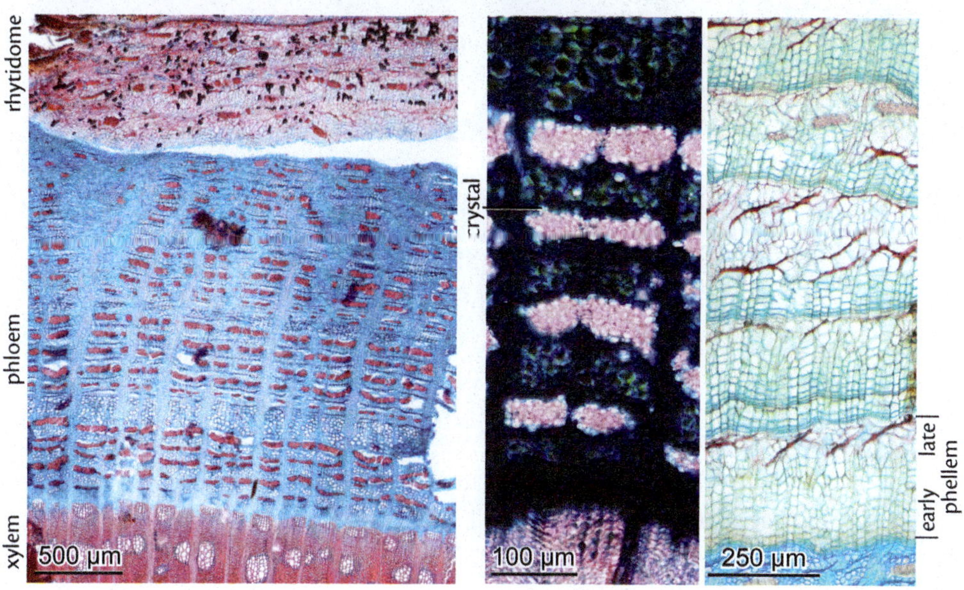

Sassafras albidum (Nutt.) Nees — Lauraceae

sassafras, Fenchelholzbaum, sassafras, sassofrasso, sassafras

Form: Upright tree, crown regular shaped, pyramidal, to 20 m tall.

Features: Leaves deciduous, alternate, long ovate or one- to three-lobed, margin entire, to 16 cm long, very aromatic, with deep orange red autumn colors. Flowers dioecious or monoecious, small, yellow green, five-petaled stars in both sexes, to 5 mm wide. Fruit an ovate, fleshy, blue black, one-seeded drupe, to 1.5 cm wide.

Habitat: Eastern North America from Maine and Ontario west to Missouri and Oklahoma, south to Texas and Florida.

On deep, moist, well-drained, acidic soils, to 1500 m.

Bark: Young smooth, olive green to brown, with scattered lenticels, with age gray brown, deeply fissured and roughly furrowed.

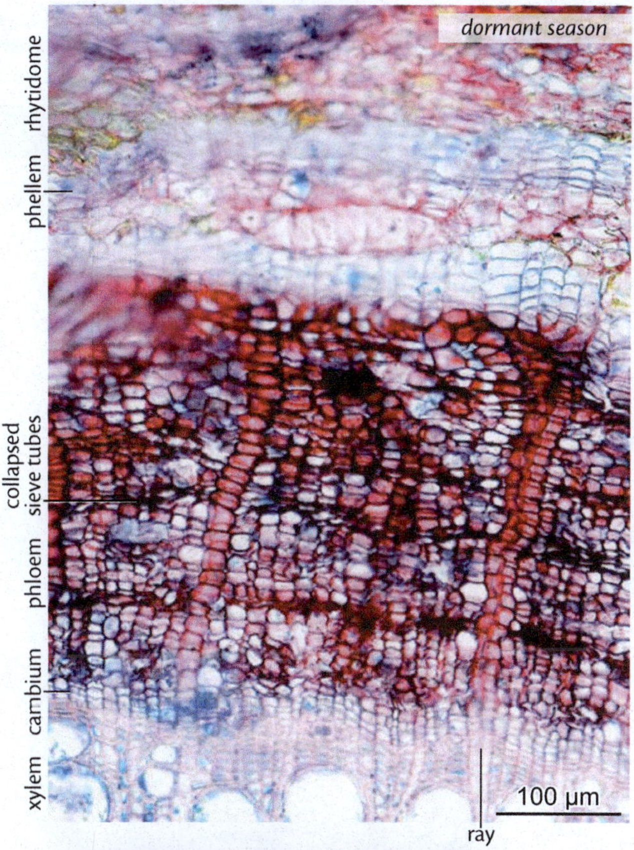

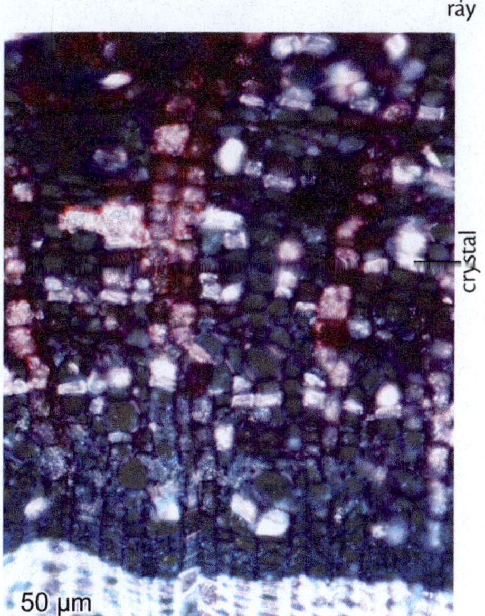

Bark anatomy: Phloem layered. Sieve tubes angular, parenchyma cells round. Sieve tubes collapsing within a short time after formation, indicated by dark tangential and radial lines. Fibers solitary, in uniseriate tangential lines. Sclereids in small groups at the outside of the phloem. Rays straight, slightly dilating. Prismatic crystals in axial parenchyma cells.

Phellem layered, cells thin-walled, rectangular and square.

Due to a marked continental or Mediterranean summer drought, the **coniferous forests of Pacific North America** are represented by a unique variety of conifers, including the tallest known trees, marked not only by many members of the pine family but also by outstanding members of the cypress family, such as the giant sequoia and coast redwood.

F. H. Schweingruber et al., *Bark Anatomy of Trees and Shrubs in the Temperate Northern Hemisphere*, https://doi.org/10.1007/978-3-030-14056-4_9

9 Coniferous forest of Pacific North America

Calocedrus decurrens (Torr.) Florin Cupressaceae

incense cedar, Kalifornische Flusszeder, calocèdre, cedro dell'incenso, cedro blanco de California

Form: Upright, straight conifer, crown conical to broad ovate, to 60 m tall.

Features: Leaves evergreen, scale-like in whorls of four, to 3 mm long, forming flattened sprays, aromatic. Flowers monoecious, yellowish, spherical, in both sexes inconspicuous, to 5 mm across. Seed cones long ovate, red brown, hanging, to 3.5 cm long.

Habitat: Western North America from Oregon south to northern Baja California, centered in the Sierra Nevada.

On moist, deep soils, to 2400 m.

Bark: Young scaly, red brown, soon deeply fissured, peeling in thin, vertical strips.

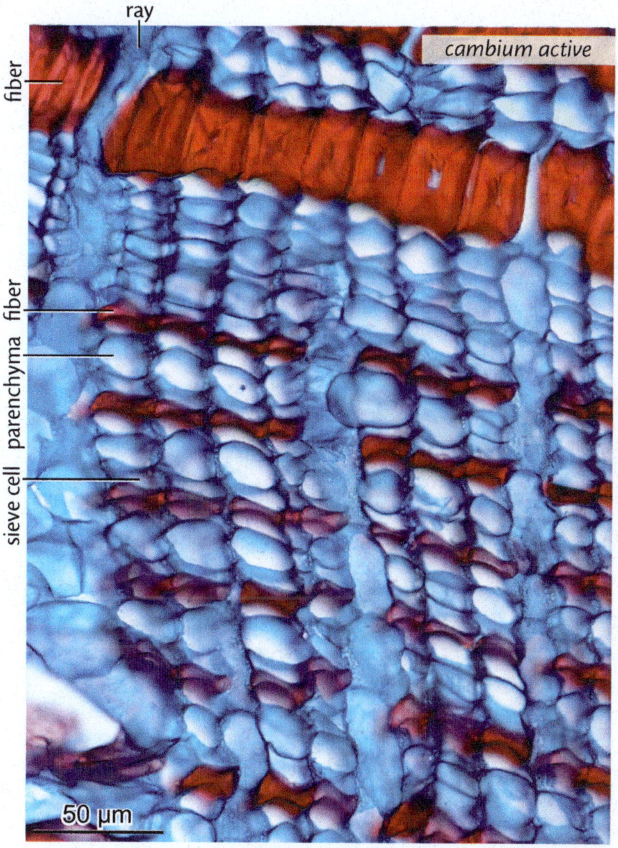

Bark anatomy: Regular tangential layers of sieve cells, fibers, and parenchyma cells.

Sieve cells thin-walled, deformed rectangular, parenchyma cells round to oval, fibers very thick-walled. Many bands of small fibers and a few bands of large, very thick-walled fibers. Sieve tubes tending to collapse. Rays remaining straight. Crystal sand in ray cells and radial intracellular layers.

Annual rings absent. Phellem cells thin-walled, irregularly shaped.

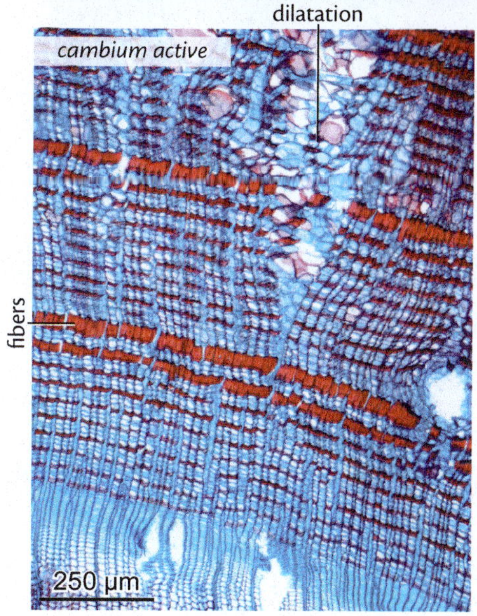

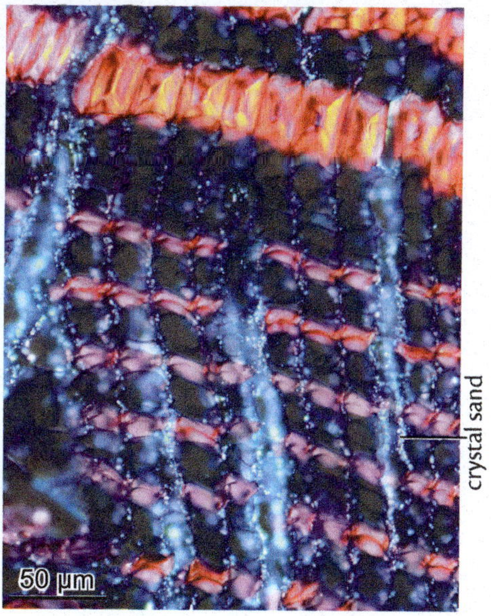

Pinus ponderosa Douglas ex C.Lawson Pinaceae

ponderosa pine, Westliche Gelbkiefer, pin ponderosa, pino giallo, pino ponderosa

Form: Upright, straight conifer, crown conical, with age turning umbellate, to 70 m tall.

Features: Needles evergreen, stiff, glossy dark green, in bundles of two or three, to 25 cm long. Flowers monoecious, both sexes globular, cone-like, to 8 mm long, male cone light brown, female cone pink to scarlet. Seed cones long, ovoid, sessile, to 15 cm long.

Habitat: Western North America from southern British Columbia south to California and Durango, east to Nebraska, West Texas and San Luis Potosí.

On dry to moist soils, a well adapted pioneer after forest fires, to 2700 m.

Bark: Young scaly, red brown or yellow orange, later deeply fissured, with darker furrows.

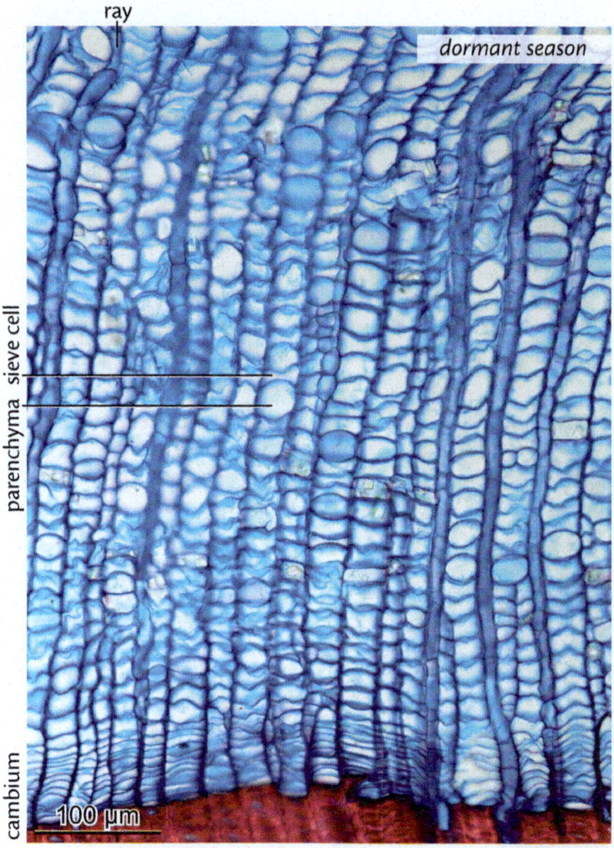

Bark anatomy: Regular tangential layers in the young phloem. Characteristic absence of fibers and isolated sclereids.

Sieve cells rectangular, arranged in tangential rows, collapsing within a few years. Parenchyma cells small and oval in the juvenile and enlarged in the adult phloem. Sclereids and fibers absent. Rays radially wavy. With elongated prismatic crystals outside of rays, mostly in axially elongated cells (crystal tubes).

Rings indistinct. Phellem irregularly layered, consisting of rectangular, thick-walled cells (sclereids) and thin-walled cells.

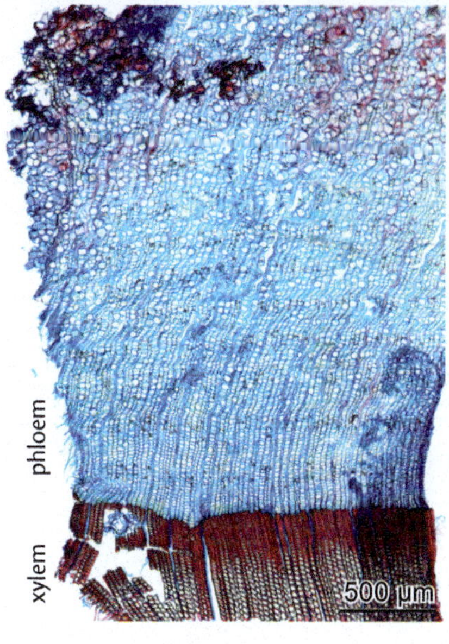

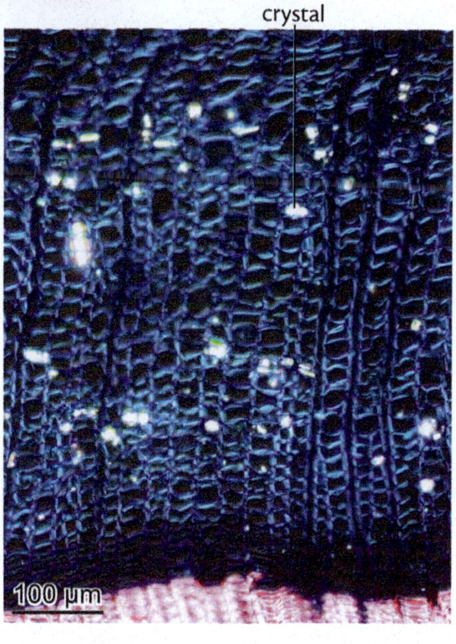

Pseudotsuga menziesii (Mirb.) Franco Pinaceae

Douglas fir, Douglasie, sapin de Douglas, abete di Douglas, abeto de Douglas

Variation: Two subspecies are recognized; subsp. *menziesii* (coastal) and subsp. *glauca* (inland).

Form: Upright, usually single-trunked conifer, crown conical, to 110 m tall.

Features: Needles flat, to 3 cm long, attached to the twig with a round disc. Flowers monoecious, female cone upright, dark red, 2–4 cm long, male cone hanging, yellow red, 2–5 cm long. Seed cones hanging, 6–10 cm long, with a long, three-pointed bract protruding prominently above each scale.

Habitat: Pacific North America from British Columbia to Northern Mexico.

Mostly on deep soils in humid temperate rainforest close to the coast to 1800 m; also on drier inland mountains to 3300 m.

Bark: Young smooth, silver gray, dotted with resinous bladders, later long fissured, cork-like, gray brown.

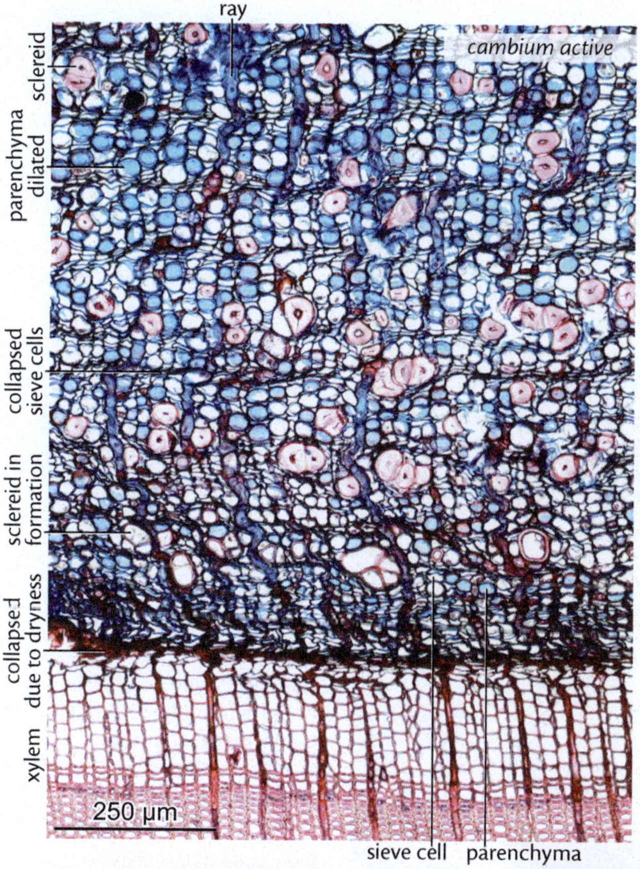

Bark anatomy: Regular tangential layers in the young phloem. Characteristic presence of isolated sclereids.

Sieve cells rectangular, arranged in 3–5-seriate tangential rows, collapsing within very few years. Parenchyma cells small and oval in the juvenile and enlarged in the adult phloem. Fibers absent. Sclereids isolated, round, very thick-walled, lignified. Rays radially wavy. Prismatic crystals small, in groups outside of sclereids.

Annual rings distinct for a few years near the cambium. Rings indistinct in the adult phloem due to sieve-cell collapse. Periderm active for several years. Phellem consisting of many layers of rectangular, thin-walled cells.

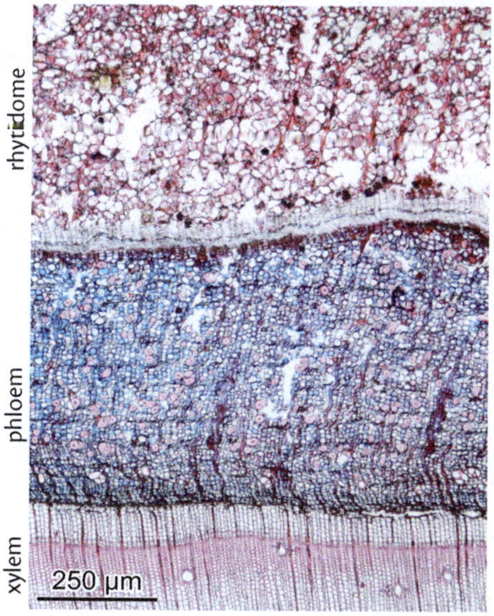

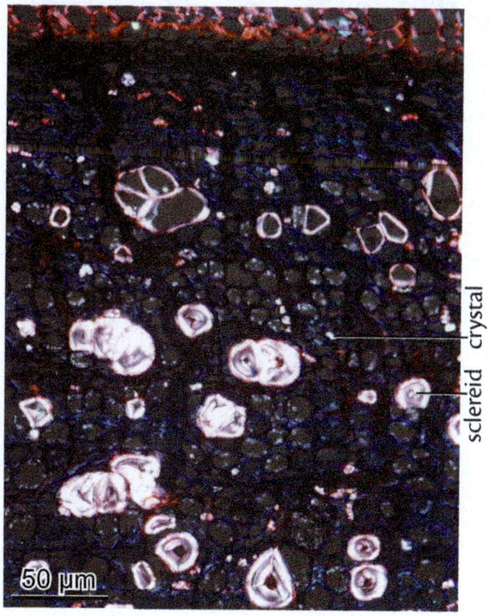

Sequoia sempervirens (D.Don) Endl. Cupressaceae

coast redwood, Küstenmammutbaum, séquoia à feuilles d'if, sequoia sempreverde, secuoya roja

Form: Upright, straight conifer, crown conical at first, broad and irregular shaped with age, to 115 m tall (the tallest known living tree worldwide).

Features: Needles evergreen, flat and linear, in two rows, to 2.5 cm long. Flowers monoecious, yellowish, spherical, in both sexes inconspicuous, to 7 mm long. Seed cones ovate to nearly globular, scaly, brown, hanging, to 3 cm long.

Habitat: Pacific North America from the Klamath Mountains in Southern Oregon south to the Santa Lucia Mountains in California.

Only close to the coast, absorbing water from maritime summer fog. On moist, deep soils, to 900 m.

Bark: Red brown or gray to dark brown, young scaly, soon fibrous, deeply fissured, peeling in long, vertical strips.

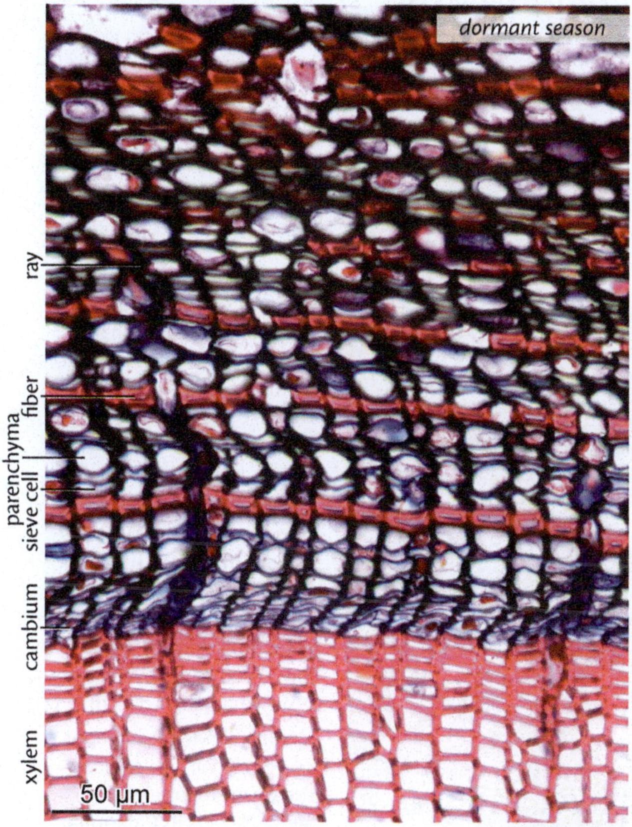

Bark anatomy: Regular tangential layers of sieve cells, fibers, and parenchyma cells.

Sieve cells thin-walled, deformed rectangular, parenchyma cells round to oval, fibers very thick-walled, rectangular. Sieve tubes tending to collapse. Rays remaining straight. Crystal sand in ray cells and radial intracellular layers.

Annual rings absent. Phellem cells thin-walled, irregularly shaped.

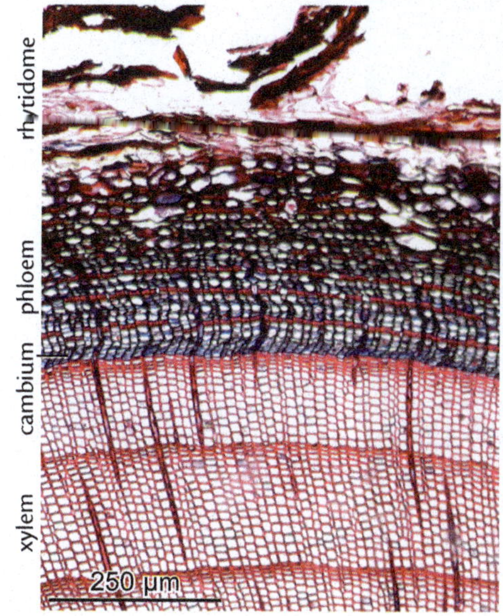

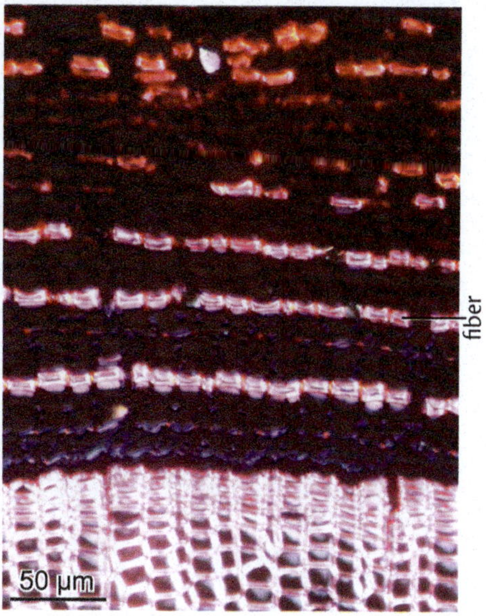

Sequoiadendron giganteum (Lindl.) J.Buchholz

Cupressaceae

giant sequoia, Riesenmammutbaum, séquoia géant, sequoia gigante, arbol del mamut

Form: Upright, straight conifer, crown conical, with age broad, irregular shaped, to 95 m tall and 8.8 m stem diameter (actually the thickest known living tree species worldwide).

Features: Leaves evergreen, pointy scales, arranged spirally in three parallel rows around the shoot, to 6 mm long. Flowers monoecious, yellowish, spherical, in both sexes inconspicuous, to 6 mm long. Seed cones ovate, scaly, brown, hanging, to 9 cm long.

Habitat: Endemic in about seventy small groves along the western slope of the Sierra Nevada in California.

On moist, but well-drained, deep soils, to 2500 m.

Bark: Red brown to cinnamon brown or gray brown to dark brown, very thick and fibrous, deeply fissured, peeling in long, vertical strips.

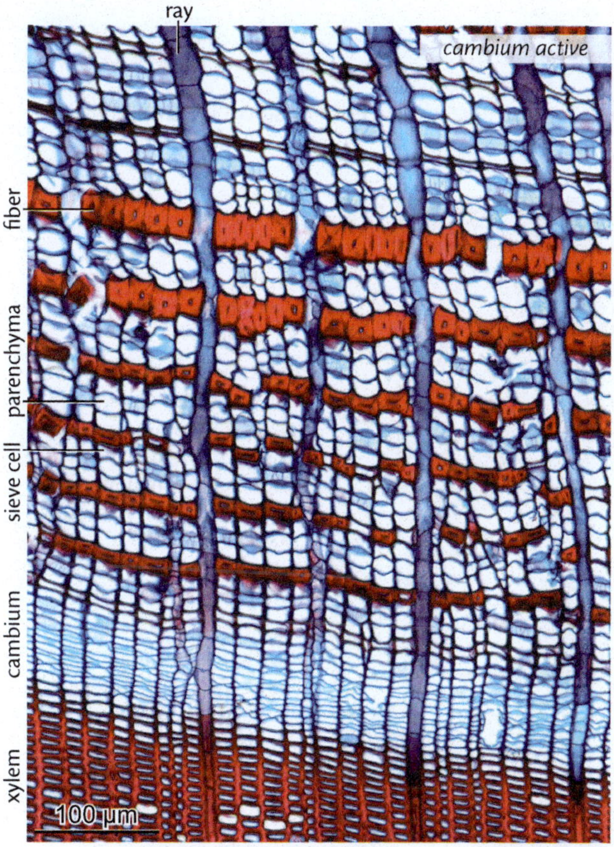

Bark anatomy: Regular tangential layers of sieve cells, fibers and parenchyma cells.

Sieve cells thin-walled, rectangular, tending to collapse. Older parenchyma cells round to oval. Fibers thick-walled, rectangular. Rays remaining straight. Crystal sand mainly in ray cells and radial intracellular layers.

Phellem cells thin-walled, irregular shaped.

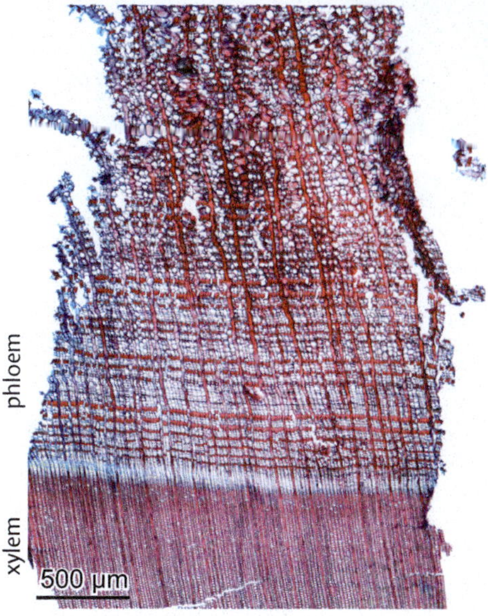

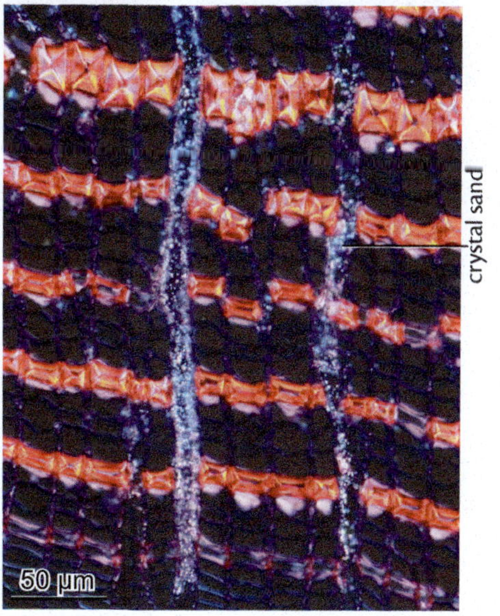

Thuja plicata Donn ex D.Don

Cupressaceae

western redcedar, Riesen-Lebensbaum, thuya géant, cedro rosso occidentale, tuya gigante

Form: Upright, straight conifer, crown small, conical, twigs mostly hanging, to 75 m tall.

Features: Leaves evergreen, flat scales, glossy green above, gray green below, aromatic, sharply pointed, to 6 mm long. Flowers monoecious, yellowish, spherical, in both sexes inconspicuous, to 5 mm long. Seed cones small, ovoid, brown, upright, to 2 cm long.

Habitat: Northwestern North America from the Prince William Sound in Alaska to the Siskiyou Mountains in Northern California along the coast, and a second area in the Rocky Mountains from British Columbia southeast to Idaho.

On moist, deep soils, in humid valleys, to 1600 m.

Bark: Red brown, young scaly, soon deeply fissured, peeling in long, vertical strips.

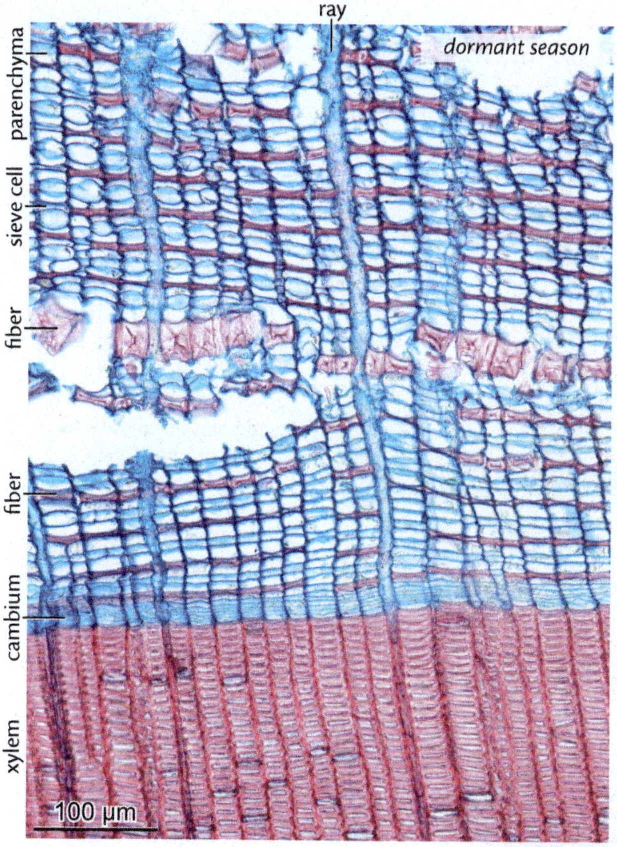

Bark anatomy: Regular tangential layers of sieve cells, fibers and parenchyma cells.

Sieve cells thin-walled, rectangular. Parenchyma cells round to oval. Fibers thick-walled, square to rectangular. Rays remaining straight, dilating in older stages. Ducts rare, large. Crystal sand in ray cells and radial intracellular layers.

Annual rings absent. Phellem cells thin-walled, regularly formed.

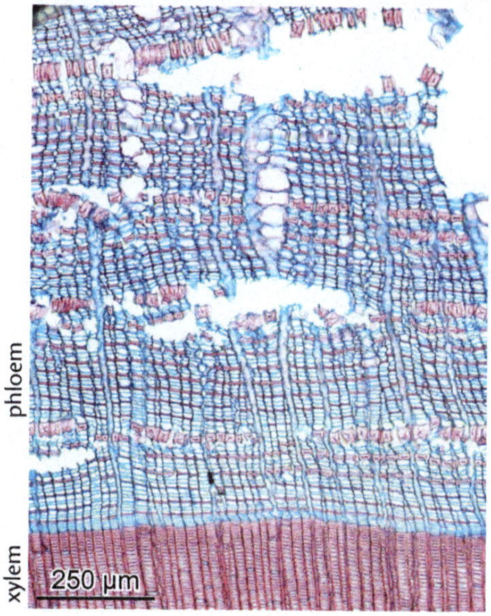

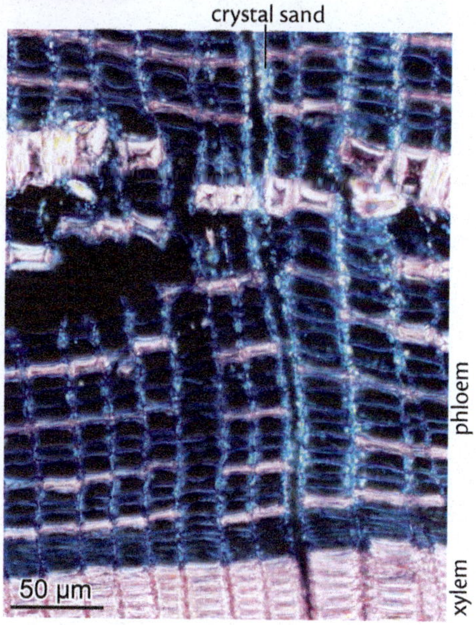

Torreya californica Torr.

Taxaceae

California torreya, Nusseibe, torreya de Californie, torreya, torreya

Form: Upright, straight conifer, crown conical, with age broad, to 40 m tall.

Features: Leaves evergreen, needle-like, linear, with a pointed tip, glossy green, to 6 mm long, aromatic. Flowers dioecious, yellowish, spherical, in both sexes inconspicuous, to 8 mm long. Fruits ovate, the hard seed is embedded in a fleshy, green arillus, to 4 cm long.

Habitat: Endemic in a few scattered locations in the Coast Ranges and the western slope of the Sierra Nevada in California; threatened species.

On dry to moist soils, at well-shaded slopes, to 1400 m.

Bark: Red brown, narrowly fissured.

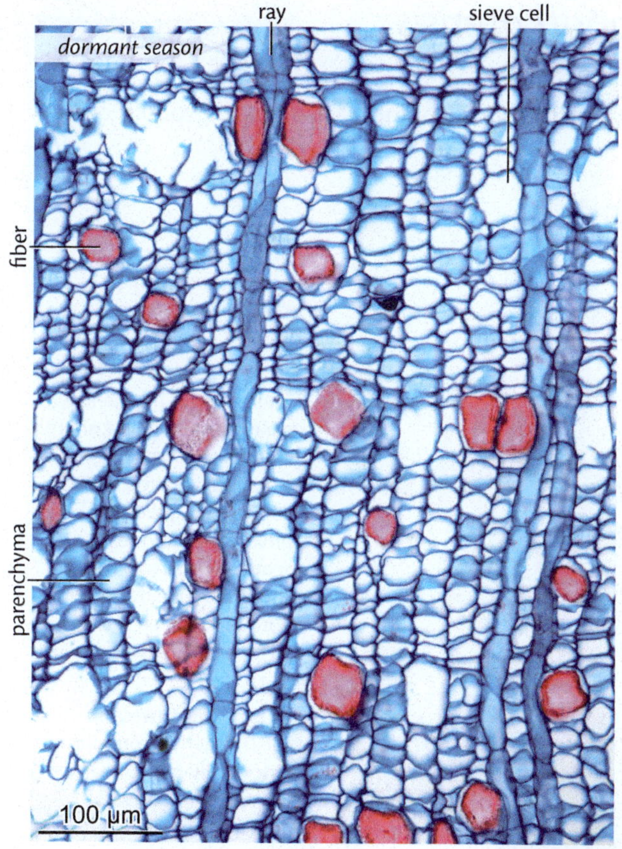

Bark anatomy: Regular tangential layers of sieve cells, isolated fibers and parenchyma cells.

Sieve cells thin-walled, rectangular. Parenchyma cells round to oval. Fibers round, isolated, unicellular, very thick-walled. Rays remaining straight. Ducts absent. Crystal sand mainly around fibers.

Annual rings absent. Rhythidome consisting of collapsed, unlignified, thin-walled cells and lignified, medium thick-walled cells.

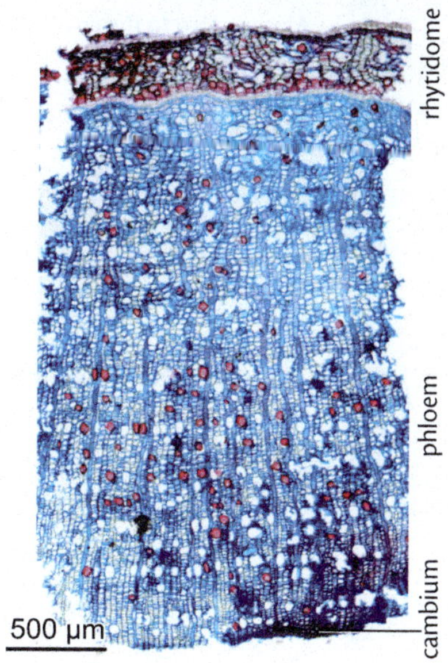

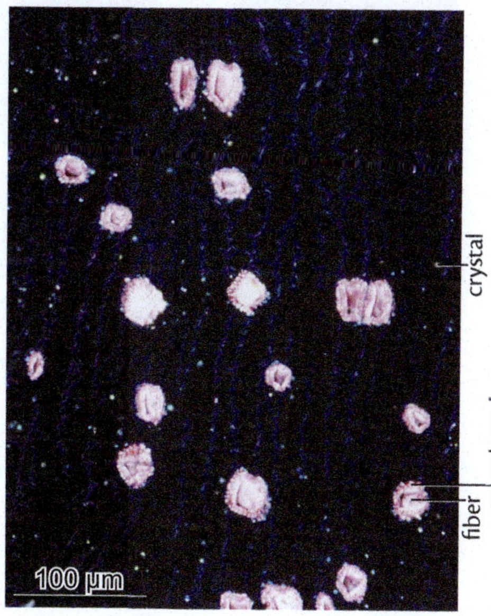

Xanthocyparis nootkatensis (D.Don) Farjon & D.K.Harder

Cupressaceae

Alaska cedar, Nutka-Scheinzypresse, cyprès de Nootka, cipresso di Nootka, cedro de Alaska

Form: Upright, straight conifer, crown conical, branches hanging, to 35 m tall.

Features: Leaves evergreen, juvenile leaves needle-like, adult leaves scale-like, sharply pointed, to 6 mm long, strongly aromatic. Flowers monoecious, yellowish, spherical, in both sexes inconspicuous, to 5 mm long. Fruit cone ovate, with protuberances on the scales, red brown, upright, to 1 cm long.

Habitat: Northwestern North America from the Prince William Sound in Alaska to the Siskiyou Mountains in Northern California.

Always close to the coast, on moist, deep soils, in humid valleys, to 2200 m.

Bark: Young red brown, scaly, soon red to silver brown, deeply fissured, peeling in long, vertical strips.

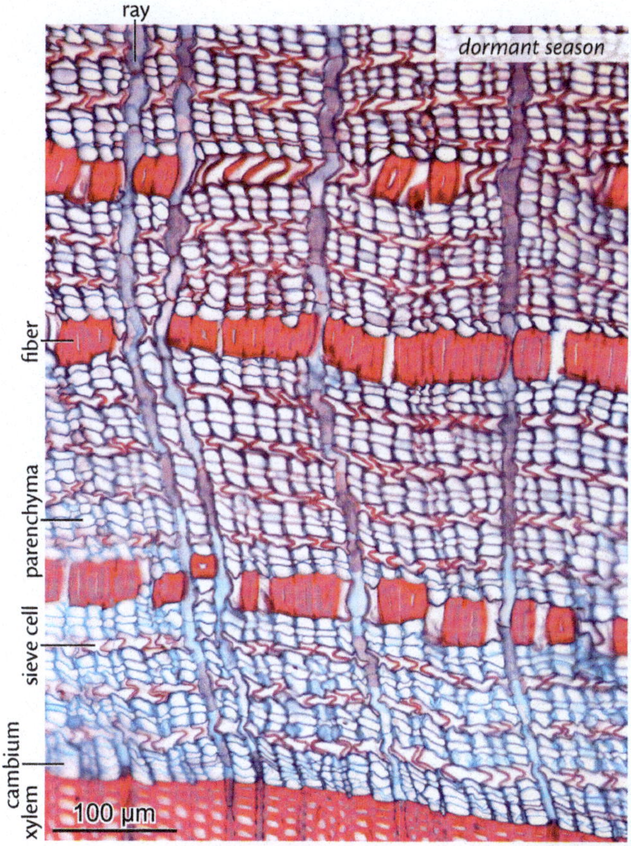

Bark anatomy: Regular tangential layers of sieve cells, fibers and parenchyma cells, characteristic for Cupressaceae.

Sieve cells thin-walled, deformed rectangular, collapsing at later stages. Parenchyma cells rectangular. Fibers very thick-walled, radially rectangular. Rays uniseriate, remaining straight, dilating in later stages. Ducts rare, large. Crystal sand in ray cells and radial intracellular layers.

Annual rings absent. Phellem cells thin-walled, irregular shaped.

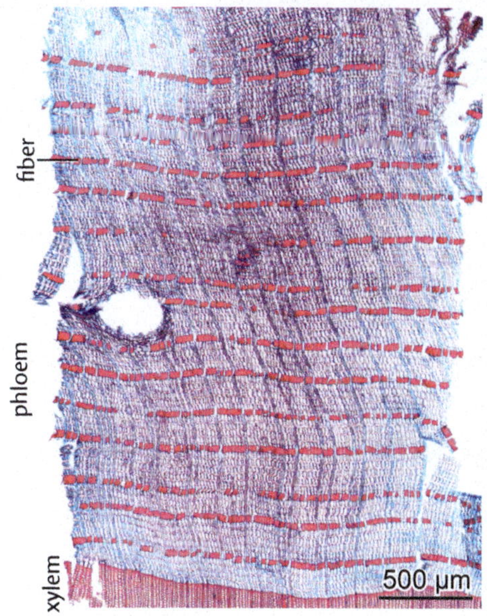

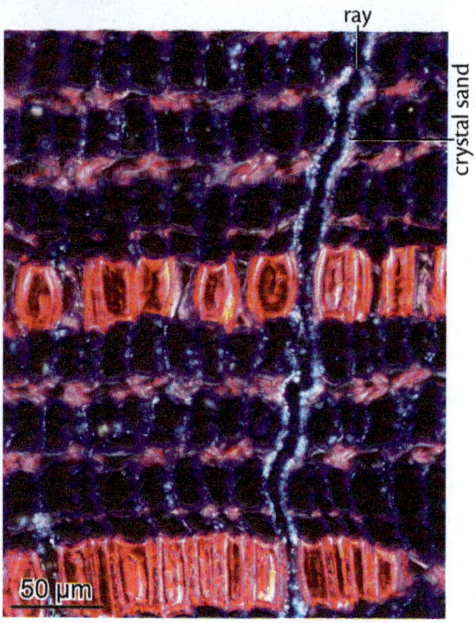

Gaultheria shallon Pursh

Ericaceae

salal, Shallon-Scheinbeere, palommier, gaultheria, gaultheria

Form: Shrub with a dense crown, to 1 m tall.

Features: Leaves evergreen, leathery, alternate, round ovate, net-veined, margin toothed, to 10 cm long. Flowers pinkish white, bell-shaped, with five petals, hanging, to 1 cm long. Fruit a fleshy, blue black, globular, berry-like capsule, to 1 cm wide.

Habitat: Pacific North America from Southeast Alaska along the coast, south to Santa Barbara in California.

On moist, acidic soils, to 700 m.

Bark: Young twigs red to green, smooth, later gray brown, fissured, peeling in long, vertical strips.

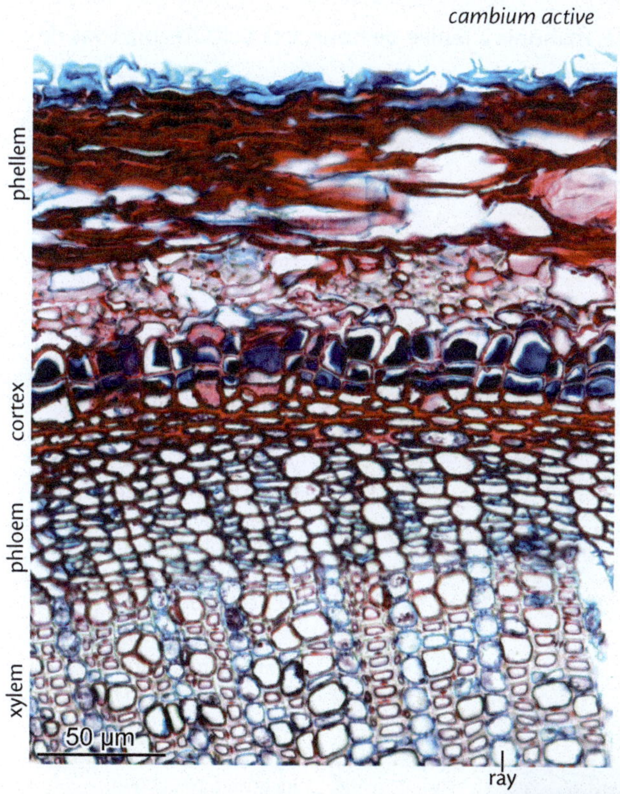

Bark anatomy: Phloem poorly structured, consisting of small sieve tubes and large parenchyma cells. Fibers and crystals absent. Ray cells large, round, radially arranged. Cortex consisting of a layer of thin- to thick-walled, irregular shaped parenchyma cells, a layer of large, square cells containing dark-staining substances and a layer of thick-walled fibers.

Phellem cells very large, irregular shaped.

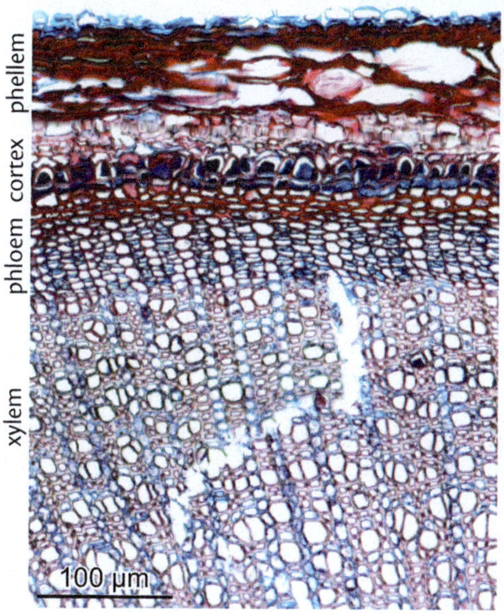

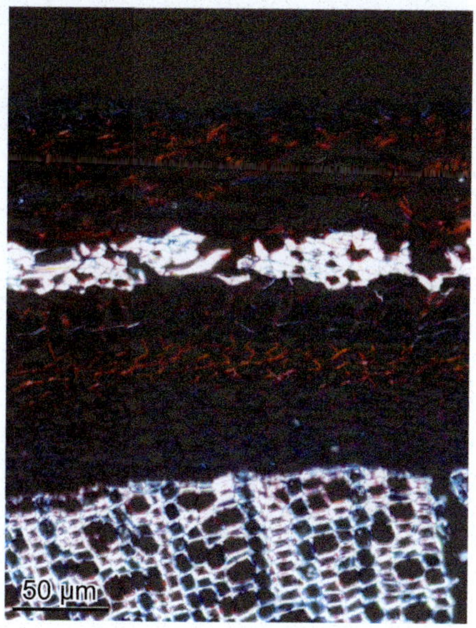

Mahonia aquifolium (Pursh) Nutt. Berberidaceae

Oregon holly grape, Mahonie, mahonia à feuille de houx, uva dell'Oregon, uvas de Oregón

Form: Shrub with a dense, broad crown, to 2 m tall.

Features: Leaves evergreen, leathery, glossy green, alternate, pinnate with 5–13 bristly toothed leaflets, to 25 cm long. Flowers yellow, with six petals, to 1.5 cm long, clustered in dense, terminal racemes. Fruit a fleshy, globular, purple black berry, with a bluish coating, to 1 cm wide.

Habitat: Pacific North America from British Columbia south to Northern California, east to Idaho.

On moist soils, in humid locations, to 2000 m.

Bark: Light gray or yellow brown, soon irregularly fissured and peeling in vertical strips.

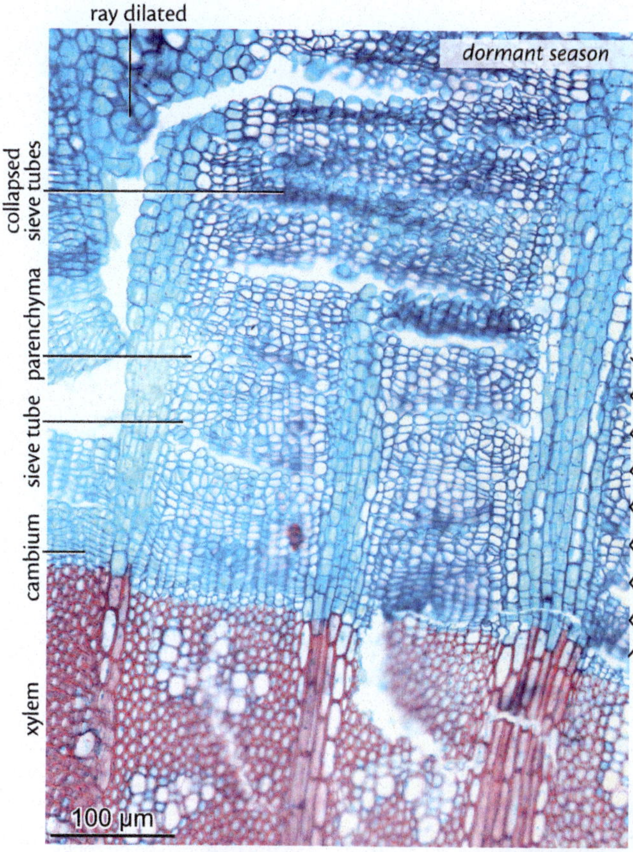

Bark anatomy: Regular tangential layers of parenchyma and sieve tubes, occasionally with sclereids.

Sieve tubes irregular shaped, companion cells small, in groups. Sieve tubes collapsing within a few years. Parenchyma cells roundish. Fibers absent. Sclereids in groups in older parts of the phloem, in irregular groups within rays. Rays large, dilating after some years. Crystals absent.

Tangential layers not corresponding with annual rings. Periderm tangentially layered, consisting of large, rectangular, thin-walled cells and small, thick-walled cells.

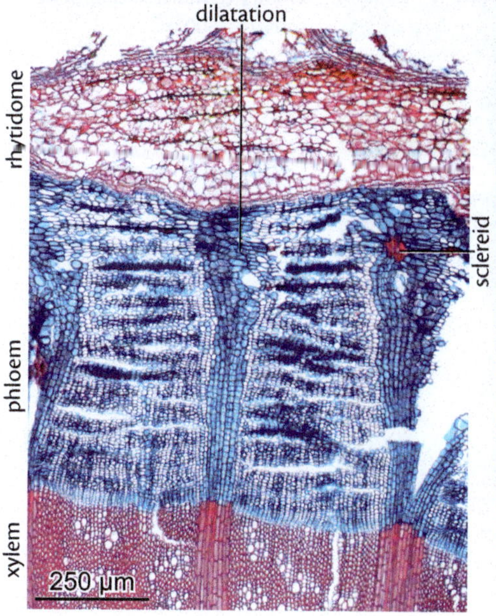

Umbellularia californica (Hook. & Arn.) Nutt. Lauraceae

California bay laurel, Kalifornischer Lorbeer, laurier de Californie, l'alloro della California, laurel de California

Form: Upright tree, crown pyramidal, to 30 m tall.

Features: Leaves evergreen, alternate, long ovate, glossy green above, leathery, heavily aromatic with a camphor scent, margin entire and slightly undulate, to 12 cm long. Flowers dioecious, creamy white, with three petals, to 6 mm long, in umbels. Fruit an ovate, fleshy, green to purple berry, to 3 cm long.

Habitat: California, from Southern Oregon along the coast southwards to San Diego, also on the western slopes of the Sierra Nevada.

On dry to moist, often rocky soils, to 1600 m.

Used as a spice.

Bark: Twigs green or yellow brown, soon turning smooth, dull, light gray, often with many small, diamond-shaped lenticels.

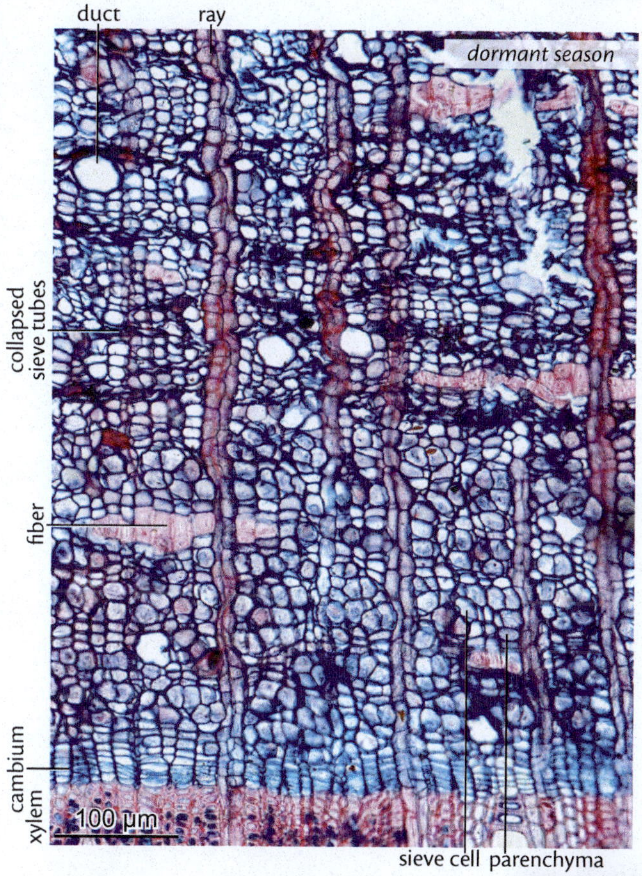

Bark anatomy: Regular tangential layers consisting of interrupted bands of fibers and zones of sieve tubes and parenchyma.

Layers of angular sieve tubes and roundish parenchyma cells alternating with 1–2-seriate, interrupted bands of thick-walled fibers. Sieve tubes collapsing early, indicated by dark bands. Rays usually straight or slightly wavy. Few rays dilating at the periphery of the phloem. With many large ducts (laticifers). Many very small acicular crystals in axial parenchyma cells and rays.

Annual ring boundaries unrecognizable. Phellem consisting of thin-walled, collapsed cork cells.

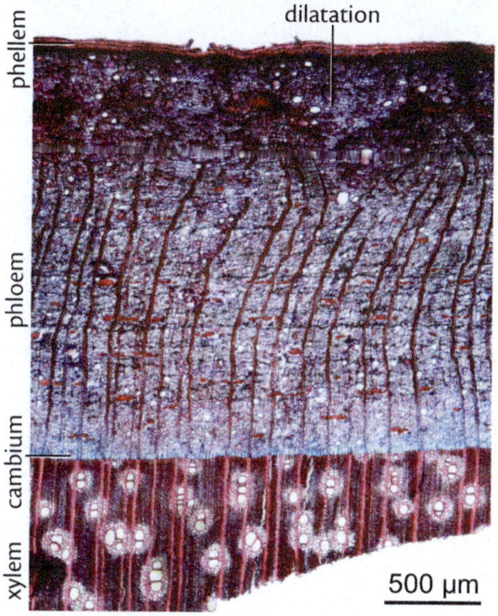

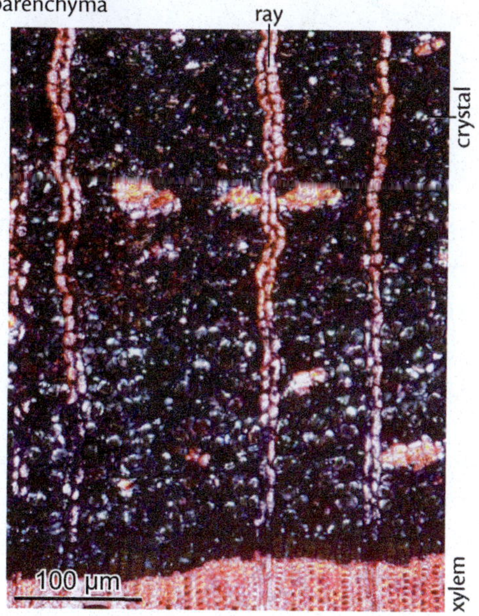

The **Mediterranean hard-leaved forest** is well represented by hard-leaved trees and shrubs, naturally often dominated by *Quercus ilex*, growing in a sometimes only narrow zone below the frost-influenced areas around the Mediterranean Basin, together with a few conifers and deciduous species such as *Platanus* x *hispanica* or *Punica granatum*. With the exception of some circum-Mediterranean species, species are often centered either in the western or eastern half of the basin.

F. H. Schweingruber et al., *Bark Anatomy of Trees and Shrubs in the Temperate Northern Hemisphere*, https://doi.org/10.1007/978-3-030-14056-4_10

10 Mediterranean hard-leaved forest

Cupressus sempervirens L. Cupressaceae

Mediterranean cypress, Mittelmeer-Zypresse, cyprès d'Italie, cipresso comune, ciprés común

Form: Conifer, often planted and typically seen in the small columnar form in Mediterranean gardens. In the wild with a broader, more irregular shaped crown, to 30 m tall.

Features: Leaves evergreen, opposite, scale-like, arranged like dense tiles around the twig, dark green, to 6 mm long. Flowers dioecious, female and male flowers in tiny yellowish or greenish globes, to 3 mm wide. Seed cones ovoid, with a berry-like, yellow to silver brown structure of coalescing scales, hanging, to 4 cm long.

Habitat: Eastern Mediterranean, native only to scattered localities in Crete, southern Turkey, Cyprus, Syria, Lebanon and Cyrenaica.

On dry to moist, often rocky soils, to 1700 m.

Bark: Thin, fibrous, gray to red brown, soon finely fissured.

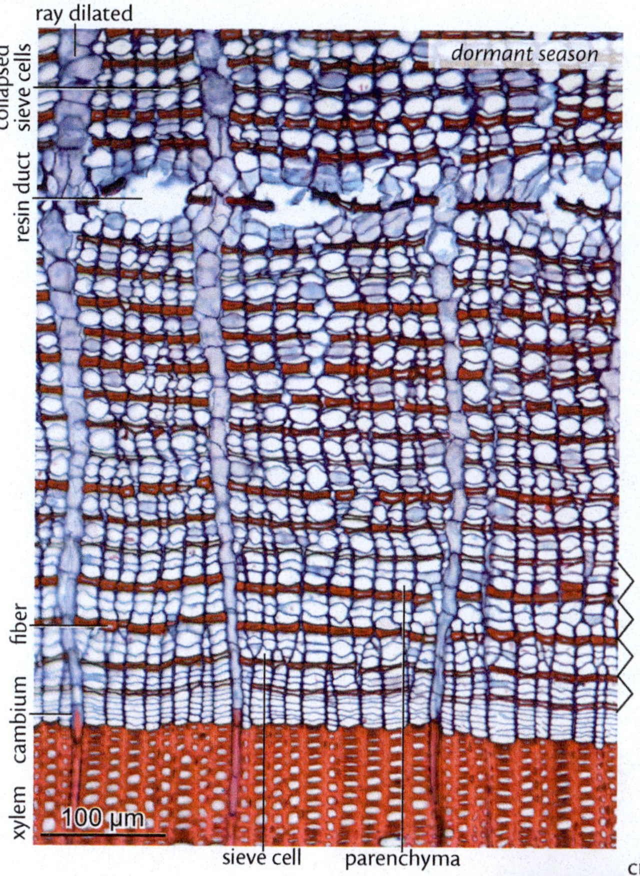

Bark anatomy: Regular tangential layers of sieve cells, fibers and parenchyma cells, characteristic for Cupressaceae.

Sieve cells thin-walled, rectangular. Parenchyma cells round to oval. Fibers thick-walled, rectangular. Sieve cells tending to collapse in later stages. Rays uniseriate, remaining straight, dilating in older stages. Ducts rare, large. Crystal sand in ray cells and radial intracellular layers.

Annual rings mostly indistinct. Ring boundaries of fast-growing trees indicated by thick-walled fibers in the early phloem. Phellem cells thin-walled, irregular shaped.

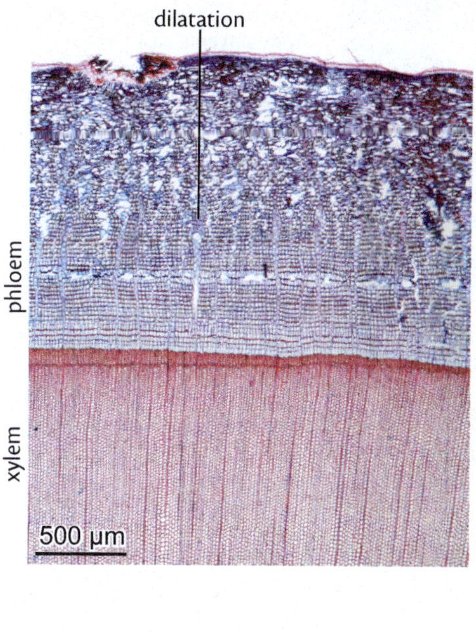

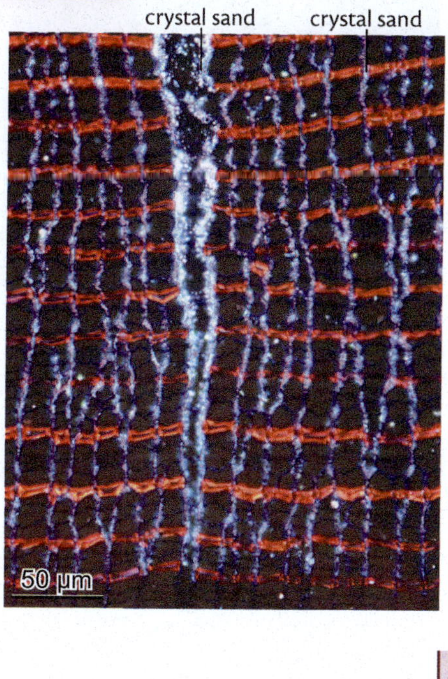

Juniperus oxycedrus L. Cupressaceae

cade, Stech-Wacholder, cade, ginepro rosso, oxicedro

Form: Conifer, usually a multi-stemmed shrub, rarely a small tree, to 14 m tall.

Features: Needles evergreen, stiff, short, prickly pointed, arranged in three rows around the twig, to 2 cm long, upper side with two white, waxy streaks, giving the leaves a grayish appearance. Flowers dioecious, female and male flowers in tiny yellowish globes, to 5 mm wide. Seed cones with a berry-like, red brown structure of coalescing scales, to 13 mm long.

Habitat: Western Mediterranean, from Portugal east to Italy, also isolated in northern Algeria.

On dry to moist, often sandy or gravelly soils, to 1500 m.

Variation: The East Mediterranean populations from Slovenia east to the Caucasus and Iran are now regarded as a separate species, *Juniperus deltoides*.

Bark: Young red brown or gray, with papery scales, later with distinct longitudinal fissures.

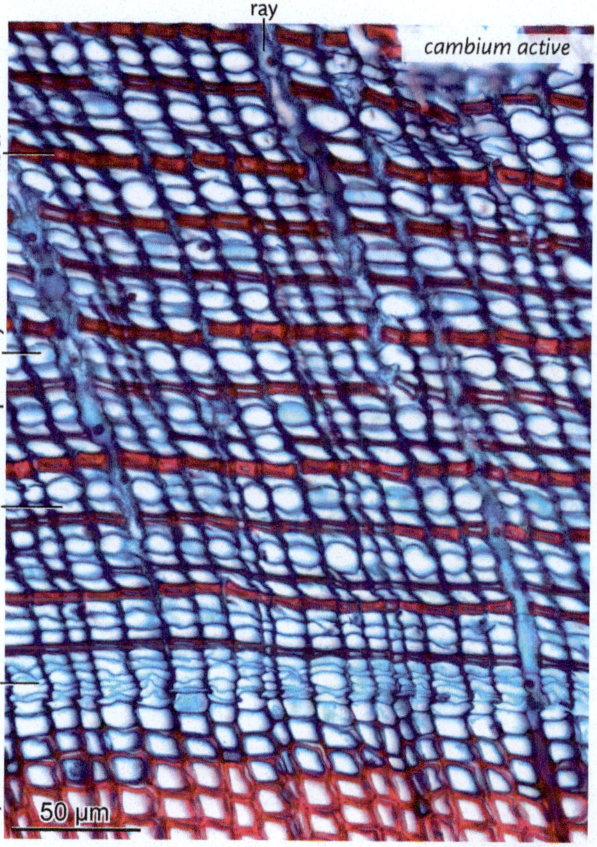

Bark anatomy: Regular tangential layers of sieve cells, fibers, and parenchyma cells.

Sieve cells thin-walled, rectangular, parenchyma cells round to oval, fibers thick-walled, square and rectangular. Sieve cells tending to collapse. Rays uniseriate, remaining straight, dilating in later stages. Rarely with large resin ducts. Crystal sand in radial intracellular layers.

Annual rings mostly indistinct. Phellem cells thin-walled, irregularly shaped.

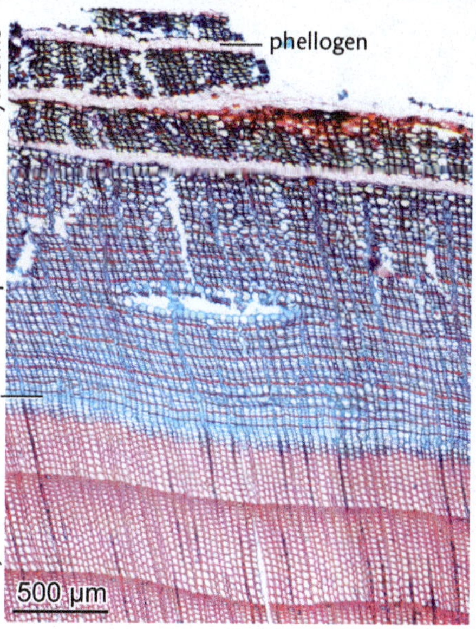

Pinus halepensis Mill. Pinaceae

Aleppo pine, Aleppo-Kiefer, pin blanc de Provence, pino d'Aleppo, pino carrasco

Form: Upright conifer, crown conical at first, with age irregular umbrella-shaped, to 25 m tall.

Features: Needles evergreen, in bundles of two, stiff, light green, to 15 cm long. Flowers monoecious, female cone upright, carmine, globular, male cone yellow, crowded together at the top of twigs. Seed cones brown, erect ovoid, later hanging, to 10 cm long.

Habitat: Western Mediterranean, centered in North Africa and in southeastern Spain to Southern France, small and isolated areas in Italy, Croatia, Greece, Turkey, Syria, Lebanon and Israel.

Usually on dry, rocky or gravelly, calcareous soils, to 2600 m.

Bark: Young smooth, silvery reddish gray, soon scaly, dark gray brown to red brown, later cracking into irregular shaped plates.

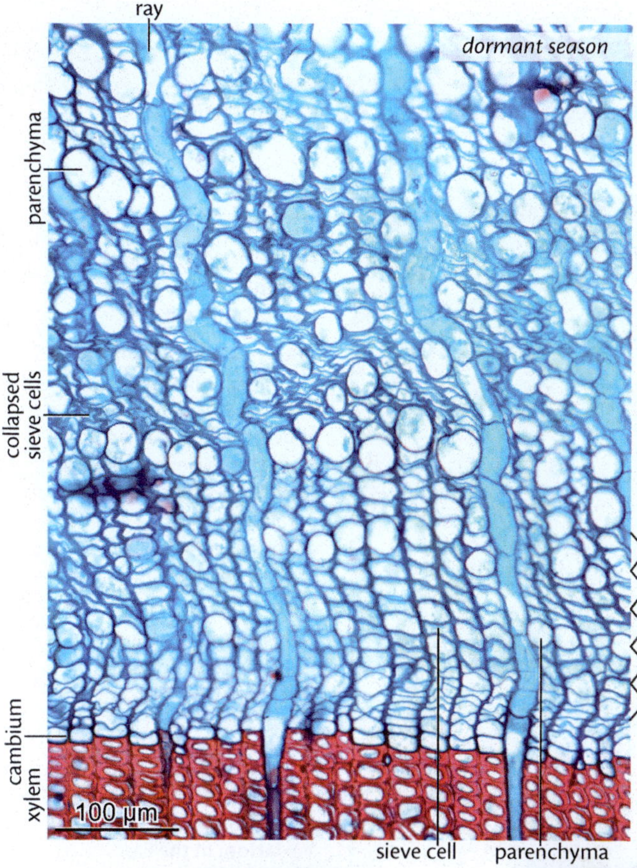

Bark anatomy: Regular tangential layers in the young phloem. Characteristic absence of fibers and isolated sclereids.

Sieve cells rectangular, arranged in 3–5-seriate tangential rows, collapsing within a few years after formation. Parenchyma cells small and oval in the juvenile and enlarged in the adult phloem. Sclereids and fibers absent. Rays radially wavy. Prismatic crystals outside of rays, mostly in axially elongated cells (crystal tubes).

Annual rings distinct for a few years near the cambium. Rings indistinct in the adult phloem due to sieve-cell collapse and dilatation of parenchyma cells. Phellem layered, consisting of rectangular, thin-walled cells.

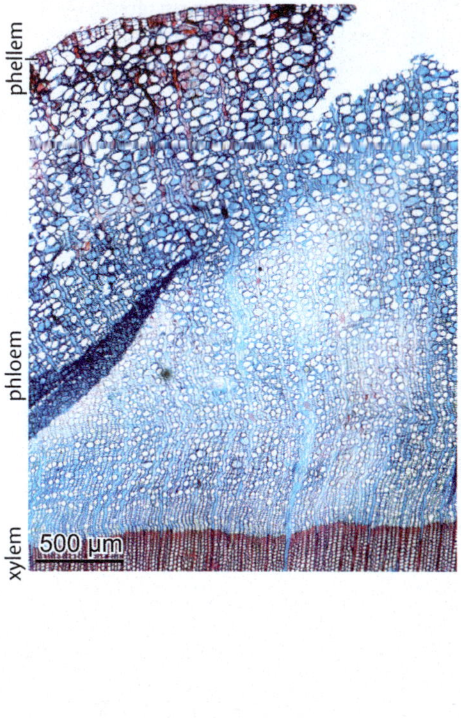

Arbutus unedo L. — Ericaceae

strawberry tree, Westlicher Erdbeerbaum, arbousier, corbezzolo, madroño

Form: Usually a multi-stemmed shrub, rarely a small tree, to 10 m tall.

Features: Leaves evergreen, alternate, long ovate, leathery, glossy dark green above, margin toothed, to 11 cm long. Flowers pinkish white, bell-shaped, with five petals, to 1 cm long. Fruit a deep red, globular berry with orange, edible flesh, to 3 cm wide.

Habitat: Mediterranean, from North Africa and Portugal, east to Turkey and Crete, isolated also in Southwest Ireland.

On dry to moist, acidic soils, to 1000 m.

Variation: In the East Mediterranean often replaced by *Arbutus andrachne*.

Bark: Young scaly, often intense red to cinnamon brown, soon shedding in long, small strips.

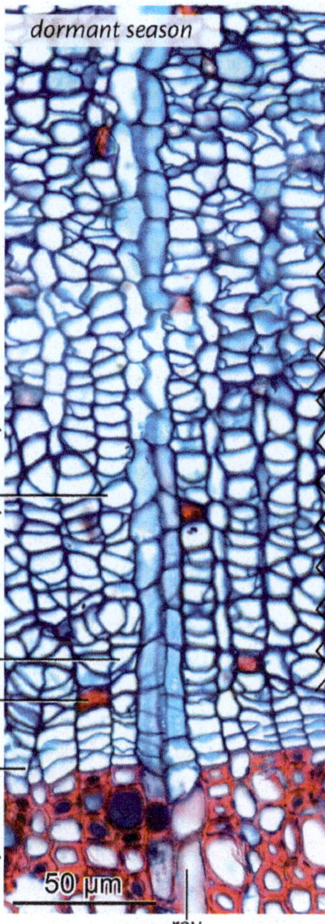

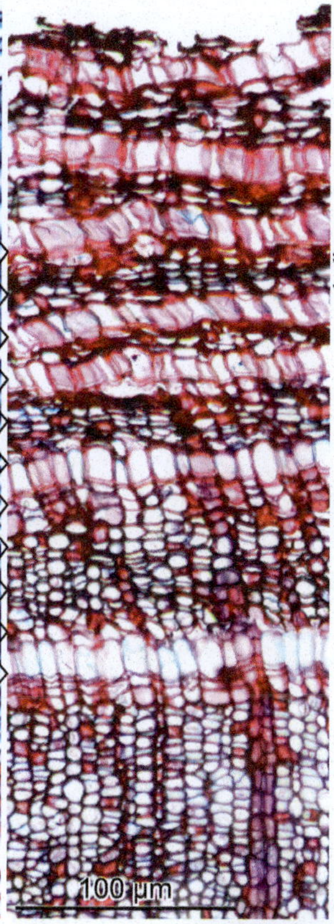

Bark anatomy: Large phloem with solitary sclereids. Sieve tubes often accompanied by relatively large companion cells and round to oval parenchyma cells. Fibers solitary or in small groups. Larger rays slightly dilating. Many prismatic crystals in axial parenchyma cells and adjacent to sclereids.

Phellem consisting of alternating layers of unstructured, small cells and uniseriate layers of large, radially elongated cells (phelloids), lignified at the cambial side.

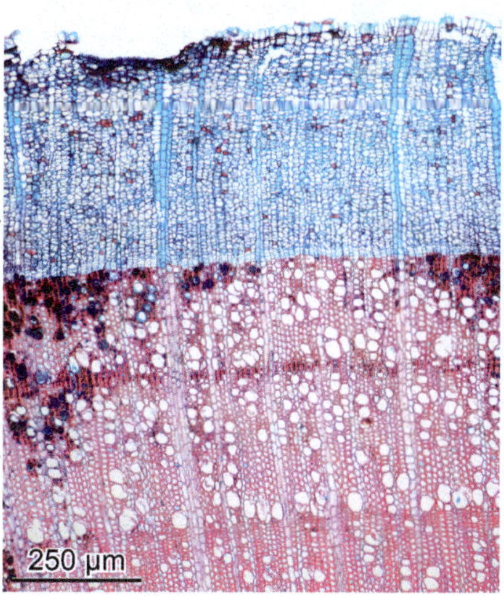

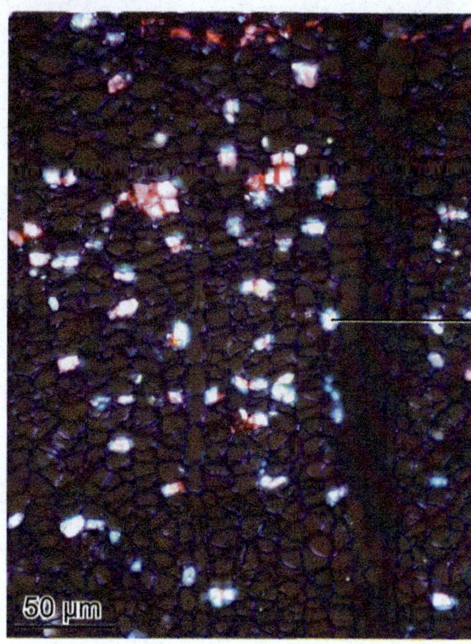

Ceratonia siliqua L. Fabaceae

carob tree, Johannisbrotbaum, caroubier, carrubo, algarrobo

Form: Tree, often with a large trunk and horizontal branches, crown broad umbellate with age, to 20 m tall.

Features: Leaves evergreen, alternate, pinnate with four to ten ovate leaflets, margin entire, to 20 cm long. Flowers cauliflorous, dioecious or monoecious, with no petals and five sepals, to 5 mm wide, male flower reddish yellow, female flower greenish, both in crowded masses in autumn. Fruit a purple brown, often curled, leathery pod with edible flesh, to 30 cm long.

Habitat: Mediterranean, from North Africa and Portugal, east to Turkey and Israel.

Growing only in the warmest locations, never far from the coast, on dry to moist, calcareous, also salty soils, to 400 m.

Bark: Dull light gray, with scattered lenticels, with age rough and often fissured.

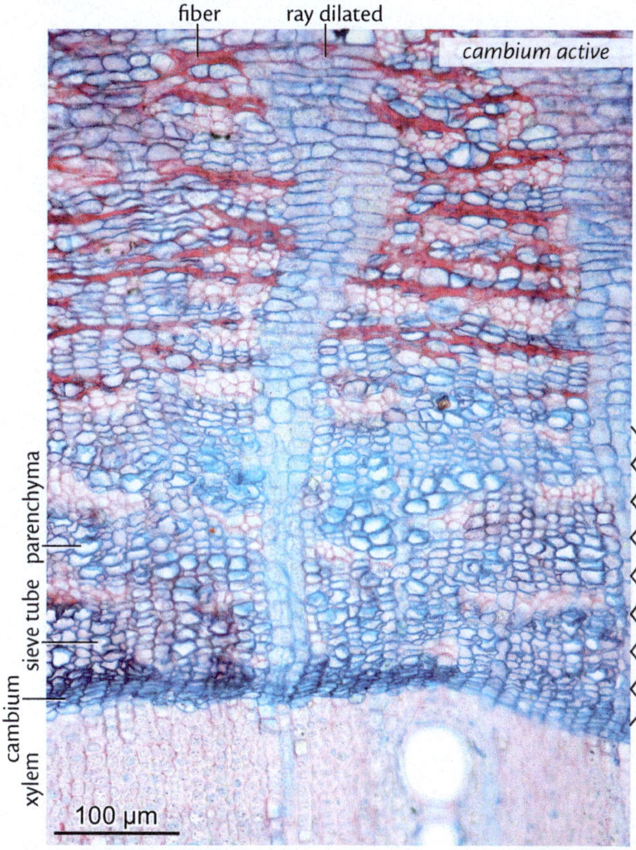

Bark anatomy: Regular tangential layers consisting of bands of small, oval fibers and zones of sieve tubes and parenchyma.

Early phloem formed by irregular shaped sieve tubes and roundish parenchyma cells. Sieve tubes collapsing within a few years, indicated by narrow, dense bands. Fibers in 2–4-seriate, laterally interrupted, tangential bands. Sclereids absent. Rays dilating. Prismatic crystals mainly within fibers and around fiber zones. Crystal druses in parenchyma cells near the cambium. Crystals absent in rays.

Periderm consisting of many layers of large, rectangular, thin- to thick-walled cells.

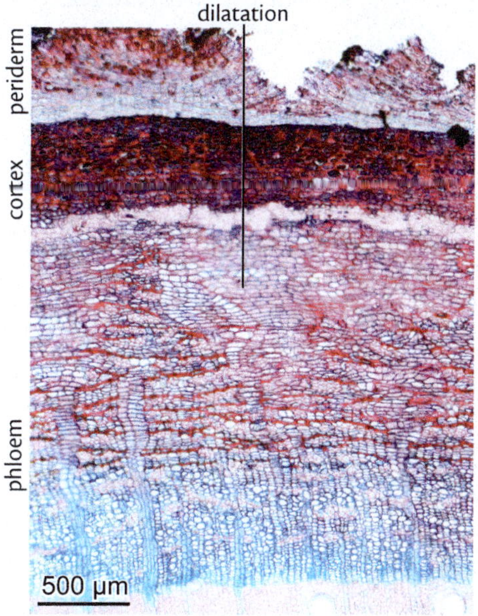

Cercis siliquastrum L. Fabaceae

Judas tree, Judasbaum, arbre de Judée, albero di giudea, ciclamor

Form: Shrub or small tree, crown globular, to 13 m tall.

Features: Leaves deciduous, alternate, kidney-shaped, margin entire, to 13 cm long. Flowers cauliflorous, magenta, typical zygomorphic pea flowers, to 1.5 cm long, huge masses of flowers forming an impressive display in spring. Fruit a brown, long ovate, dry pod, to 10 cm long.

Habitat: Mediterranean, from North Africa and Portugal, east to Turkey and Israel.

Growing only in warm locations, on dry to moist, calcareous soils, to 400 m.

Bark: Young smooth, dull, gray brown, with small, scattered lenticels, with age rough, cracking into small, rectangular plates.

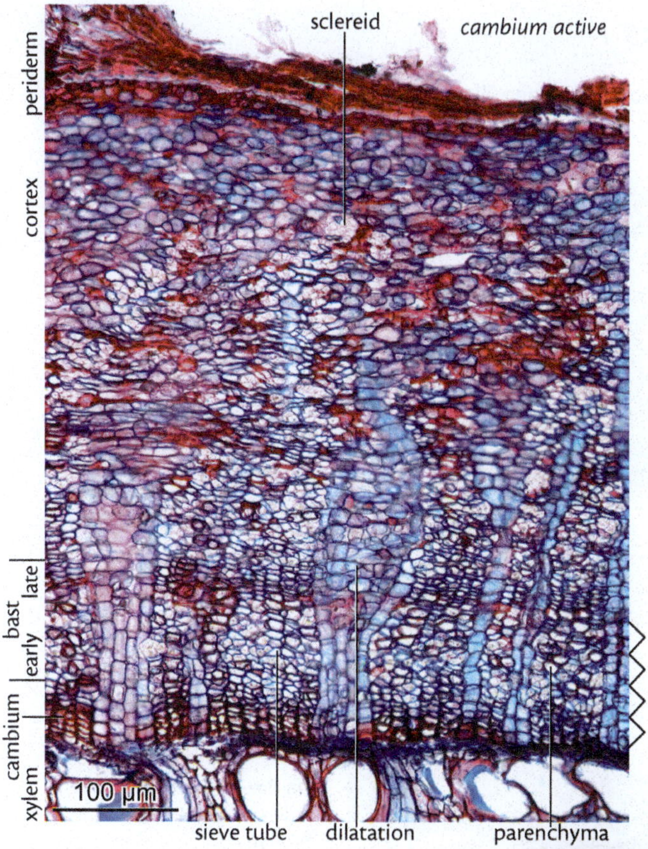

Bark anatomy: Soft tissue of sieve tubes and parenchyma with many rays.

Early phloem formed by large, angular sieve tubes and round to oval parenchyma cells. Late phloem formed by a zone of smaller cells. Fibers absent. Sclereids rare, in very small groups. Large rays frequently dilating within a short time. Many prismatic crystals in parenchyma cells, rarely in rays.

Phellem layered, cells thin-walled and rectangular.

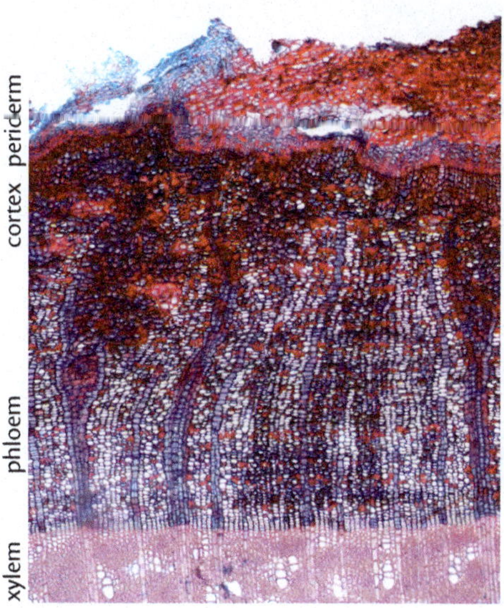

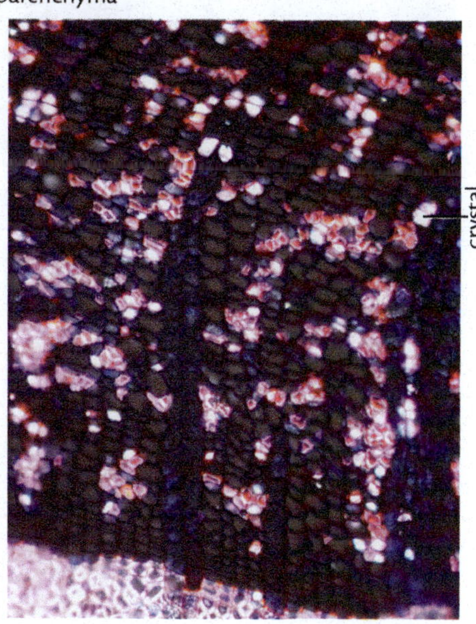

Cistus laurifolius L. Cistaceae

laurel-leaved rock rose, Lorbeerblättrige Zistrose, ciste à feuilles de laurier, cisto maggiore, jara de estepa

Form: Shrub, twigs erect, to 2 m tall.

Features: Leaves evergreen, opposite, lanceolate, glossy green, with aromatic, sticky resin, margin undulated entire, to 9 cm long. Flowers with five large, white, papery petals and many yellow stamens, to 5 cm across. Fruit a brown, five-valved, dry capsule, to 4 cm long.

Habitat: Mediterranean, at scattered localities from Morocco and Portugal, east to Greece and Turkey.

On dry, acidic, often rocky soils, in warm locations, to 1900 m.

Bark: Young smooth, green, soon turning red to cinnamon brown, peeling in long, thin strips.

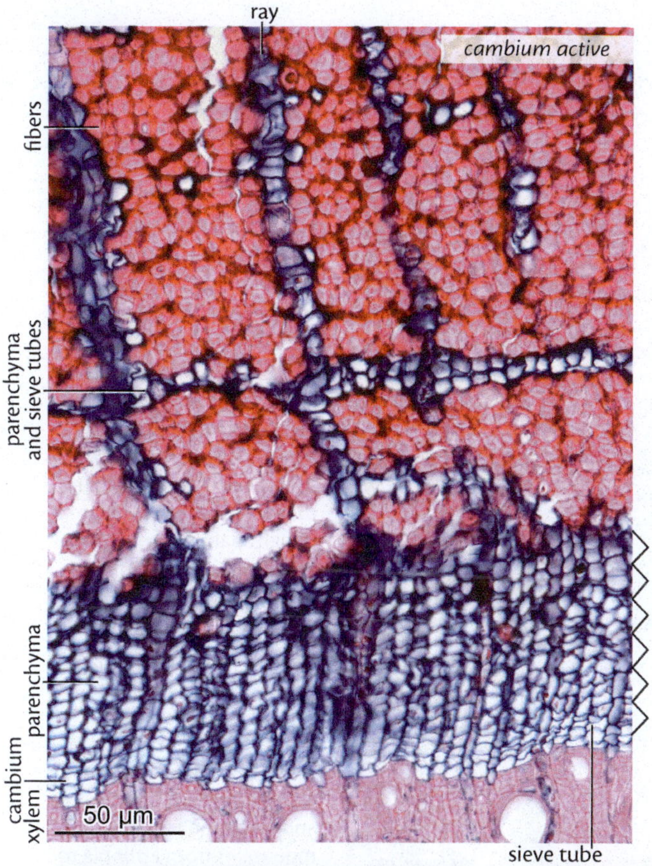

Bark anatomy: Regular tangential layers consisting of large bands of fibers and small band of sieve tubes and parenchyma zones.

Sieve tubes and parenchyma difficult to distinguish in the earlywood zone. Fibers in large, layered tangential bands. Prismatic crystals around the fiber bands. Few crystal druses in parenchyma cells.

Annual ring boundaries usually corresponding with fiber bands.

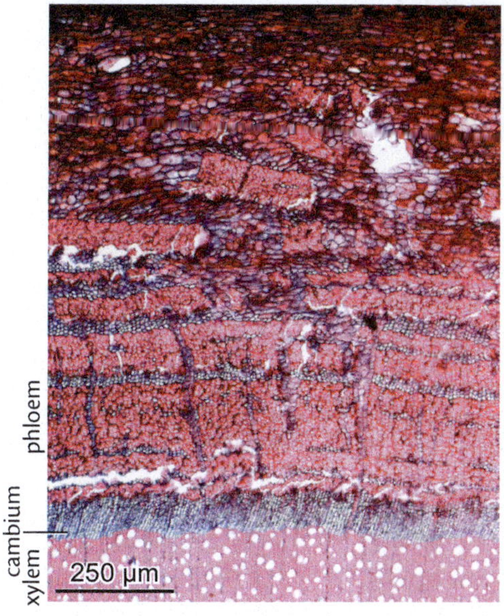

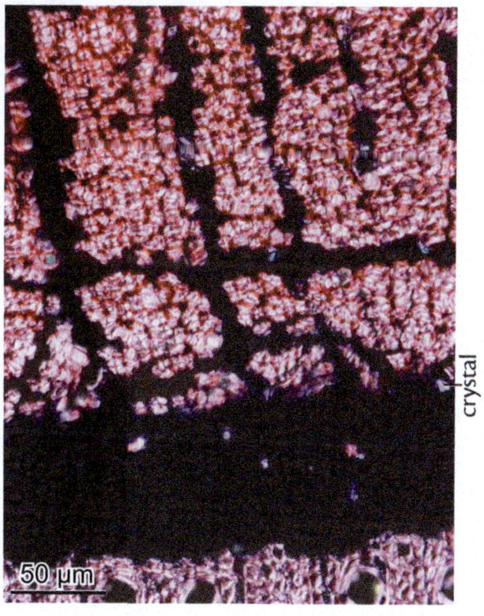

Erica arborea L.

Ericaceae

tree heath, Baumheide, bruyère en arbre, erica arborea, brezo blanco

Form: In cultivation usually a shrub with erect twigs, to 6 m tall. In its native African and Canarian range also a tree, to 20 m tall.

Features: Leaves evergreen, arranged in whorls, small, needle-like, to 6 mm long. Flowers with five white petals, bell-shaped, hanging, to 4 mm long, crowded in upright panicles, to 40 cm long. Fruit a brown, dry capsule, hidden in the wilted crown, to 2 mm long.

Habitat: Mediterranean and Africa, at scattered localities from the Canary Islands, Morocco and Portugal, north to the Southern Alps and Bulgaria, east to Greece and Turkey, in East African mountains from Ethiopia to Uganda and the Cameroon Highlands.

On dry, acidic, often rocky soils, to 3700 m.

Bark: Young felty hairy, red brown, soon peeling in long, thin strips.

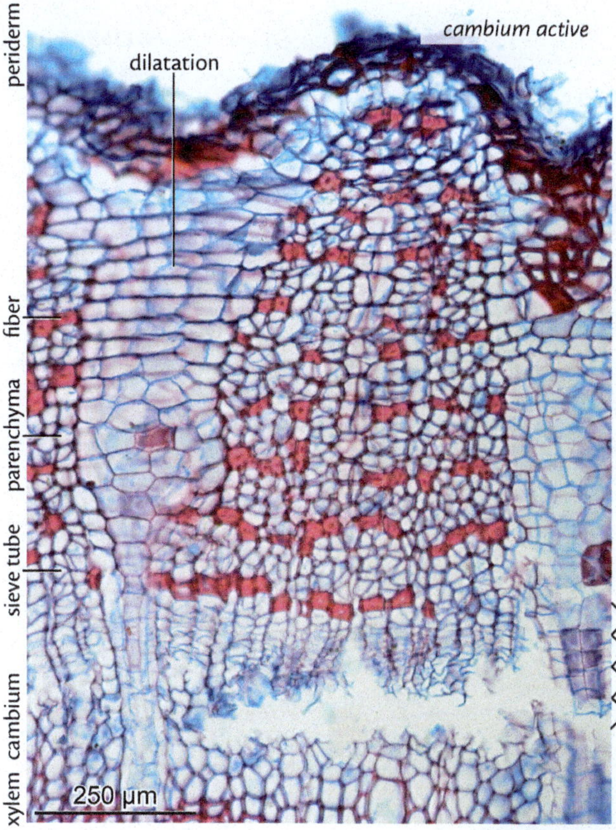

Bark anatomy: Phloem with fibers in uniseriate tangential lines.

Sieve tubes and parenchyma cells difficult to distinguish. Fibers in uniseriate tangential rows. All rays dilating. With very few prismatic crystals.

Phellem not observed.

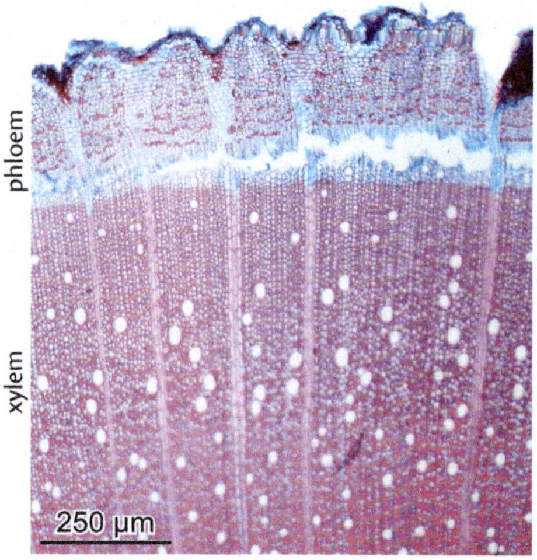

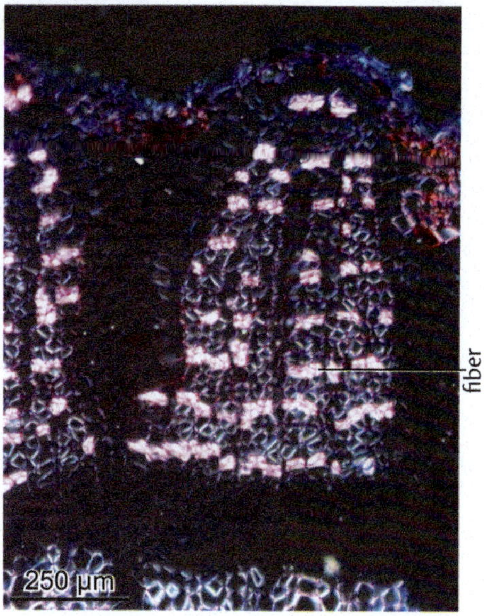

Ficus carica L.

Moraceae

fig tree, Echte Feige, figuier, fico comune, higuera

Form: Shrub or small tree, crown irregular shaped, low and broad, to 10 m tall.

Features: Leaves deciduous, alternate, deeply three- to five-lobed, rough, margin slightly toothed, to 20 cm long, with white latex. Flowers dioecious, inconspicuous, greenish, hidden inside of the future fruit base, to 2 mm long. Fruits fleshy, edible, spurious, pear-shaped, turning from green to black blue, to 8 cm long.

Habitat: Originally only in the South Caucasus and Northern Iran; for a long time widely planted and escaped in the whole Mediterranean and other warm regions.

On dry to moist, calcareous soils, to 1700 m.

Bark: Smooth, gray brown, with many small lenticels.

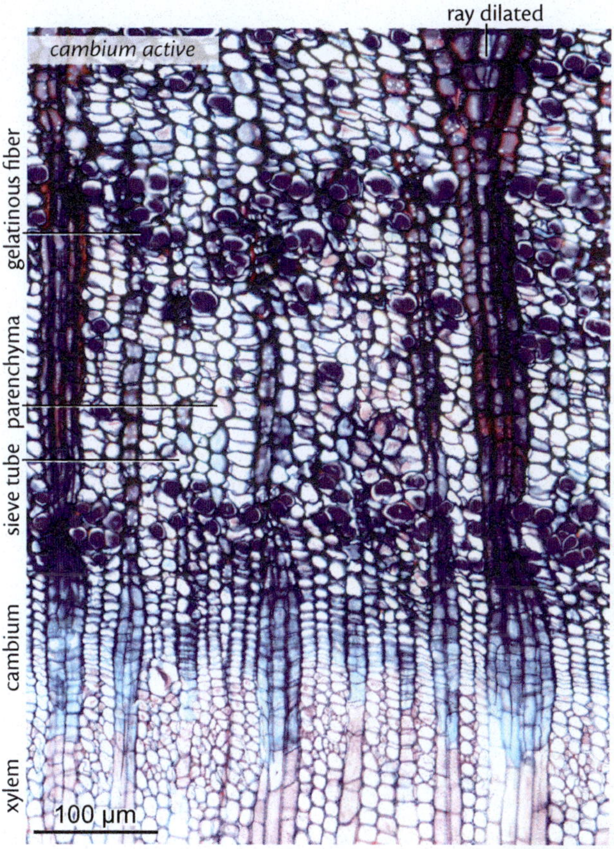

Bark anatomy:
Homogeneous tissue of sieve tubes and parenchyma, without any lignified elements.

Sieve tubes angular, accompanied by round to oval parenchyma cells. Fibers isolated or in small groups, often gelatinous. Rays often intensively dilating. Large prismatic crystals in parenchyma cells of the dilatations.

Phellem layered, consisting of irregular zones of large, thin-walled cells, filled with dark-staining substances.

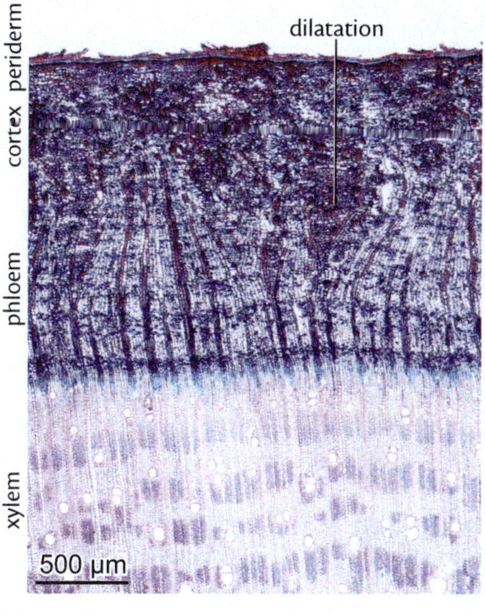

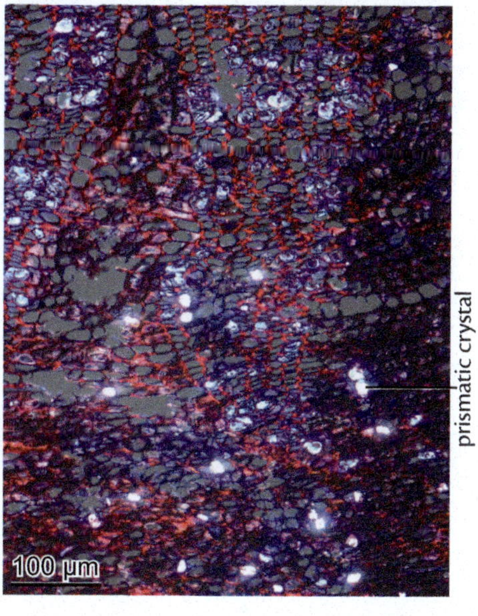

Laurus nobilis L.

Lauraceae

bay tree, Echter Lorbeer, laurier vrai, alloro, laurel

Form: Tree, crown upright, pyramidal, to 15 m tall.

Features: Leaves evergreen, alternate, long ovate, leathery, glossy green above, aromatic, margin entire and slightly undulate, to 12 cm long. Flowers dioecious, greenish to creamy white, with four petals, to 6 mm long. Fruit an ovate, fleshy, glossy black berry, to 1.5 cm wide.

Habitat: Mediterranean, from North Africa and Portugal east to Turkey and Israel; widely grown as a spice.

On moist soils, to 800 m.

Bark: Smooth, light gray, with many small, horizontal lenticels.

Bark anatomy: Unlayered phloem of sieve tubes and parenchyma and a dense belt of fibers in the cortex.

Sieve tubes angular, collapsing within a short time after formation. Parenchyma cells roundish. Fibers solitary. Sclereids in groups, forming a tangential band. Rays frequently dilating. Many small acicular crystals and a lot of crystal sand in ray cells.

Phellem consisting of a few layers of square to rectangular, thick-walled cells.

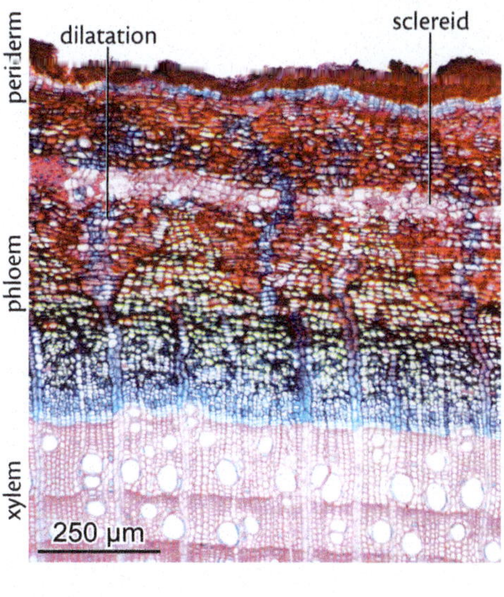

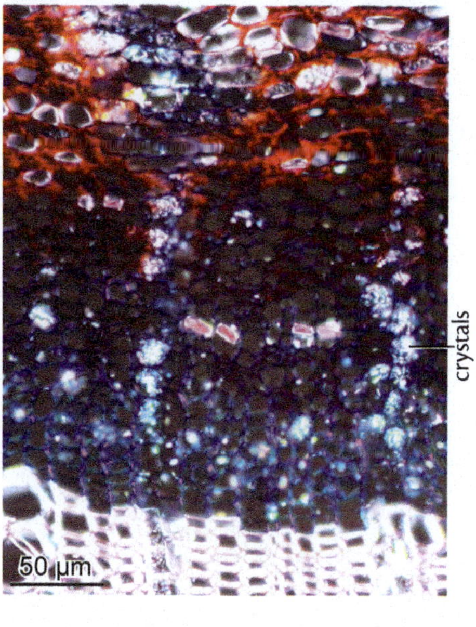

Myrtus communis L.

Myrtaceae

myrtle, Myrte, myrte, mirto, arrayán

Form: Shrub, twigs erect, to 5 m tall.

Features: Leaves evergreen, opposite, ovate to lanceolate with a pointed tip, leathery, strongly aromatic, margin entire, to 6 cm long. Flowers solitary and long-stalked, with five white petals and many white, conspicuous stamens as long as the petals, to 1.5 cm long. Fruit a blue black, fleshy berry, to 1.5 cm wide.

Habitat: Mediterranean, from Morocco and Portugal, east to Greece and Turkey, also along the coast of the Black Sea.

On moist, mostly acidic soils in warm locations, to 1000 m.

Bark: Light to cinnamon brown, with many small fissures.

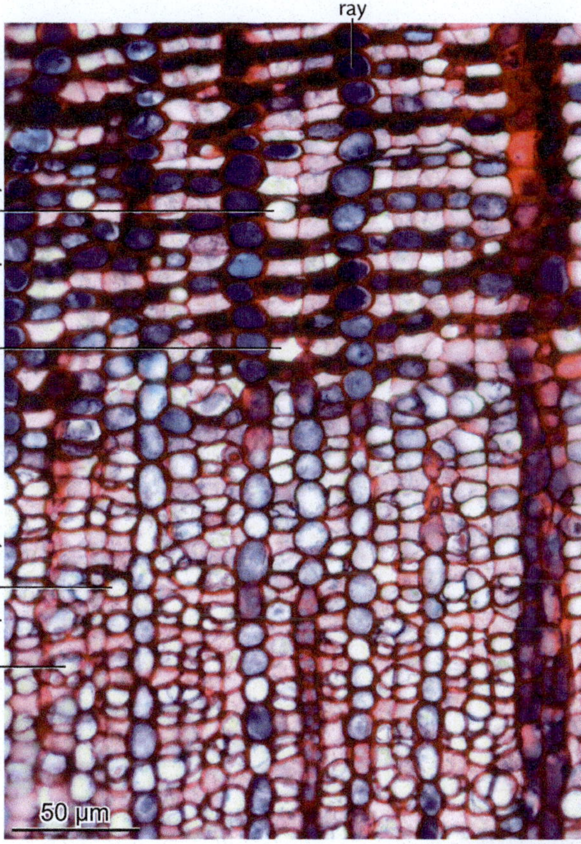

Bark anatomy: Finely layered structure of sieve tubes and parenchyma.

Sieve tubes rectangular, accompanied by small companion cells and round to oval parenchyma cells. Sclereids rare, isolated, large. Rays straight in the beginning, bent in outer parts of the phloem, hardly differentiated from axial tissue of sieve tubes and parenchyma. Rays dilating occasionally. Many prismatic crystals in long axial cell chambers.

Phellem consisting of layers of large, square, thin-walled cells and small, rectangular, thick-walled cells (phelloids).

dormant season

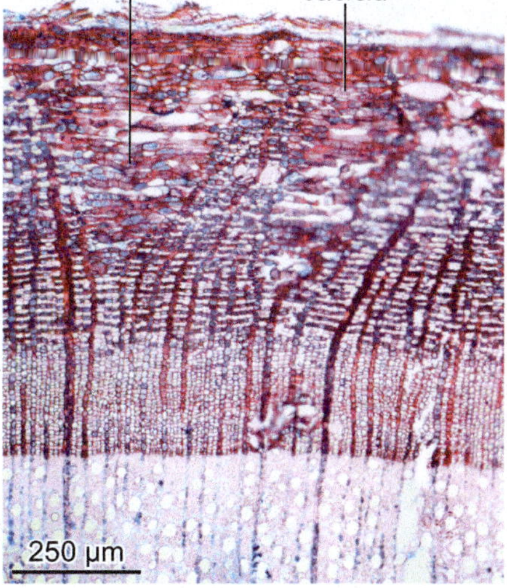

Nerium oleander L.

Apocynaceae

oleander, Oleander, laurier-rose, oleandro, adelfa

Form: Shrub, twigs erect, to 5 m tall.

Features: Leaves evergreen, opposite, lanceolate with a pointed tip, leathery, side veins parallel, margin entire, to 12 cm long. Flowers with five magenta to pink or white petals, in terminal spikes, to 6 cm long. Fruit a long, linear, brown follicle, to 12 cm long.

Habitat: Mediterranean, from Morocco and Portugal, east to Greece, Turkey and Western Asia; widely planted and locally invasive.

On dry to moist, gravelly or sandy riparian soils, close to the sea, to 300 m.

Bark: Dull light gray, with many small lenticels.

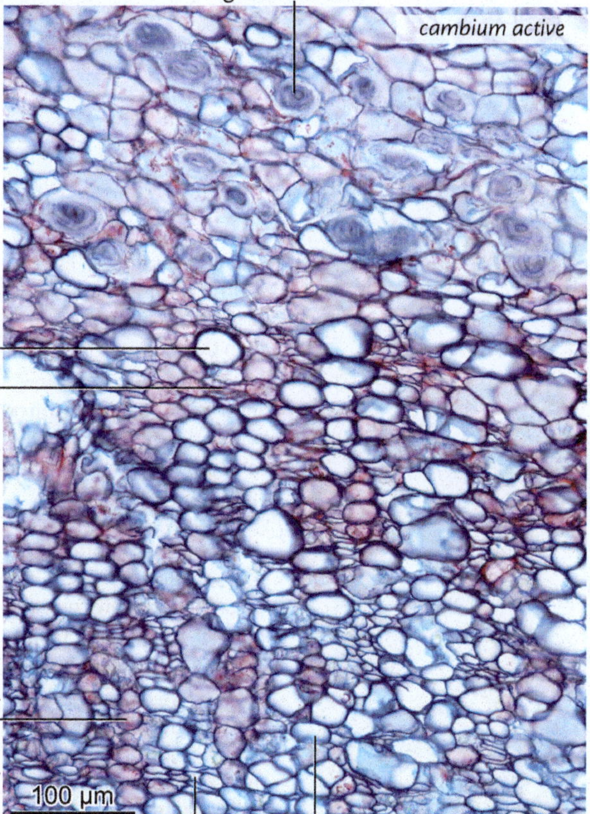

Bark anatomy: Irregularly structured tissue of sieve tubes and parenchyma with some round groups of sclereids.

Sieve tubes angular, accompanied by round to oval parenchyma cells. Gelatinous fibers in the transition between cortex and phloem. Dilatations absent. Laticifers mainly in the phloem. Many small crystal druses in axial parenchyma cells and rays.

Periderm consisting of 2–3-seriate rows of unlignified, square to rectangular cells.

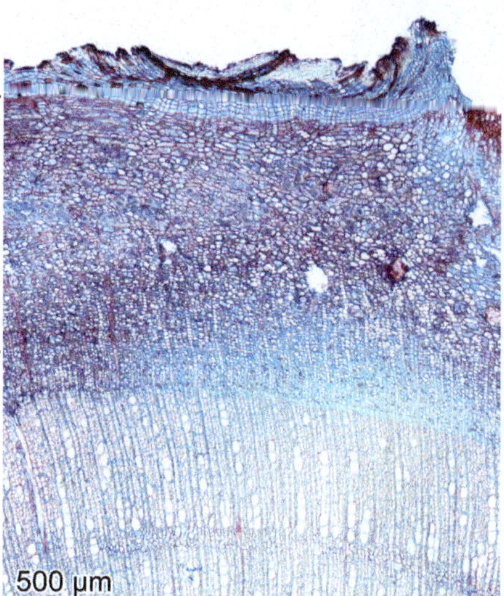

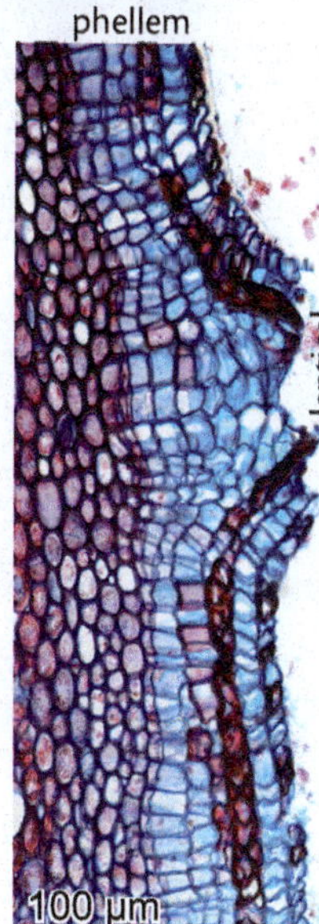

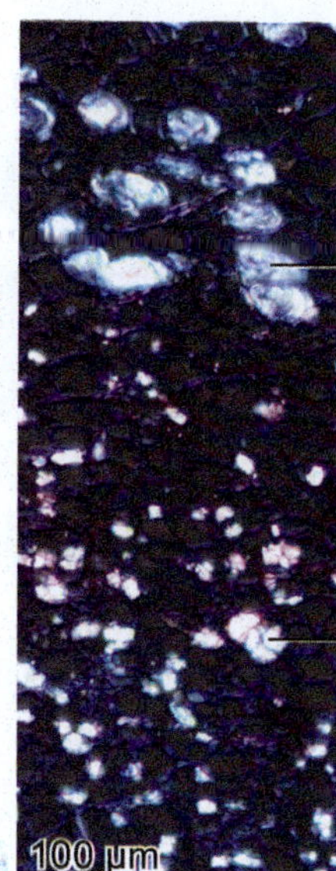

Olea europaea L. Oleaceae

olive tree, Olivenbaum, olivier, olivo, olivo

Form: Tree or shrub, often multi-stemmed, crown broad, irregular shaped, to 10 m tall.

Features: Leaves evergreen, opposite, long ovate, leathery, gray to silvery green, margin entire, to 8 cm long. Flowers dioecious or monoecious, yellowish white, with four small petals, to 4 mm long, in terminal panicles, to 5 cm long. Fruit an ovate to nearly globular, fleshy, green to black drupe, to 4 cm long.

Habitat: Mediterranean, from North Africa and Portugal east to Turkey, Syria and Israel; cultivated forms widely planted as an oil fruit and ornamental.

On dry to moist soils, to 900 m.

Bark: Young smooth, dull, light to green gray, with many small lenticels, later light brown gray, cracked into many small, square plates. Old stems often gnarled and with many hollows.

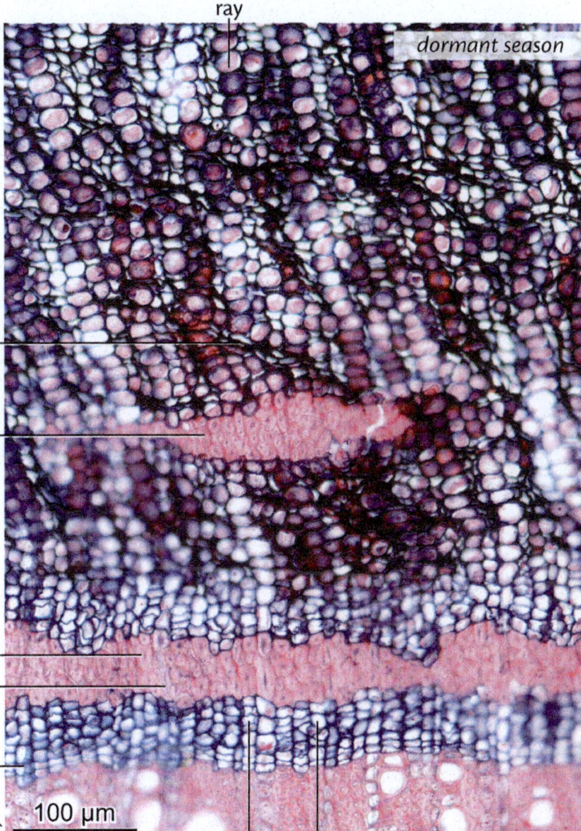

Bark anatomy: Tangentially structured tissue with many large, oval groups or bands of fibers.

Sieve tubes angular, collapsing within a short time after formation. Parenchyma cells roundish. Fibers in tangentially arranged groups. Sclereids rare, surrounding fiber groups or in rays. Rays slightly wavy, dilatations rare and only at later stages. Crystal sand in axial parenchyma cells and in rays, arranged in radial lines.

Phellem consisting of many layers of rectangular cells.

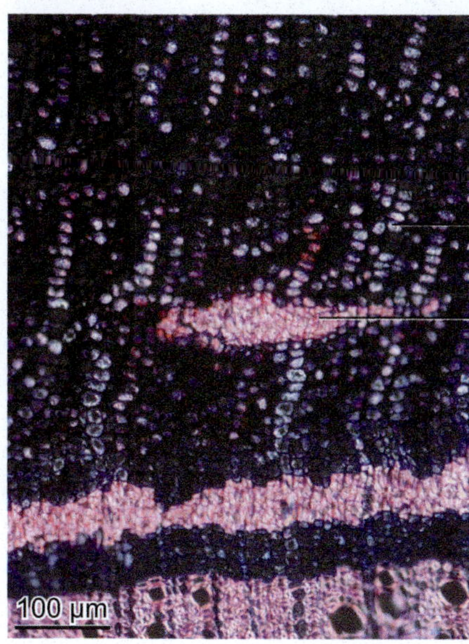

Phillyrea latifolia L. Oleaceae

green olive tree, Breitblättrige Steinlinde, filaire à large feuille, ilatro latifoglie, labiérnago prieto

Form: Shrub or small tree, often multi-stemmed, crown broad, irregular shaped, to 10 m tall.

Features: Leaves evergreen, leathery, opposite, ovate, margin regularly toothed, to 7 cm long. Flowers yellowish white, with four small petals, to 5 mm long, in terminal panicles, to 5 cm long. Fruit a globular, fleshy, blue black drupe, to 1 cm long.

Habitat: Mediterranean, from North Africa and Portugal east to Turkey.

On dry to moist soils, to 800 m.

Bark: Young dull light gray, with many small lenticels, later light brown gray, cracking into many small, square plates.

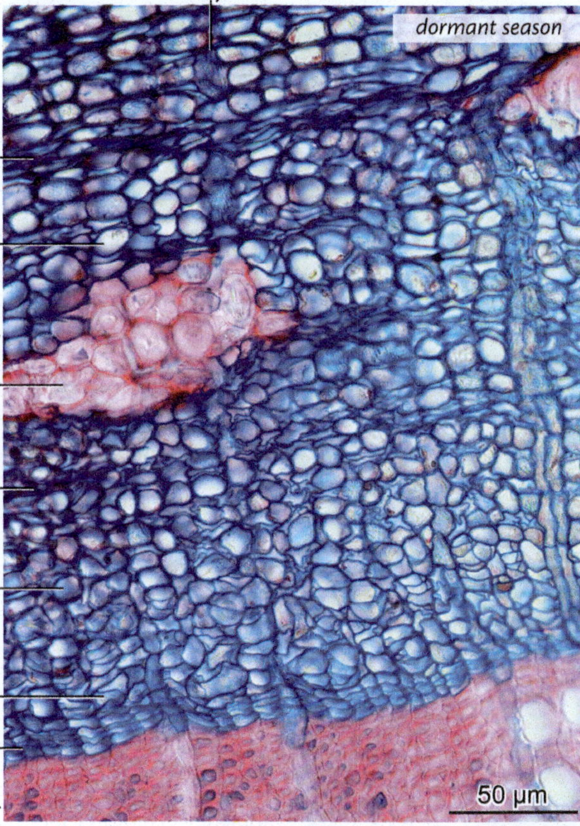

Bark anatomy: Tangentially structured tissue with many round to large oval groups of fibers.

Sieve tubes angular, collapsing within a short time after formation, indicated by irregular lines. Parenchyma cells roundish. Fibers absent. Sclereids in large, irregular shaped groups. Rays straight, dilatations rare. Small acicular crystals in most axial parenchyma cells and rays.

Phellem consisting of many rows of thin-walled, unlignified cells.

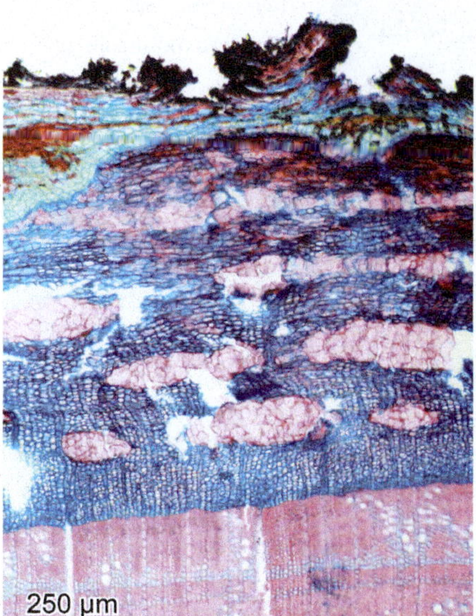

Pistacia lentiscus L. Anacardiaceae

evergreen pistache, Mastixstrauch, arbre au mastic, lentisco, mata charneca

Form: Shrub or a small tree, crown densely branched, to 8 m tall.

Features: Leaves evergreen, alternate, pinnate with 8–12 long ovate leaflets, margin entire, to 5 cm long. Flowers dioecious, with five short petals, to 4 mm across, male flower with deep red stamens, female flower greenish, both in dense, upright panicles. Fruit a red, fleshy, globular drupe, maturing to black, to 4 mm long.

Habitat: Mediterranean, from the Canary Islands and North Africa and Portugal, east to Turkey and Israel; sometimes cultivated for their mastic resin.

On dry to moist soils, to 1000 m.

Bark: Dull, gray brown or light gray, dotted with many small lenticels.

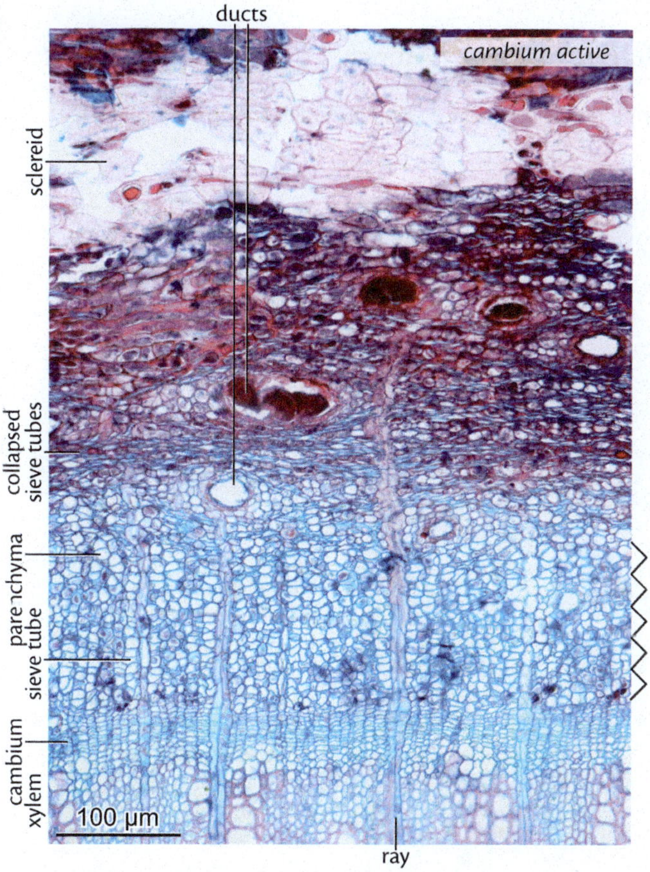

Bark anatomy: Poorly structured phloem with irregular groups of sclereids and ducts.

Sieve tubes angular, irregularly distributed. Parenchyma cells roundish. Fibers absent. Sclereids in tangential oval groups, forming radial lines. Rays bent, occasionally dilating. Ducts frequent in the phloem, partially filled with dark-staining substances. Crystals absent.

Phellem consisting of many rows of thin-walled, unlignified cells.

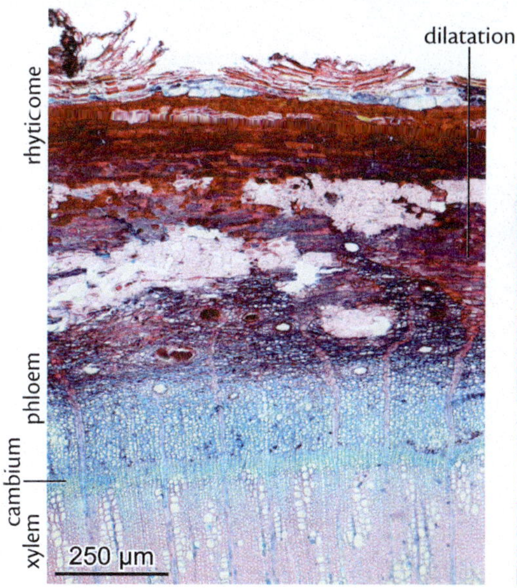

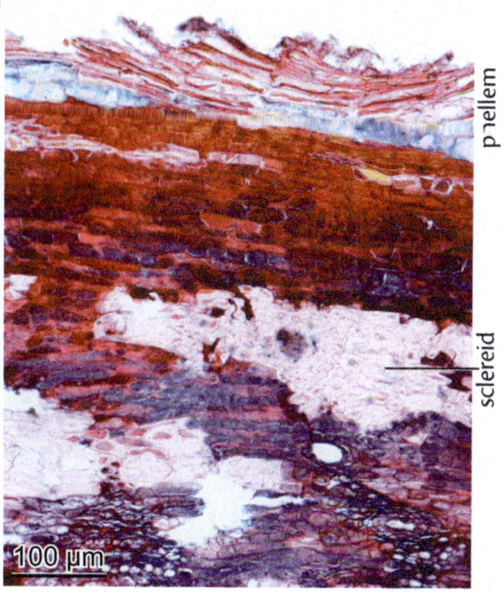

Platanus x hispanica Mill. ex Münchh. Platanaceae

London plane, Ahornblättrige Platane, platane commun, platano ibrido, plátano

Variation: The London plane is a hybrid between the East Mediterranean and Western Asiatic *Platanus orientalis* and the North American *P. occidentalis*.

Form: Tree, crown conical at first, later broad, irregular shaped, to 45 m tall.

Features: Leaves deciduous, alternate, long-stalked, palmately three- to seven-lobed, maple-like, margin with a few, sharp teeth, to 25 cm long. Flowers dioecious, long-stalked, in a globular ball, male flower green yellow, female flower scarlet, to 3.5 cm wide. Fruit a cluster of one-seeded, long-haired nuts in a long-stalked, globular, brown ball, to 3.5 cm wide.

Habitat: *Platanus orientalis* is native to the East Mediterranean and Western Asia, from Southern Albania and Greece east to Turkey and India. *Platanus occidentalis* is native to Eastern North America from Ontario to Texas and Tamaulipas.

In the wild, both parent species grow on riverine soils, to 1500 m; the hybrid is widely planted.

Bark: Young dull, light gray, soon shedding in large, irregular shaped plates, creating trunks with a conspicuous light gray to yellow or greenish gray pattern.

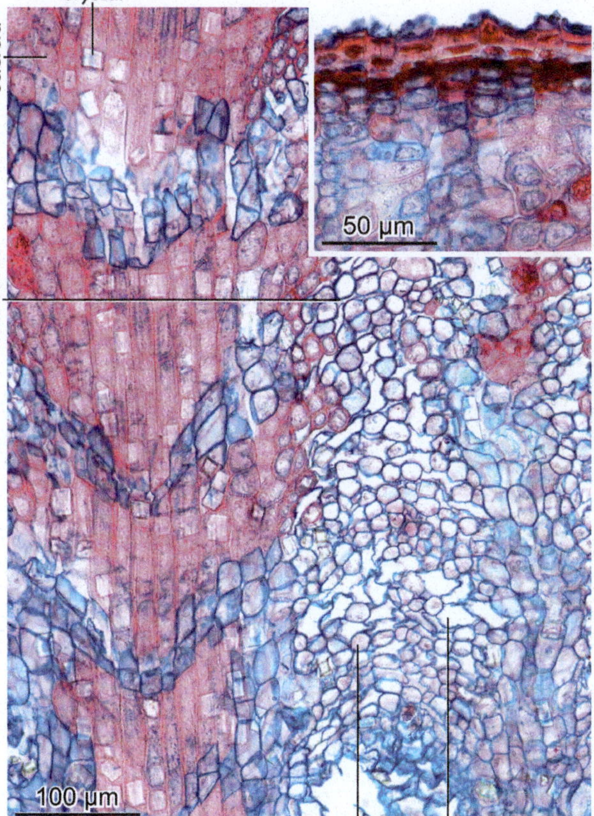

Bark anatomy: Intensively lignified, radially structured tissue.

Sieve tubes angular, collapsing within a short time after formation. Parenchyma cells roundish. Fibers absent. Bands and groups of sclereids in tangential waves. Sclereids relatively thin-walled. Rays straight, occasionally dilating. Large rays formed by wing-shaped groups of sclereids, tangentially interrupted by a few unlignified parenchyma cells. Large prismatic crystals mainly in rays.

Periderm consisting of a few layers of thin-walled cells and a thin row of intensively sclerified cells.

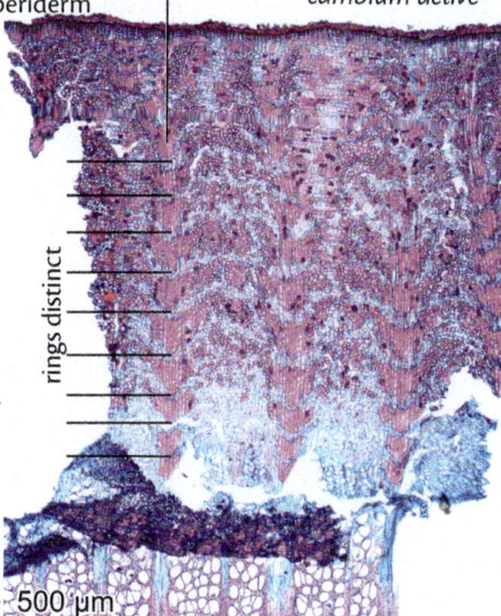

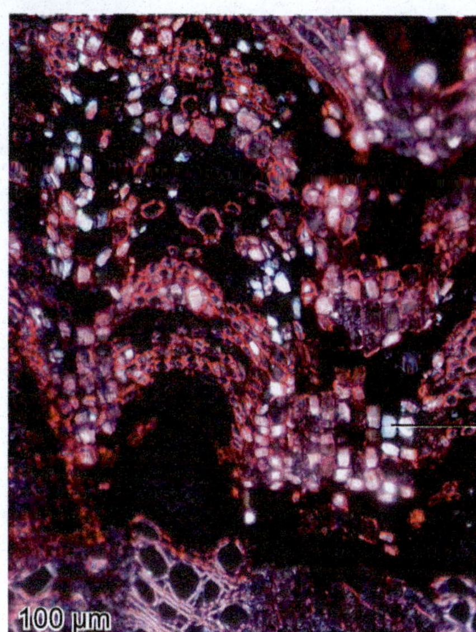

Punica granatum L.

Lythraceae

pomegranate, Granatapfel, grenadier, melograno, granado

Form: Shrub or small tree, crown broad, irregular shaped, spiny, to 10 m tall.

Features: Leaves deciduous, opposite, crowded at the top of twigs, long ovate, margin entire, to 8 cm long. Flowers red, with three to seven large, papery petals, overlapping each other like tiles, to 8 mm long. Fruit a globular berry, to 8 cm long, with a hard, yellow to red skin, inside fleshy, with many seeds embedded in a scarlet, edible pulp.

Habitat: East Mediterranean and Western Asia, from Turkey to India, with an obscure natural distribution due to long-time cultivation; widely planted as an ornamental and for fruits.

On dry to moist soils, to 2000 m.

Bark: Smooth, greenish or light gray, soon with many vertical lenticels.

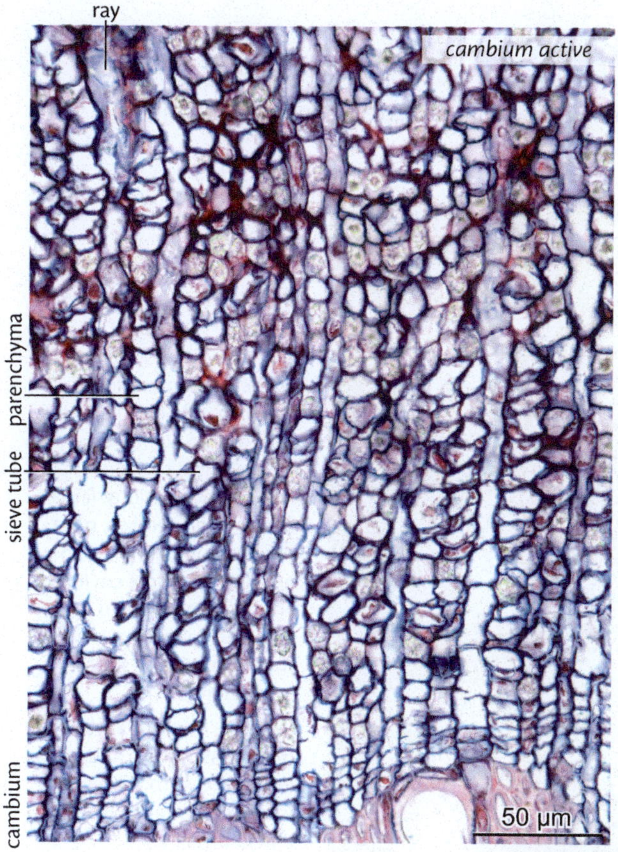

Bark anatomy: Poorly structured tissue.

Sieve tubes angular, often collapsing within a short time after formation. Parenchyma cells roundish. Fibers and sclereids absent. Rays straight, without dilatations. Small acicular crystals in most axial parenchyma cells and rays.

Phellem consisting of many rows of rectangular, thin-walled, unlignified cells.

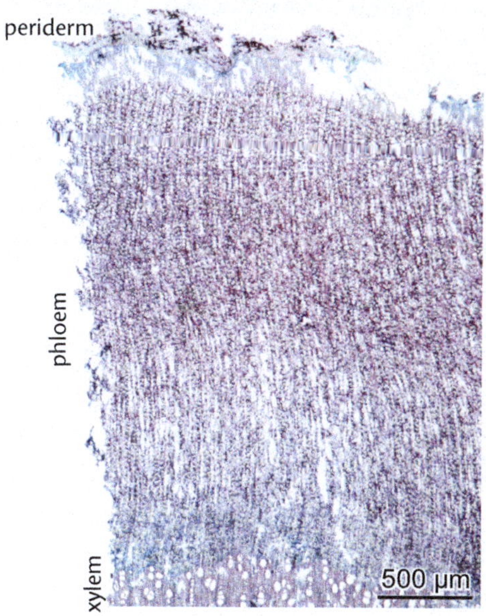

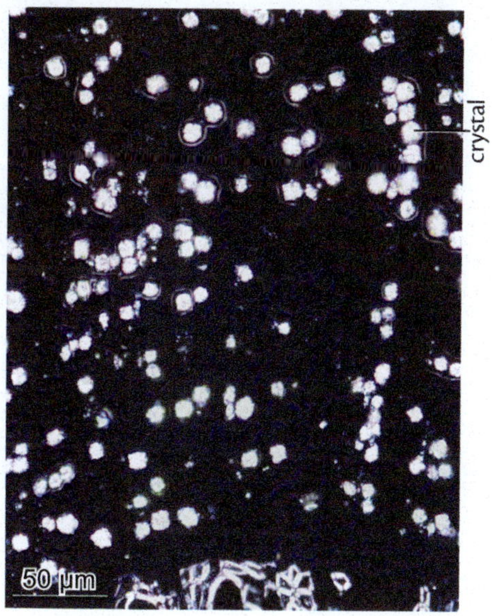

Quercus coccifera L. Fagaceae

kermes oak, Kermeseiche, chêne des garrigues, quercia spinosa, coscoja

Form: Shrub or a small tree, crown very dense, irregular shaped, to 8 m tall.

Features: Leaves evergreen, leathery, alternate, long ovate, margin undulated, with four to eight sharp teeth on each side, to 6 cm long. Flowers monoecious, female flower small, reddish, male flowers in long, thin, yellow green catkins, to 8 cm long. Fruit an ovate, brown nut, sitting deeply in a cup with thorny, spreading scales, to 4.5 cm long.

Habitat: Mediterranean, from North Africa and Portugal east to Turkey and Israel, with a gap in Northern and Central Italy.

On dry, often stony soils, on warm to hot slopes, to 1200 m.

Bark: Young dull, light gray, with many small lenticels, with age cracking into vertical fissures.

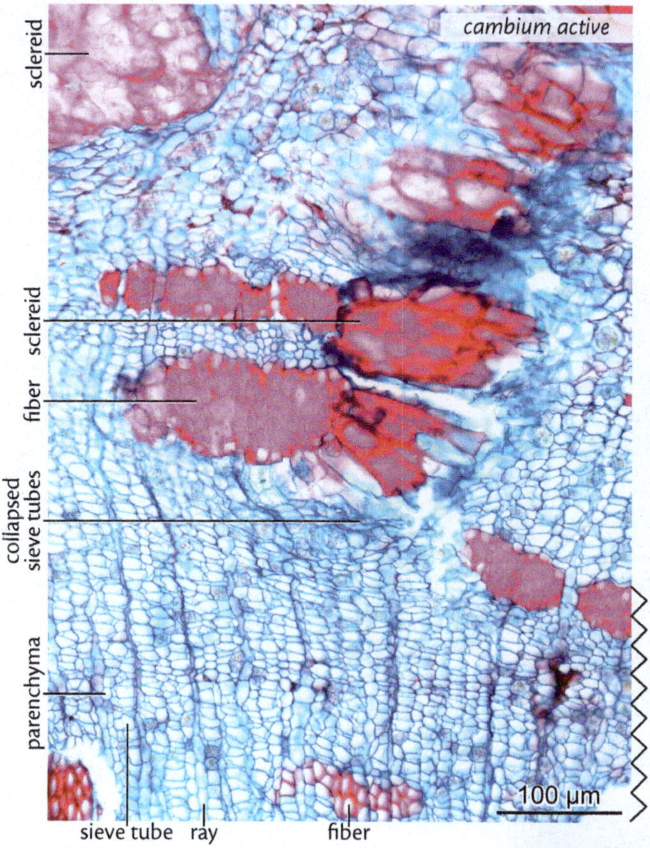

Bark anatomy: Indistinctly tangentially layered structure, sieve tubes and parenchyma cells radially arranged.

Sieve tubes angular, collapsing within a short time after formation. Parenchyma cells round. Groups of fibers alternating with the dominant zones of sieve tubes and parenchyma. Sclereids in round groups. Rays straight. Dilatations towards the periphery. Prismatic crystals surrounding fiber bands and within sclereids. Many crystal druses in the parenchymatic zones.

Phellem consisting of layers of rectangular, thin- to thick-walled cells.

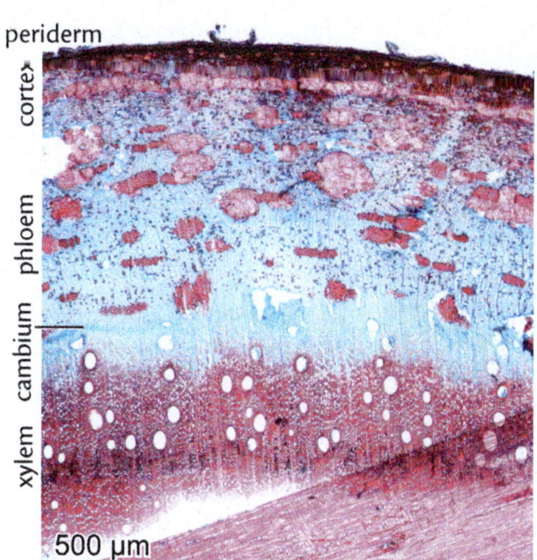

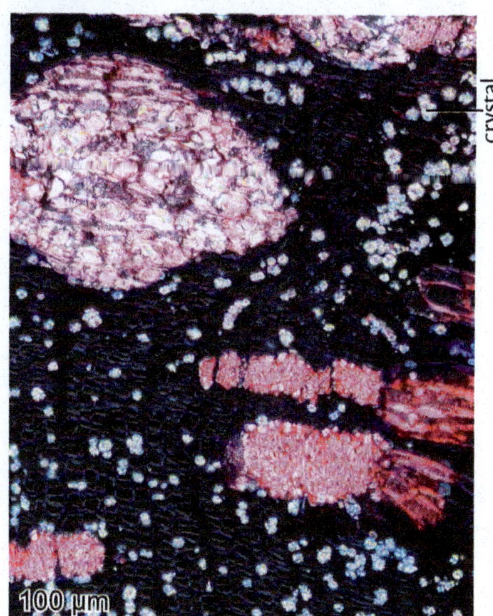

Quercus ilex L.

Fagaceae

holm oak, Steineiche, chêne vert, leccio, encina

Form: Tree, crown large, irregular shaped, to 25 m tall.

Features: Leaves evergreen, leathery, alternate, from elliptic to ovate and nearly round, margin entire, to 9 cm long. Flowers monoecious, female flower small, reddish, male flowers in long, thin, yellow green catkins, to 8 cm long. Fruit an ovate or nearly globular, brown nut, sitting in a cup with small scales, to 3.5 cm long.

Habitat: Mediterranean, the most dominant tree in the Western and Central Mediterranean Basin, before human impact forming large, often almost pure stands.

On dry to moist, often stony soils, to 2000 m.

Variation: Subspecies *ilex* from Northern Spain and western France east to Greece and western Anatolia, subsp. *rotundifolia* in Spain, Portugal and North Africa.

Bark: Young dull, smooth, light gray, with age cracking into small, square plates.

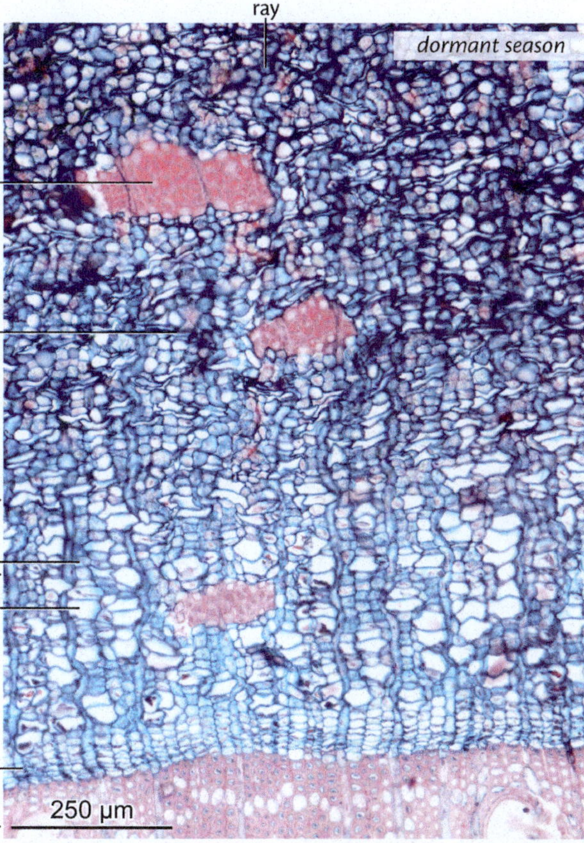

Bark anatomy: Indistinctly tangentially layered structure, sieve tubes and parenchyma cells radially arranged.

Sieve tubes angular, collapsing within a short time after formation. Parenchyma cells round. Groups of fibers alternating with the dominant zones of sieve tubes and parenchyma. Sclereids in round groups and dominating large rays. Rays large and straight or uniseriate and wavy. Large rays sometimes dilating. Prismatic crystals surrounding the fiber bands, crystal druses in the parenchymatic zones.

Phellem consisting of layers of rectangular, thin-walled cork cells.

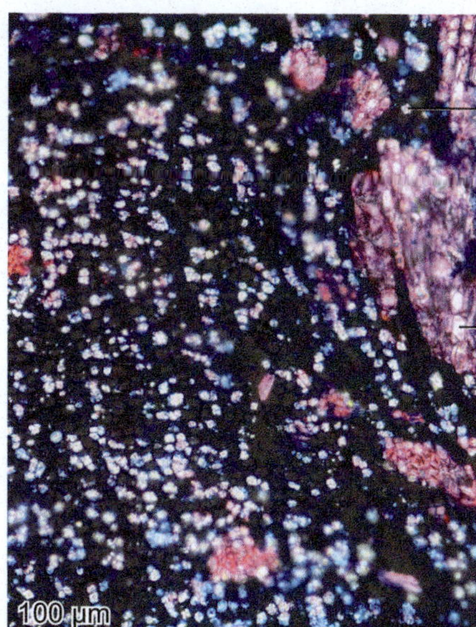

Quercus suber L. Fagaceae

cork oak, Korkeiche, chêne-liège, quercia da sughero, alcornoque

Form: Tree, crown globular to irregular, to 25 m tall.

Features: Leaves evergreen, leathery, alternate, long ovate, glossy green above, margin with four to seven small teeth on each side, to 7 cm long. Flowers monoecious, female flower small, reddish, male flowers in long, thin, yellow green catkins, to 7 cm long. Fruit a long ovate, brown nut, sitting half in a cup with small scales, to 4.5 cm long.

Habitat: Western Mediterranean, centered in Portugal, western Spain, northern Morocco and Algeria, with scattered stands in Southern France, Italy and Tunisia; important source for natural cork.

On dry to moist, acidic soils, to 1500 m.

Bark: Young dull, light to brown gray, soon cracking into thick, bulging, corky fissures, the target of cork harvesting.

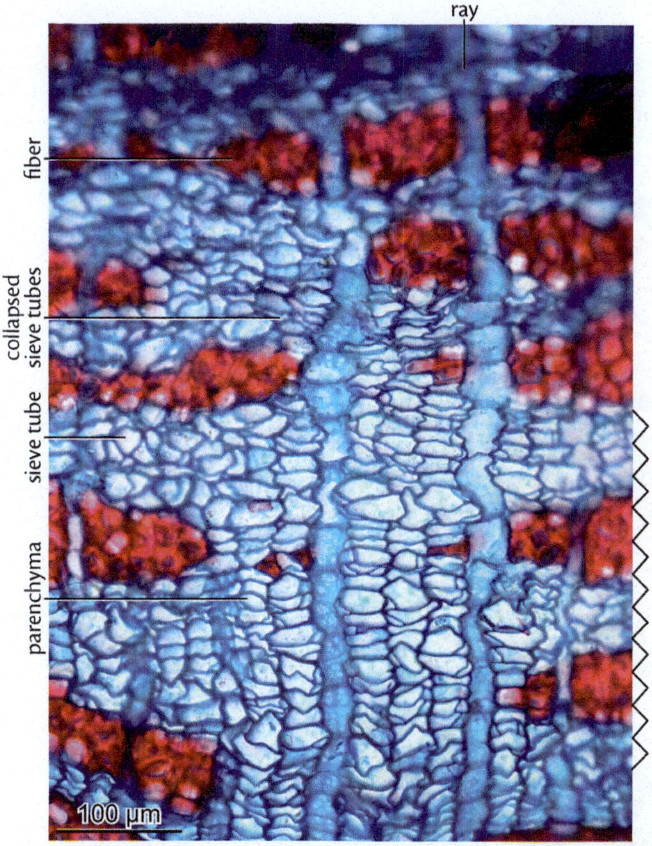

Bark anatomy: Tangentially layered structure of sieve tubes and parenchyma, fiber bands and sclereids.

Sieve tubes angular, collapsing within a short time after formation. Parenchyma cells round. Fiber bands alternating with zones of sieve tubes and parenchyma. Sclereids in round groups and dominating large rays. Rays large and straight or uniseriate and wavy. Large rays sometimes dilating. Prismatic crystals surrounding the fiber bands, crystal druses in the parenchymatic zones.

Phellem consisting of layers of rectangular, thin-walled cork cells.

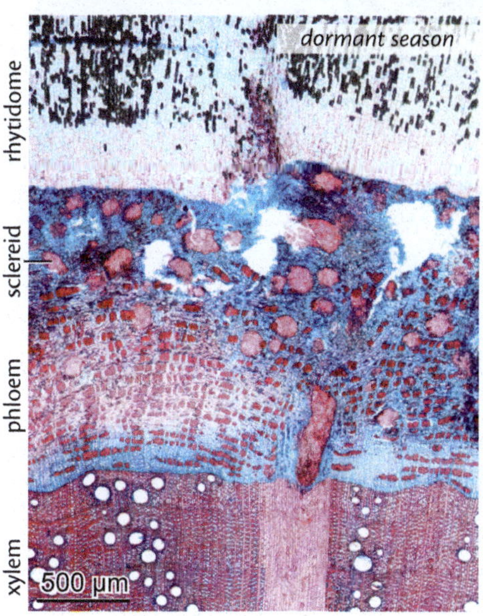

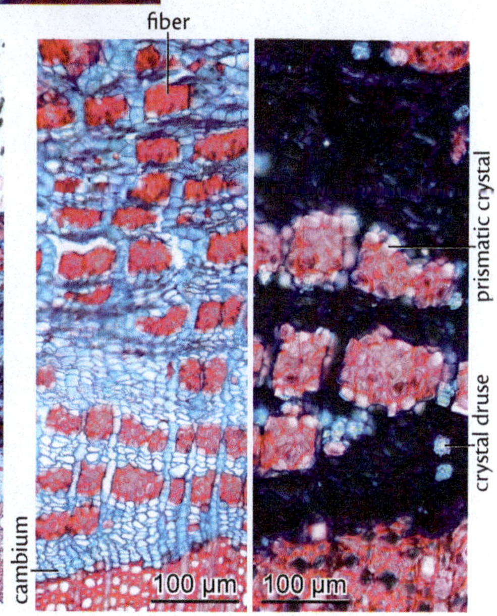

Spartium junceum L. Fabaceae

Spanish broom, Pfriemenginster, spartier à tiges, ginestra odorosa, retama de olor

Form: Shrub, twigs flexible, erect and unbranched, usually without leaves, to 3 m tall.

Features: Leaves deciduous, alternate, linear, margin entire, to 3 cm long, soon aborted to conserve water. Flowers intensely yellow, typical zygomorphic and odorous pea flowers, to 2.5 cm long. Fruit a black brown, long ovate, slightly bent, dry pod, to 10 cm long.

Habitat: Mediterranean, from North Africa and Portugal east to Turkey, Crimea, the Caucasus and Israel; widely planted and invasive.

On dry to moist, often loamy or sandy soils, to 1200 m.

Bark: Young smooth, dull, dark gray green, with age greenish brown, with small fissures.

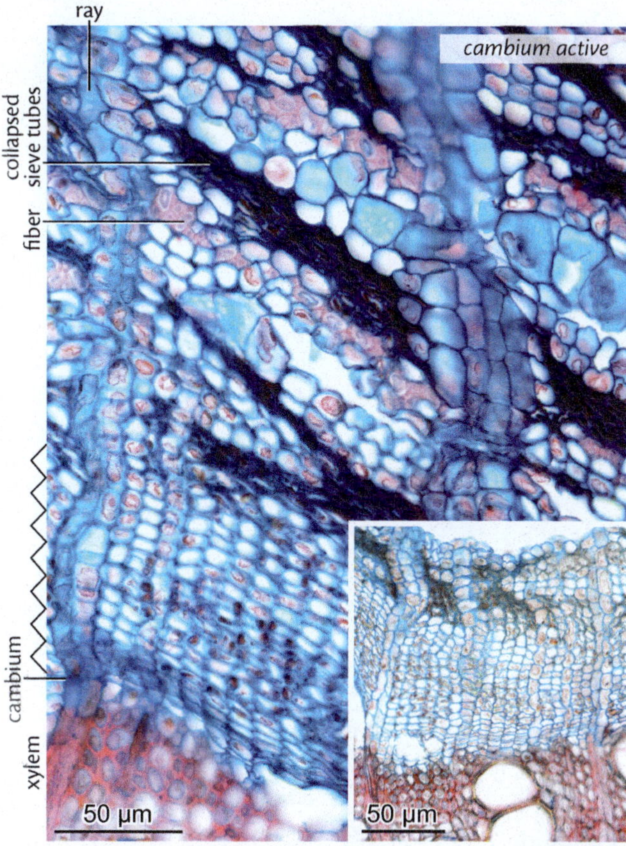

Bark anatomy: Tangentially layered structure.

Sieve tubes angular, collapsing within a short time after formation, indicated by dark tangential lines. Sieve tubes and companion cells alternating with bands of oval parenchyma cells and 2–3-seriate tangential fiber bands. Sclereids in groups at the periphery of the phloem. Rays straight, often tangentially interrupted by fibers. Dilatations absent. Many acicular crystals, mainly in axial parenchyma cells and rays.

Phellem consisting of layers of square to rectangular, thick-walled cells, filled with dark-staining substances.

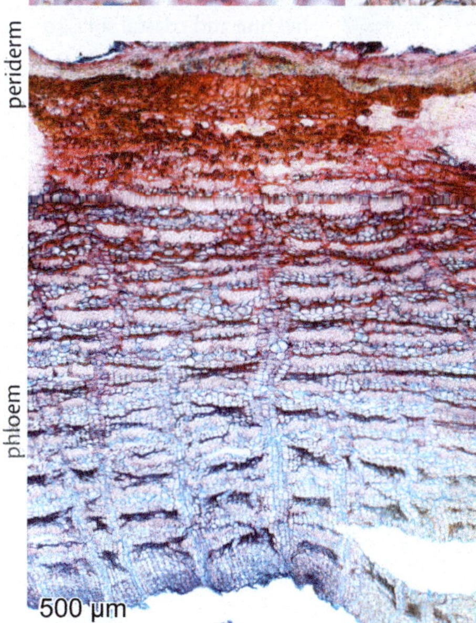

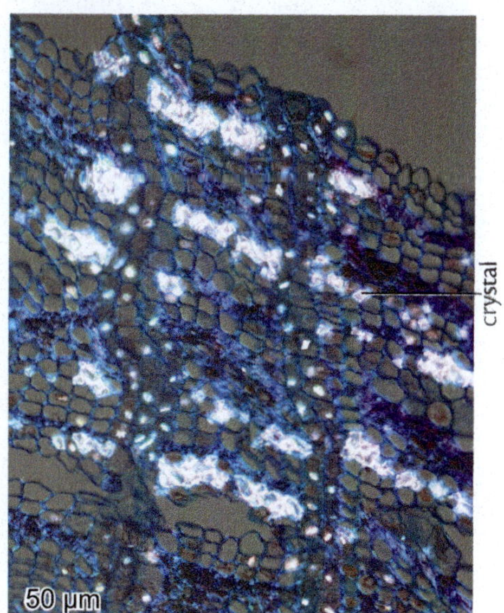

Tamarix gallica L.

Tamaricaceae

French tamarisk, Französische Tamariske, tamaris commun, tamerice comune, taray

Form: Shrub or small tree, crown dense, irregular shaped, twigs strongly ramified, to 10 m tall.

Features: Leaves evergreen, gray to blue green, linear scales, tile-like overlapping, to 1 mm long. Flowers pink to white, with five petals, to 3 mm long, in dense panicles. Fruit a light pink, three-valved, dry capsule, to 4 mm long.

Habitat: Western Mediterranean, from North Africa and Portugal east to Sicily and Dalmatia; also planted and locally invasive.

On dry to moist, often salty, riverine and coastal soils, to 800 m.

Bark: Young smooth, dull or glossy, red brown or gray, with age with horizontal lenticels and some fissures.

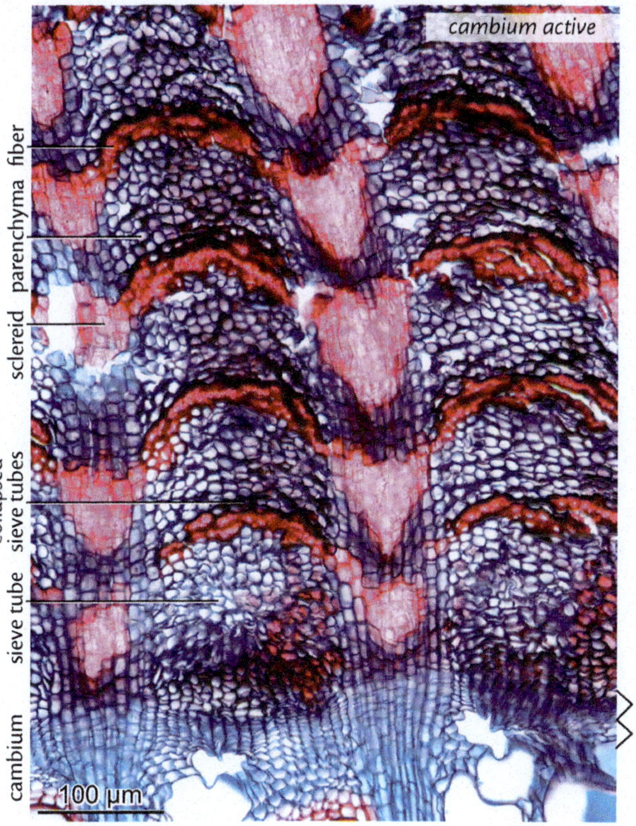

Bark anatomy: Tangentially arc-like layered structure.

Sieve tubes angular, collapsing within a short time after formation, indicated by dark tangential lines. Sieve tubes and companion cells alternating with bands of oval parenchyma cells and 2–3-seriate tangential fiber bands. Large rays dominated by groups of sclereids, periodically interrupted by unlignified parenchyma cells. Rays straight, dilatations absent. Crystals absent.

Rhythidome consisting of irregular layers of thin-walled, lignified or unlignified cells.

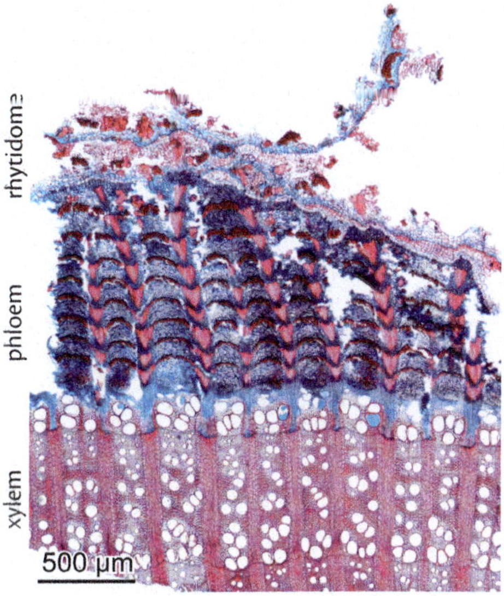

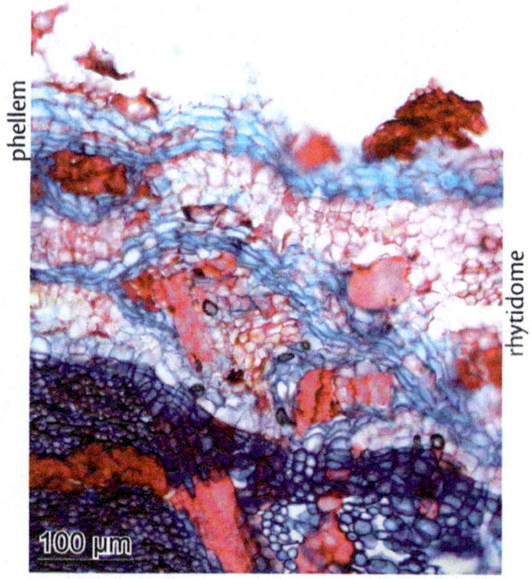

References and recommended reading

Macroscopic bark and tree descriptions

Alam M (2011) Trees and Shrubs of Afghanistan. *Rossolis, Lausanne, Switzerland*

Blamey M, Grey-Wilson C. (1993) Mediterranean wild flowers. *HarperCollins, London, UK*

Blanca G, Cabezudo B, Cueto M, Lopez CF, Torres CM (2009) Flora Vascular de Andalucia Oriental, Volume I–IV. *Consejeria de medio ambiente, Junta de Andalucia, Sevilla, Spain*

Britton NL, Brown A (1990) An illustrated Flora of the Northern United States and Canada, Volume I–III. *Dover Publications, New York, USA*

Brockmann F, Merrilees R (1979) Trees of North America. *Golden Press, New York, USA*

Bärtels A, Schmidt P (2014) Enzyklopädie der Gartengehölze. *Ulmer, Stuttgart, Germany*

Chiappini M (1985) Flora e paesaggi vegetali della Sardegna. *Della Torre, Cagliari, Italy*

Del Favero (2008) I boschi delle regioni meridionali e insulari d'Italia. *Cleup, Padova, Italy*

Fang J, Zhiheng W, Zhiyao T (2009) Atlas of Woody Plants in China, Volume I–IV. *Higher Education Press, Beijing, China*

Farjon A (2010) A Handbook of the World's Conifers. *Brill, Leiden, Netherlands*

Fennana M, Ibn Tattou M, El Oualidi J (2007) Flore pratique du Maroc, Volume I–III. *Institut Scientifique, Série Botanique no. 40, Rabat, Morocco*

Fielding J, Turland N, Matthew B (2005) Flowers of Crete. *Kew Publishing, Richmond, UK*

Glatz F, Balogh M, Péterfi A (2006) Magyarorszag tajainak növenyzete és allatvilaga. *Mta Tarsadalomkutato Kozpont, Budapest, Hungary*

Grey-Wilson C, Cribb P (2011) Flowers of Western China. *Kew Publishing, Richmond, UK*

Griffin J, Critchfield W (1972) The distribution of forest trees in California. *USDA Forest Service Research paper no. 82, Washington D.C., USA*

Grisa S, Benglowskii B, Carimsakow S, Carraro G (2008) Tipologia lesow Kirgiskoi Respubliki. *Gosadarstwenoe agenturo po ochruzhenjuscheni sredj, Bischkek, Kyrgistan*

Haeupler H, Muer T (2007) Bildatlas der Farn- und Blütenpflanzen Deutschlands. *Ulmer, Stuttgart, Germany*

Holubec V, Krivka P (2006) The Caucasus and its flowers. *Loxia, Praha, Czechia*

Jeanmonod D, Gamisans J (2007) Flora Corsica. *Edisud, Aix-en-Provence, France*

Jovanovic B (1986) Prirodna potencijalna vegetacija Jugoslavije. *Do Paralele, Ljubljana, Slovenia*

Klötzli F, Dietl W, Marti K, Schubiger-Bossard C, Walther G (20110) Vegetation Europas. *Ott, Bern, Switzerland*

Knapp R (1973) The vegetation of Africa. *Fischer, Jena, Germany*

Krishen P (2013) Jungle trees of Central India. *Penguin, Harayana, India*

Larsson JY, Sögnen SM (2003) Vegetasjon i norsk skog. *Landbruksforlaget, Oslo, Norway*

Lauber K, Wagner G, Gygax A (2018) Flora Helvetica – Illustrierte Flora der Schweiz. *Haupt, Bern, Switzerland*

Lesur L (2017) Arboles de México. *Trillas, Ciudad de México, México*

Li Z, Shi l, Wang W (2007) Plants of Mount Emei. *Baitong, Beijing, China*

Lopez Gonzalez G (2004) Guia de los arboles y arbustos de la Peninsula Ibérica y Baleares. *Ediciones Mundi-Prensa, Madrid, Spain*

Mayer H (1984) Wälder Europas. *Fischer, Stuttgart, Germany*

Mayor M, Diaz T (1977) La flora Asturiana. *Ayalga, Gjion, Spain*

Miyawaki A, Landolt E (1988) Contributions to the knowledge of flora and vegetation of Japan. *Geobotanisches Institut der ETH Zürich, volume 98, Zürich, Switzerland*

Mossberg B, Stenberg L (2003) Den nya nordiska Floran. *Wahlström & Widstrand, Stockholm, Sweden*

Mucina L, Grabherr G, Wallnöfer S (1993) Die Pflanzengesellschaften Österreichs, Volume I–III. *Fischer, Jena, Germany*

Muer T, Sauerbier H, Calixto Cabrera F (2016) Die Farn- und Blütenpflanzen der Kanarischen Inseln. *Margraf, Weikersheim, Germany*

Munz P, Keck D (1968) A California Flora and supplement. *University of California Press, Berkley, USA*

Nowak A, Nobis M, Nowak S, Nobis A, Kusza G (2011) Atlas of Vascular Plants of Tajikistan. *Institute of Botany, Jagellonian University, Opole, Poland*

Oberdorfer E (1992) Süddeutsche Pflanzengesellschaften, Volume I–IV. *Fischer, Jena, Germany*

F. H. Schweingruber et al., *Bark Anatomy of Trees and Shrubs in the Temperate Northern Hemisphere*, https://doi.org/10.1007/978-3-030-14056-4

Parker T (2008) Trees of Guatemala. *The Tree Press, Austin, USA*

Pfadenhauer J, Klötzli F (2014) Vegetation der Erde – Grundlagen, Ökologie, Verbreitung. *Springer Spektrum, Berlin, Germany*

Pignatti S (2003) Flora d'Italia, Volume I–III. *Edagricole, Bologna, Italy*

Pils G (2006) Flowers of Turkey. *Pils, Linz, Austria*

Polunin O (1987) Flowers of Greece and the Balkans. *Oxford Press, Oxford, UK*

Polunin O, Stainton A (1986) Flowers of the Himalaya. *Oxford University Press, Delhi, India*

Quosja X (1992) Flora e Shqiperise, Volume I–III. *Akademia e shencave e republikes se Shquiperise, Tirana, Albania*

Rameau JC, Mansion D, Dumé G, Gauberville C (2008) Flore forestière Française, Volume I–III. *ENGREF, Agroparistech, Paris, France*

Rzedowski J (1988) Vegetacion de Mexico. *Editorial Limusa, Ciudad de México, México*

Schönfelder I and P (1994) Kosmos Atlas Mittelmeer- und Kanarenflora. *Kosmos, Stuttgart, Germany*

Silic C (1984) Endemicne Biljke. *Priroda Jugoslavije, Svjetlost, Beograd, Serbia*

Speta E, Rakosy L (2010) Wildpflanzen Siebenbürgens. *Plöchl, Freistadt, Austria*

Steiger P (2010) Wälder der Schweiz – Vielfalt der Waldbilder und Waldgesellschaften in der Schweiz. *Ott, Bern, Switzerland*

Steiger P (2016) Esche, Espe oder Erle? Pflanzenportraits aller wildwachsender Gehölze Mitteleuropas. *Ott, Bern, Switzerland*

Storrs A and J (1998) Trees and shrubs of Nepal and the Himalayas. *Pilgrims Book House, Kathmandu, Nepal*

Tekin E (2005) Türkiye en güzel yaban çiçekleri. *Basimevi, Istanbul, Turkey*

Tison JM, Jauzein P, Michaud H (2014) Flore de la France méditerranéenne continentale. *Naturalia, Conservatoire botanique national méditerranéen de Porquerolles, France*

Vaucher H (1990) Baumrinden. *Ferdinand Enke Verlag, Stuttgart, Germany*

Walter H (1974) Die Vegetation Osteuropas, Nord- und Zentralasiens. *Fischer, Stuttgart, Germany*

www.efloras.org Flora of China

Yasaka H, Azegami C, Hisiyama C (1985) Woody plants of Japan. *Yama-kei, Tokyo, Japan*

Microscopic bark descriptions and preparation techniques

Angyalossy V, Pace MR, Evert RF, Marcati CR, Oskolski AA, Terrazas T, Kotina E, Lens F, Mazzoni-Viveiros SC, Angeles G, Machado SR, Crivellaro A, Rao KS, Junikka L, Nikolaeva N, Baas P (2016) IAWA List of Microscopic Bark Features. *IAWA Journal* 37 (4): 517–615.

Echrich W (1995) Funktionelle Pflanzenanatomie. *Springer, Berlin, Heidelberg, New York*

Evert RF (2006) Esau's plant anatomy. Meristems, cells and tissues of the plant body. Their structure, function and development. 3rd edition. *Wiley Interscience, John Wiley & Sons. Inc. Hoboken, New Jersey, USA*

Crivellaro A, Schweingruber FH (2013) Wood, bark and pith anatomy of Eastern Mediterranean trees and shrubs. *Springer, Heidelberg, New York, Dordrecht, London*

Esau K (1969) The phloem. *Gebrüder Bornträger, Berlin, Stuttgart, Germany*

Gärtner H, Lucchinetti S, Schweingruber FH (2015) A new sledge microtome to combine wood anatomy and tree-ring ecology. *IAWA Journal* 36 (4): 452–459.

Gärtner H, Schweingruber FH (2013) Microscopic preparation techniques for plant stem analysis. *Verlag Dr. Kessel, Remagen, Germany*

Grosser D (2007) Die Hölzer Mitteleuropas: Ein mikrophotographischer Lehratlas. *Verlag Dr. Kessel, Remagen, Germany*

Holdheide W (1951) Anatomie mitteleuropäischer Gehölzrinden. In Freund H (ed.).Handbuch der Mikroskopie in der Technik, Volume 5. *Umschau Verlag, Frankfurt am Main, Germany*

Ravaomanalina BH, Crivellaro A, Schweingruber FH (2017) Stem anatomy of *Dalbergia* and *Diospyros* species from Madagascar. *Springer International Publishing*

Schweingruber FH (1978) Microscopic wood anatomy. *Swiss Federal Institute of Forestry Research, Birmensdorf, Switzerland*

Schweingruber FH (1990) Anatomy of European woods. *Paul Haupt Verlag, Bern, Stuttgart*

Schweingruber FH (2005) Xylem Database *https://www.wsl.ch/dendropro/xylemdb/*

Schweingruber FH, Börner A, Schulze E-D (2103) Atlas of stem anatomy in herbs, shrubs and trees, Volume 1–2. *Springer Verlag Berlin, Heidelberg, Germany*

The manufacturer's authorised representative in the EU is Springer Nature Customer Service Centre GmbH, Europaplatz 3, 69115 Heidelberg, Germany. If you have any concerns regarding our products, please contact ProductSafety@springernature.com

Printed and bound by CPI Group (UK) Ltd, Croydon, CR0 4YY
28/09/2026
02224078-0006